Living with Nature, Cherishing Language

Justyna Olko · Cynthia Radding
Editors

Living with Nature, Cherishing Language

Indigenous Knowledges in the Americas Through History

The Americas
Research Network

palgrave
macmillan

Editors
Justyna Olko
University of Warsaw
Warsaw, Poland

Cynthia Radding
University of North Carolina
Chapel Hill, NC, USA

ISBN 978-3-031-38738-8   ISBN 978-3-031-38739-5   (eBook)
https://doi.org/10.1007/978-3-031-38739-5

The preparation and publication of this book were funded by the European Union's Horizon 2020 research and innovation programme under the Marie Skłodowska-Curie grant agreement No 778384. It was developed within the project "Minority Languages, Major Opportunities. Collaborative Research, Community Engagement, and Innovative Educational Tools (COLING). The Americas Research Network (ARENET), a signatory to the COLING project, hosted several authors of this book on research secondments and facilitated the participation of Indigenous scholars in this book. Research on this book was supported under 2018-2023 science funding program allocated for the implementation of the co-financed international project 3928/H2020/2018/2.

Cover illustration: © Kevin Schafer/Getty Images

This Palgrave Macmillan imprint is published by the registered company Springer Nature Switzerland AG
The registered company address is: Gewerbestrasse 11, 6330 Cham, Switzerland

Paper in this product is recyclable.

Contents

Abbreviations

COCIHP	*Coordinadora de Organizaciones Campesinas e Indígenas de la Huasteca Potosina* (Coordinating Committee of the Rural and Indigenous Organizations of the Huasteca Potosina)
EIA	The U.S. Energy Information Administration
INEGI	*Instituto Nacional de Estadística y Geografía* (The National Institute of Statistics and Geography)
PEMEX	*Petróleos Mexicanos* (Mexican Petroleum)
PROCEDE	*Programa de Certificación de Derechos Ejidales y Titulación de Solares* (The Program for Certification of Ejido Rights and Titles of Lots)

List of Figures

List of Tables

1

Living with Nature Across Time, Space and Cultural Perspectives: Introduction

Justyna Olko and Cynthia Radding

Throughout human history, the past has always been part of the present, but today the awareness that past is not just past and that the renderings of the past very much shape the world in which we live today is perhaps more pervasive than ever before. Thus, as historians we have become increasingly conscious of the fact that "history is not only constructed in the present – our understanding of it also constructs the sort of present we live in. We live with the legacies of the past" (Carr & Lipscomb, 2021, p. 13). This process works both ways, however, as the past shapes our present and present concerns and agendas inform our searches and readings of the past. "Historians will always be busy—not because the past changes, but because the questions we ask of it change as conditions in our own times change" (Crosby, 2001, p. xi). This has

J. Olko (✉)
University of Warsaw, Warsaw, Poland
e-mail: jolko@al.uw.edu.pl

C. Radding
University of North Carolina, Chapel Hill, NC, USA
e-mail: radding@email.unc.edu

© The Author(s) 2024
J. Olko and C. Radding (eds.), *Living with Nature, Cherishing Language*,
https://doi.org/10.1007/978-3-031-38739-5_1

been the case of environmentally informed history that from its beginnings in the 1970s has been driven by and reflected concerns of our world and its disruptive relationships with nature. These concerns, shared by the authors of this book, have inspired historians to probe into complex ecological processes of different time periods. While grounded strongly in the study of the past, this book consciously crosses the frontiers and methods of history and related social sciences, and it addresses the major concerns of our time in different regions of the Americas. These are not limited, however, to environmental challenges, but extend to the cultural survival, autonomy, and resilience of Indigenous people, vitality of their languages, and recognition of traditional knowledge among their own communities and by the larger societies in which they live.

Over nearly a century environmental history has developed as an inherently interdisciplinary field with strong methodological ties to anthropology, geography, and human ecology. Environmental history has developed the conceptual frameworks for bringing space and nature into the woven tapestries of events in both the recent and ancient past, thus challenging conventional methods for building historical narratives and designing socioeconomic analyses and cultural interpretations. Focused on the reciprocal relations between different social and ethnic communities and the natural environment, it is centered in the materiality of soil textures, vegetation biomes, wildlife distribution, and stream flow as these relate to human settlement patterns, technologies, and cultural practices. Environmental scholars working in different world regions have developed common themes dealing with agrarian and pastoral systems, built urban and rural spaces, water management, cultural adaptations for living in tropical and arid environments, deforestation, and reforestation. Environmental studies offer a unique explanatory framework by weaving together nature and culture, power and negotiation, adaptation and innovation, to explain the different ways that human societies craft the landscapes in which they live in the course of both external and internal conflicts and the reciprocal processes of adaptation and innovation (Radding, 2019, p. 58).

The conceptual frameworks developed to address environmentally informed histories point to different emphases and visions of historical change. Environmental historians who work on both the recent and remote past and present scenarios of disruption and decline contrasted with the resilience and regeneration of ecological biomes and human communities (Dean, 1995; Radding, 2022; Denevan, 2002). The tension between human disruption of natural environments and the creative adaptation of Indigenous societies provides the guiding theme for the second chapter in this book, "Flexible borders, permeable territories..." by Prządka-Giersz, Giersz, and Chyla. A related contrast in the crafting of environmental histories concerns the values ascribed to *wilderness* and to lived spaces in different natural surroundings (Cronon, 1996; White, 2004). In this volume, Radding's chapter on "The *Yoreme* creation of *itom ania*..." moves away from the notion of untouched "wilderness" to emphasize culturally produced landscapes in the context of colonialism. Environmental historians often work within the tension between the material relations of production and the cultural meanings ascribed to nature and to the landscapes that distinct human communities have created (Ingold, 2000). This tension produces creative contrasts in nearly all the chapters in this book, particularly in the contributions by Olko, "*Ihuan yehhuan tlacuauh tlamauhtiah in ichcapixqueh...,*" by Dexter-Sobkowiak, "*Amo kitlapanas tetl!..,*" and De la Cruz Cruz, "*Tlaneltoquilli tlen mochihua ica cintli.*" At the same time, these approaches, grounded in distinct methodological and theoretical tools and representing different positionalities, provide complementary insights and build a multi-angled and diachronic account of complex entanglements between Indigenous peoples and their environs.

Environmental histories of Latin America have centered their inquiries in the areas of dense pre-contact populations and centers of Iberian settlement toward the ecological and imperial borderlands of riverine tropics or semi-arid steppes and grasslands. Environmental studies of the Mesoamerican and Andean core regions and their regional spheres of influence have focused on Indigenous systems of food production and water management. Indigenous technologies associated with the *chinampas* of the Valley of Mexico and the raised fields that cultural geographers and archaeologists have found throughout the Andes and the interior lowlands of South America, illustrate the imprint of complex societies on the natural environments in which they lived. In this volume, Chapter 2, "Flexible Borders, permeable territories...," underscores both the necessity of adaptation to a regional environment of periodic droughts and the innovative technologies that supported numerous settlements and complex social structures in the ecological borderland between the Andean cordillera and the coastal plains.

Environmental historians working on Latin America in both early and modern periods have focused on Indigenous systems for gathering and cultivating diverse species of plants, enriching soils, harvesting water, and building terraces and irrigation canals. Working in interdisciplinary teams together with local communities, scholars have researched the origins and distribution of cultigens, especially the evolution of maize (*Zea maiz*), manioc (*Manihot sp.*), and quinoa (*Chenopodium quinoa*) as well as the kinds of plants that have co-evolved with human societies, e.g., amaranthus, agaves, pitahayas, nopales, tepary beans, and other species that occur in both cultivated and wild varieties. In the extended agroecological hearths of Mesoamerica, the antiquity of Oaxacan horticultural traditions is well known, as captured in the colonial Mixtec legal strategies analyzed in Chapter 4, "Ñudzahui Custom, Contracts, and Territoriality...," and the religious festivals associated with the agrarian cycle and interpreted in Chapter 11 of this volume, "Nakua nukuu ini Ñuu Savi..."

Mesoamerican environmental history is rooted in its agricultural traditions; nevertheless, hunting remained an important part of the Mesoamerican heartland and in its extended frontier regions. Hunting and gathering were essential components of Indigenous subsistence strategies, and they sustained their ritual cycles. The symbolic meanings of deer, jaguars, coyotes, and eagles emerge from the pictorial codices and dance cycles related to the cosmologies and oral traditions of distinct ethnic peoples in different ecological settings extending from Central America to northern Mexico. Terracing, cultivation, and defending community forests (*monte*) emerge as long standing poles for understanding the production of space and the environmental dynamics of community life in Mesoamerica and its borderlands, as is illustrated by Chapter 3, "Ihuan yehhuan tlacuauh tlamauhtiah in ichcapixqueh," and Chapter 5, "The *Yoreme* creation of *itom ania* in northwestern Mexico." For the Andean highlands of South America the concept of an archipelago of ecological niches distinguished by altitude and subsistence lifeways has supported the resilience of traditional knowledge systems (Murra, 2002). In the interior of the continent, the riverine borderlands followed the major tributaries of the Paraguay and Río de la Plata basins. The peoples living in these lowlands combined their riverine resources with the grasslands and tropical forests that supported foraging and horticultural economies (Radding, 2019, p. 59–61). Their environmental and linguistic traditions persisted even through the altered circumstances created by the colonial missions and European pastoral and commercial economies, as shown in Chapter 6, " Gender Disparities in Guaraní Knowledge..."

In these diverse ecological settings, Indigenous communities in both the past and recent times have built distinct societies related historically to their natural environments and to the wider polities in which they move. The resilience of their identities and their defense of territories and resources are intimately linked to their living languages and the cultural traditions they express. Striving to understand deeper connections between nature, cultural knowledge, and local languages, we build on the "anthropological experience of culture," recognizing that "culture functions as a synthesis of stability and change, past and present, diachrony and synchrony" (Sahlins, 1985, p. xvii, 114). This is clearly seen in Chapter 3, "*Ihuan yehhuan tlacuauh tlamauhtiah in ichcapixqueh*," Chapter 9 "The Interrelation between Language, History and Traditional Ecological Knowledge...," Chapter 10 "Cenotes and placemaking in the Maya world...," Chapter 11, and Chapter 12 "*Tlaneltoquilli tlen mochihua ica cintli...*" highlighting the cultural continuity anchored in dynamic relationships with the environment, natural resources and community knowledge, which persist despite long-term acculturation processes and the sense of loss. The volume has also been inspired by the methods and insights from linguistic anthropology that views social interactions as mediated by language, both spoken and written, verbal and nonverbal (Ahearn, 2012, p. 3). As seen in Chapter 7 "Combining Visions of Well-Being...," and Chapter 8 "*Amo kitlapanas tetl!*," the sense of ethnic belonging, protection of the natural environment and language practices mutually constitute and reinforce each other in communities' collective perceptions and actions. With its strong focus on environmental and socio-cultural history of the Native people, our approach is fully congruent with the tradition of New Philology that has fruitfully explored colonial history through Indigenous sources, demarginalizing local perspectives and putting in the spotlight conceptualizations of experience communicated and encoded in Indigenous languages (Lockhart, 1991, 1992; Restall, 2003). Expanding on the methods of this important vein of scholarship, Chapters 3, 4, and 5 unlock the potential of complementary Indigenous- and Spanish-language sources for recovering sociocultural microhistories of local communities, their collective knowledge, and actions. By integrating synchronic and diachronic perspectives on Indigenous relationships with nature, we follow the view of history as "the broadest and most flexible of the 'disciplines,'" recognizing that "disciplinary rigidities, orthodoxies, and preconceptions are often the greatest hurdle" (Lockhart, 2000, p. X).

Hence, central to this book is a vivid and critical dialogue not only between different time periods and regions, but also distinct disciplines, methodological approaches and epistemologies. Such a cross-disciplinary and comparative perspective illuminates distinct but counter-balancing conceptualizations of nature and Indigenous environmental strategies. Our approach stems from a multidisciplinary research project funded by the European Union under Horizon 2020, *Minority Languages, Major Opportunities. Collaborative Research, Community Engagement and Innovative Educational Tools*, including scholars and students whose work contributed to this international effort; to enrich the volume further, we invited expert contributors who are not directly involved in the project. In terms of methodological approach, the book aims to produce a dialogue among academics, Indigenous scholars, and local communities to compare and mutually enrich their perspectives on environment, knowledge, identities, and well-being in remote and recent historical periods. Therefore, the contributions to this book are based on original research in the fields of anthropology, history, archaeology, linguistics, and cultural studies. Its authors bring together academics who work on both colonial and contemporary sources and Native scholars who are speakers of Indigenous and minority languages. The participation of two Indigenous scholars and activists (Nahua and Mixtec) has made it possible to include Indigenous research approaches and the languages themselves. Together, the authors are testing new, collaborative, and decolonizing approaches to studying linguistic-cultural heritage, past and present, in its essential relationship to environmental knowledge. By its nature then, our volume is not only rooted in the readings of historical sources, but it also brings together archaeological data, oral history and oral traditions, linguistic research as well as ritual life and different testimonies of agency of Native communities. This synergy of sources and approaches makes it possible to understand more fully the roles of heritage languages in the processes of decolonization of knowledge and research practices as well as the creation of more culturally sensitive epistemological tools. In particular, the contributors offer new insights on environmental/territorial rights, claims, and protection strategies as well as on the links between environment, traditional knowledge, and well-being. Tracing the close integration of nature and culture through historical processes of environmental change in different regions of the Americas, we together address the question of how local knowledge(s) today are nurtured through ancestral languages and oral traditions. We are convinced that posing such questions can be especially fruitful through a comparative historical approach. It facilitates our understanding of processes across time and space along with their causal mechanisms (Schutt, 2006) and deriving "lessons from past experiences that speak to the concerns of the present" (Mahoney & Rueschemeyer, 2003, p. 9). Unlike many published monographs, this is not a policy-directed book on environmental management by Indigenous peoples, but a historically and culturally oriented approach to the production of humanly crafted environments in different time periods and regions. It is grounded in exploring the significance and the roles of local languages and traditional knowledge in the processes of environmental management, and, more broadly, in culturally sensitive ways of interacting with nature, while acknowledging a broad and varied scope of transformations over time. Consciously bridging temporal and geographical constraints, we explore the deep connections between environment, language, and cultural integrity from early modern times to the present. In terms of the geographical spread we focus both on North and South America to showcase the similarities and differences in a wide range of Native American cultures, their responses and adaptations to highly diversified geographic biomes as well as common challenges to the continuity and sustainability of Indigenous communities on both continents today. Thus, our volume fills a unique space: while there is a substantial body of publications using Indigenous-language texts and oral traditions in both anthropology and history, especially in Nahuatl, Mixtec, and Maya, there are few to none comparative edited books that bring together the issues of environment, language, and cultural integrity in Latin American historical and cultural spheres.

The environment as we understand it is nature worked upon by historical and modern communities in culturally sensitive and, sometimes, transformative ways. In this sense, the environment embraces the concept of territory and place linked to the sense of identity and belonging that in turn connects to a holistic sense of well-being. While we set up this broad and strongly cultural definition in our invitation to approach the topics from different angles and perspectives in a collective manner, we also invited the authors to define their own views of the environment/nature as their positionalities, research tools, and data inform them. The authors underscore the ways that local knowledge is expressed and transmitted through ancestral languages, as local communities and Indigenous peoples adapt to change and initiate innovations in their ways of living and relating to their environments. This collaborative approach has made it possible to explore the reciprocal and necessary relations between language/culture and environment; how they can lead to sustainable practices, how language is reproduced, recovered, and maintained in this relationship; how environmental knowledge and sustainable practices toward the environment are reflected in local languages, local sources, and local socio-cultural practices. In accordance with the developing paradigms of environmental history, we also account for the role of nature in historical processes that help to shape cultural and social developments, including ethnic identities. We recognize that "explorations of the various ways in which climate, soils, forests, mountains, rivers, and animals act as 'co-creators of histories'" (Mosley, 2006, p. 917). Accordingly, the recognition of both human and natural agency in historical and contemporary processes, constituting a fundamental paradigm underlying Indigenous worldviews and production of traditional knowledge, has the strong potential for transcending traditional epistemological boundaries in the academy.

Our book is structured in two complementary parts that organize the chapters in broad early modern and modern periods, without imposing a rigid chronology. Part I, "Environment and the Knowledge of the Ancients," brings together five chapters across the broad expanse of the Americas from the Andes to northwestern Mexico. Combining the research methods and interpretive arts of archaeology, documentary history, oral history, and ethnographic field work, the authors of this section of the book develop their views of the sources and meanings of *knowledge* among diverse Native peoples of the Americas in the remote and recent past. Chapter 2, "Flexible borders, permeable territories and the role of water management 'in territorial dynamics in pre-Hispanic and early Hispanic Peru," co-authored by Patrycja Prządka-Giersz, Miłosz Giersz and Julia M. Chyla, places the natural environment of the river valleys flowing from the western cordillera of the Andes to the arid coastal plain at the center of their archaeological research on the complex societies that flourished and declined during the historical phases preceding European contact and the early developments of Spanish colonialism. Water management provides the key to understanding technological advancement, aesthetics, and societal organization that permitted ancient and early modern Andean societies to adapt to the cyclical droughts related to the ENSO phenomenon and to political conflict arising from imperial claims to power among Andean urban centers and European invaders. The authors weave a narrative firmly rooted in over two decades of archaeological and ecological research, in collaboration with local communities, at the same time that they draw meaningful comparisons between the ancient and early modern histories of the Andes and the environmental exigencies that present-day communities face in the region.

Chapter 3, "*Ihuan yehhuan tlacuauh tlamauhtiah in ichcapixqueh.*

And the shepherds are inspiring great fear'. Environment, control of resources and collective agency in colonial and modern Tlaxcala," integrates Justyna Olko's experience in the rich Nahua language sources of Mexican archives with her fieldwork among living Tlaxcalan communities of central Mexico. She explores "complex battlegrounds," in which the Tlaxcaltecah have been actively resisting and counteracting the loss of their control over land, environment, and various kinds of resources, protecting essential assets and components of their well-being. Attempting to offer a *longue durée* view of Tlaxcalan history grounded in specific microhistorical insights, Olko outlines connections between past and present facets of the same longer narrative. This story embraces the struggles against the intrusion of Spanish settlers, loss and degradation of land, access to forests and keeping key traditional practices and ways of healing as fundamental components of local well-being, all in the context of progressing and often catastrophic climatic, environmental, socioeconomic, and demographic transformations. Such microhistories of resilience and resistance reveal how collective threats, be they social, economic, environmental or climatic, generated collective responses, grounded in the ability to maintain and mobilize networks for action. Olko shows how weaving together common threads of Tlaxcalan local history across longer periods of time and different places helps us understand how apparently remote historical processes continue to unfold to this day and matter for the present.

Chapter 4, "Ñudzahui Custom, Contracts, and Territoriality in Eighteenth-Century Oaxaca," by Yanna Yannakakis develops the main themes for this book in the Ñudzahui (Mixtec) region of highland Oaxaca, in southeastern Mexico. Yannakakis illustrates the close historical relationships between humanly crafted environments and territoriality through the legal practices of local custom ("usos y costumbres") and Mixtec adoption and adaptation of the colonial instruments of written contracts. Ñudzahui communities responded to varied policies of the Spanish Crown to privatize Native commons through a genre of social contract (*societas*) rooted in Roman Law, and which shaped medieval and early modern Spanish political thought and contract theory. Rather than regulate economic transactions between parties with antagonistic interests, these consensual contracts instantiated legally binding ties centered on partnership for a common purpose. The contracts' emphasis on bonds of obligation allowed Native communities to canalize relationships rooted in the *yuhuitayu*, the Ñudzahui expression of socio-political and territorial organization, into new forms, thereby modifying their jurisdiction, reconfiguring relations with their neighbors, and establishing new social hierarchies during a period of mounting demographic, economic, and political pressures. These laws of obligation, rooted in medieval practices of *ius commune* (European common law), allowed Native authorities to move strategically between agreement and conflict, maintaining a difficult balance between social harmony and exploitation.

In Chapter 5, "The *Yoreme* creation of *itom ania* in northwestern Mexico: histories of cultural landscapes," Cynthia Radding carries the themes of territory, dwelling in nature, and Indigenous strategies for negotiating with colonial authorities to the northern borderlands of New Spain. Radding affirms that language plays a fundamental role in our interpretation of the rich archival sources that allow us to comprehend the deeply rooted knowledge base that Indigenous peoples developed from the material and spiritual worlds through which they moved in seasonal patterns of migration and dwelling. Moreover, she shows that language is an essential part of the living histories we construct in collaboration with the Indigenous peoples that maintain their traditions in a radically transformed ecological region through ritual cycles and collective memories that derive their meaning from the natural world. This chapter offers new readings of land titles for the colonial provinces of Ostimuri and Sinaloa. Its analysis of changes in land tenure and use that are documented in these archival sources foregrounds ecological conditions and cultural meanings through the dual lenses of environmental and ethnohistorical perspectives. It privileges Indigenous knowledge of landforms, biological species, and the cultural values that the communities of this region ascribed to the physical features and the territorial extension of the spaces they inhabited and defended. It seeks to highlight the parallel production of oral and written sources and, thus, to suggest points of intersection in the languages and modes of communication that are inferred from both colonial documents and ethnographic registers. Finally, its purpose is to contribute a historical analysis that is useful for the Yoreme communities in their present-day defense of their territory and its resources through both documentary and cartographic evidence.

Barbara A. Ganson, in Chapter 6, "Gender Disparities in Guaraní Knowledge, Literacy, and Fashion in the Ecological Borderlands of Colonial and Early Nineteenth-Century Paraguay," brings the concept of gender to her historical analysis of Guaraní knowledge of the natural environment and their selective adoption of European skills and technologies. Centered on education in the mission towns built by both Guaraní villagers and Jesuit missionaries, beginning in the seventeenth century, Ganson shows how the Guaraní sustained significant elements of their spiritual and material cultures beyond Jesuit tutelage through the mid-nineteenth century. Her focus on education includes formal literacy in reading, writing, and mathematics, as imparted in mission schools, artisan skills for both men and women, and the oral traditions that conserved healing knowledge based in the tropical forests across the generations. As she shows in this chapter, gender differences among the Guaraní in the mission towns were made visible by differences in dress, following both Indigenous and Iberian norms of social hierarchies. The primary sources that support Ganson's research include extensive archival documents, published Jesuit histories like *The Spiritual Conquest* by Antonio Ruiz de Montoya (1639/2017), edited and annotated by Ganson, and her own field notes from interviews carried out among the Avá-Chiripá communities on the upper Paraná river. The passages included in this chapter from Guaraní language texts enrich this discussion of language, nature, and enduring communities through the crucible of colonialism and the social and gendered inequalities that it reinforced. The Guaraní voices that emerge from this chapter cross the threshold from the colonial regime to the Paraguayan nation-state.

Part II, "Language, environment, and well-being: contemporary challenges," groups together the contributions that focus on the threats faced by Indigenous communities today with regard to the transmission of traditional knowledge and practices, securing environmental resources and maintaining balanced relationships with nature. One of the most widely shared challenges is an increasing intergenerational gap in the transmission of linguistic, cultural, and environmental knowledge, which brings about multiple negative consequences for Indigenous communities, eroding their sense of individual and collective well-being. As argued by Gregory Haimovich, who conducted field research in local communities in present-day Tlaxcala in Mexico and analyzed the narratives of his interlocutors, their emic conceptualizations of well-being include the vital role of environment, cultural knowledge, and heritage language use. An indispensable background of this Chapter 7 "Combining Visions of Well-Being through the Generational Gap: The Views of Tlaxcala Old and Young on Environment, Tradition and Language" is provided by profound social and economic transformations that deeply affected Native communities in the second half of the twentieth century and to the present: aggressive assimilation pressures, widespread shift to Spanish and monolingualism, economic marginalization of Indigenous communities as well as environmental degradation. Haimovich shows that when approaching community perspectives on their identity and cultural attitudes, it is vital to include the views represented by different age groups of community members. In his research a sense of distrust between the older and the younger generations became quite salient, which hindered them from establishing more effective forms of supportive exchange and collaboration when dealing with the sense of cultural and environmental losses. Among the elderly community members, the disappearance of sources of clean water or reduction of maguey cultivation evoked similarly strong emotions as the loss of their heritage language and they emphasized the connection between the two processes. While they were convinced that the youth were not able to experience a similar sense of loss, since they were born after the devastating process accelerated, in his analysis Haimovich points out that there are in fact many points of convergence in the discourses of both generations that can potentially offer bridges to reducing the perceived gaps and foster intergenerational collaboration. Importantly, this common ground refers to the fundamental components of well-being that link the natural environment and traditional knowledge to the ancestral language and the sense of identity.

The inherent connections between heritage languages and nature are further explored in Elwira Sobkowiak's Chapter 8 "*Amo kitlapanas tetl!*: Heritage language and the defense against fracking in the Huasteca Potosina, Mexico," focusing on an impressively biodiverse region of the Huasteca Potosina in Mexico, also forming an important locus of cultural-linguistic diversity. Despite its ecological potential, the region has been exposed to numerous environmentally damaging projects, the most recent of which involves a highly contaminating method of hydraulic fracturing (fracking) to extract shale natural gas. Looking closely at Indigenous resistance to this project, she explores the relationship between two locally spoken and endangered languages, Nahuatl and Ténck, the defense of Native land and natural environment. Sobkowiak convincingly argues that the anti-fracking movement in the Huasteca reveals the perceived similarity between environmental degradation and cultural-linguistic loss. Interestingly, in the local protests both heritage languages became the symbols of their fight against fracking, and the means of Indigenous mobilization and solidarity. Moreover, the increased visibility of Nahuatl and Ténck in the linguistic landscape started to be perceived as a protective shield for local communities, awakening more positive attitudes toward ancestral languages, still commonly associated with the sense of shame and poverty as the widespread shift to Spanish continues at an alarming pace. This study convincingly pinpoints the deeper, relational connection between heritage languages, nature, and environmental resources fundamental for the survival, sustainability and well-being of Indigenous communities. Therefore, as Sobkowiak aptly proposes, this important link should inform future linguistic-cultural revitalization efforts, integrating them with ongoing struggles for environmental protection as inseparable pillars of local sustainability.

Indigenous languages are not only powerful tools for collective action and resistance, but also provide the means of cultural survival in more nuanced and durable ways: they act as vital reservoirs and carriers of traditional knowledge passed through the generations and they reflect complex historical experiences of their speakers in the face of present and future threats. As shown by Ebany Dohle in Chapter 9 "The Interrelation between Language, History and Traditional Ecological

Knowledge within the Nahuat-Pipil context of El Salvador," the role of the heritage language can be particularly vital in the community that suffered the horror of ethnocide, which resulted in profound historical traumatization and the disruption of the transmission of traditional knowledge. In her study of the complex interactions between ancestral language, migration, environment, and violence experienced by the Nahuat-Pipils, she argues that their suffering and marginalization has also resulted in the preservation of traditional ecological knowledge as a fundamental asset for the survival of the group. Thus, being able to identify plants and name them in the heritage languages helps to maintain the relationship with the land and secure their sustenance, while lexical resources provide useful means for understanding and communicating necessary information about the plants of economic and socio-cultural importance. Despite the exposure to trauma, sense of loss, and pressures toward assimilation, the environmental knowledge encoded within the Nahuat-Pipil language is still necessary and productive within this community as it struggles to maintain balanced relationships with land and to respond not only to the environmental crisis, but also to new forms of violence and (re)colonization.

The transmission of traditional ecological knowledge is central to the team-led, multidisciplinary project carried out by Khristin N. Montes, Dylan J. Clark, Patricia A. McAnany and Adolfo Iván Batún Alpuche with school children for participatory research and community engagement around the unique ecological resource of the *cenotes* in the Yucatán peninsula of Mexico. The Chapter 10 "Cenotes and placemaking in the Maya world: biocultural landscapes as archival spaces," they contributed to this volume rests on the conviction that the health and sustainability of the Yucatec Mayan language is intimately connected with environmental conservation and the value placed on Maya cultural heritage. McAnany and her colleagues received funding from the National Geographic Society to promote the cultural heritage, healthy ecology, and conservation of Yucatec *cenotes* (water-bearing solution sinkholes in a limestone substrate). Linked to *re-patrimonialization* of Indigenous landscapes, this project focuses on middle-school students in nine small communities of eastern Yucatán to learn how students relate to *cenotes*, which historically were a key part of the Yucatec Maya sacred landscape and cosmology, as well as a vital source of life-sustaining water. As students presented their thoughts and concerns about the history and ecological health of *cenotes*, a series of workshops produced instructional workbooks that have helped to prepare local teachers to bring experiential learning about the cultural heritage and conservation of *cenotes* into their classrooms. The heritage of *cenotes* includes their archaeological study, their conspicuous presence in two of the four extant Maya codices from Postclassic times, and the central role of *cenotes* in oral histories and mythic narratives in Maya communities. Indigenous knowledge systems in Yucatán accord well with western scientific ideas about the Great Maya Aquifer that underlies the peninsula and through which *cenotes* are interconnected. Thus, the pollution of even one *cenote* is cause for great concern for the health of the aquifer, and environmental contaminants pose a real threat to the ecosystems of *cenotes*. The environmental science of *cenotes* includes student and teacher-organized water quality testing programs in order to encourage citizen science and inspire young people to take the lead in conserving the biocultural resources of their communities. The fertile common ground encountered during this project suggests that Indigenous and Western perspectives may be effectively braided together to work toward environmental and heritage conservation.

In Chapter 11 "*Nakua nukuu ini Ñuu Savi: Nakua jíno, nakua ka'on de nakua sa'on ja kuatyi Koo Yoso*. Memory and cultural continuity of the Ñuu Savi People: Ancestral knowledge, language and rituals around Koo Yoso deity," Omar Aguilar Sánchez brings a similar sensitivity to the inseparable linkages between language, communal ceremonies, and environmental well-being through his profound knowledge of *Sahan Savi* language and ecological knowledge in the communities of the Mixteca Alta of Oaxaca, Mexico. His chapter aims to reintegrate the cultural memory of *Koo Yoso* in Ñuu Savi (People of the Rain) and to show its diverse meanings from antiquity to the present, written first in Sahan Savi and secondly in English. This diachronic study is supported by the cultural continuity through the Sahan Savi-Mixtec language, showing how the Plumed Serpent is key for the union and identity of the communities and between them and the landscape. *Koo Yoso* is the Mixtec Quetzalcoatl, one of the most important Ñuu Savi deities in the

Mesoamerican pantheon. *Koo Yoso* is represented in pictorial manuscripts as the creator of humanity and founder of rituals and communities. The Plumed Serpent plays an important role today in the well-being of *Ñuu Savi*, despite the effort by friars and Spaniards to suppress the Mesoamerican religion in colonial times. *Ñuu Savi* communities say that storms (*tachi-savi*) start when *Koo Yoso* moves into another house (lakes); in addition, *Koo Yoso* is associated with fertility and abundance. This is commemorated every year by the Mixtec community of Santo Tomás Ocotepec, where the festival has taken on new meanings through the experiences of transnational migration. Today many migrants in México or USA return to Ocotepec to celebrate *Koo Yoso*, to see their families again, and to speak the Sahan Savi language.

In Chapter 12 "*Tlaneltoquilli tlen mochihua ica cintli ipan tlalli Chicontepec: tlamantli chicahualiztli ipan tochinanco.* Ceremonial practices relating to corn in the region of Chicontepec: local aspects of wellbeing," Eduardo de la Cruz conveys the perspective of an active community member on the ritual cycle of maize, a sacred plant and a basic food staple for the Nahuas of Chicontepec, Veracruz. Written in both Nahuatl and English, this chapter explains the details of the ritual cycle that continues to structure the life and work of the community and synchronize it with the cycle of nature. De la Cruz argues that maize ceremonies are not only crucial for securing good harvests and favorable weather, but also transmit fundamental knowledge about social harmony, reciprocity, and the concept of good living, *cualli nemiliztli*. The rituals reveal and perpetuate the principles of *cualli nemiliztli* reflected in the daily practices of harmonious coexistence and respect for all beings. He argues that the prayers conveyed in the Nahuatl language intimately link the lives of the sacred plant and people, as in ancient myths that have shaped the relationships with divine beings, nature, and other humans. However, the continuity of this way of life is at risk as some of the ritual practices are abandoned and become forgotten. The message of the author is therefore directed not only to academic readers, but, above all, it becomes a call for action for the Nahuas of Chicontepec: the further loss of religious practices and traditional knowledge will result in the destabilization of communities, decline of the sense of well-being, and the risk of severing life-sustaining bonds with the environment.

The present book is much more than a collection of related case studies that weave together important threads in environmental history, traditional knowledge, heritage language use, and present challenges faced by Indigenous communities. Perspectives from specific places that bridge past and present offer contextualized, focused, and deepened insights into local sociocultural and natural ecosystems. In turn, they make it possible to form a macroscale picture of Indigenous universe(s) that can challenge simplified generalizations, an ever present threat in historical and anthropological research. While the longevity of colonization processes and their contemporary legacies are widely acknowledged, tangible connections between the pre-conquest developments, the history of the Indigenous peoples under colonial regimes and contemporary communities are seldom foregrounded in the existing scholarship. Reconstructing the long duration of Indigenous history through the richness of its local trajectories is, for us, an essential part of the decolonizing research paradigm that recognizes different forms of neo-colonization and violence affecting many Native groups today. Neo-colonization is particularly salient in the context of ongoing struggles for the preservation of territory, environment, and natural resources, often paired with the fight for linguistic-cultural survival and integrity. Native communities continue to be exposed to state-sanctioned extractivism of their resources and to criminal violence (Makaran & López, 2018). One of its obvious sources are the cartels' struggles to gain control over resource-rich Native regions, propitious for the cultivation of marijuana and opium poppies and providing strategic trafficking corridors (Ley et al., 2019).

Looking at the past is often a key for understanding the challenges of the present, as we increasingly recognize the need for social justice and solidarity with historically oppressed peoples. We hope that stimulating a fruitful dialogue between Western academics and Indigenous scholars helps unveil the implications and potential for exploring deeper connections between sustainable relationships with nature, place-based knowledge, language vitality, and the sense of well-being. The awareness of this relationality is crucial for counteracting dispossession, (neo)colonial appropriation, and the loss of cultural diversity.

References

Ahearn, L. (2012). *Living language: An introduction to linguistic anthropology*. Wiley.

Carr, H., & Lipscomb, S. (2021). Prologue: Ways in. In H. Carr & S. Lipscomb (Eds.), *What is history, now? How the past and present speak to each other* (pp. 3–16). Weidenfeld & Nicolson.

Cronon, W. (1996). *Uncommon ground: Rethinking the human place in nature*. W.W. Norton.

Crosby, A. W. (2001). Foreword. In T. Myllyntaus & M. Saikku (Eds.), *Encountering the past in nature essays in environmental history* (pp. xi–xvi). Ohio University Press.

Dean, W. (1995). *With broadax and firebrand*. University of California Press.

Denevan, W. M. (2002). *Cultivated landscapes of native Amazonia and the Andes: Triumph over the soil*. Oxford University Press.

Ingold, T. (2000). *The perception of the environment: Essays in livelihood, dwelling, and skill*. Routledge.

Ley, S., Mattiace, S., & Trejo, G. (2019). Indigenous resistance to criminal governance: Why regional ethnic autonomy institutions protect communities from Narco rule in Mexico. *Latin American Research Review, 54*(1), 181–200. https://doi.org/10.25222/larr.377

Lockhart, J. (1991). *Nahuas and Spaniards: Postconquest Central Mexican history and philology*. Nahuatl Studies Series 3. Stanford University Press.

Lockhart, J. (1992). *The Nahuas after the conquest: A social and cultural history of the Indians of Central Mexico*. Stanford University Press.

Lockhart, J. (2000). *Of things of the Indies, essays old and new in early Latin American history*. Stanford University Press.

Mahoney, J., & Rueschemeyer, D. (2003). *Comparative historical analysis in the social sciences*. Cambridge University Press.

Makaran, G., & López, P. (2018). *Recolonización en Bolivia: neonacionalismo extractivista y resistencia comunitaria*. Universidad Nacional Autónoma de México, Centro de Investigaciones sobre América Latina y el Caribe,

Bajo Tierra Ediciones. https://rilzea.cialc.unam.mx/jspui/handle/CIALC-UNAM/L53

Mosley, S. (2006). Common ground: Integrating social and environmental history. *Journal of Social History, 39*(3), 915–933.

Murra, J. (2002). *El mundo andino: Población, medio ambiente, y economía.* Pontificia Universidad Católica del Perú.

Radding, C. (2019). Crafting landscapes in the Iberian borderlands of the Americas. In A. Danna, L. Rojo, & C. Radding (Eds.), *The Oxford handbook of borderlands of the Iberian world* (pp. 58–82). Oxford University Press.

Radding, C. (2022). *Bountiful deserts: Sustaining indigenous worlds in Northern New Spain.* University of Arizona Press.

Restall, M. (2003). A history of the new philology and the new philology in history. *Latin American Research Review, 38*(1), 113–134.

Sahlins, M. (1985). *Islands of history.* University of Chicago Press.

Schutt, R. K. (2006). *Investigating the social world: The process and practice of research.* Sage.

White, R. (2004). From wilderness to hybrid landscapes: The cultural turn in environmental history. *The Historian, 66*(3), 557–564.

Part I

Environment and the Knowledge of the Ancients

2

Flexible Borders, Permeable Territories and the Role of Water Management in Territorial Dynamics in Pre-Hispanic and Early Hispanic Peru

Patrycja Prządka-Giersz, Miłosz Giersz, and Julia M. Chyla

Introduction

The Peruvian Andes are well known for their ethnic and highly compressed ecological diversity (Sandweiss & Richardson, 2008). This diversity has shaped the degree and kind of human mobility and interactions that played a crucial role in social and political developments for millennia (Roosevelt, 1999). Ancient and modern Andean societies faced and continue to face various crises of climatic, social, political, or economic nature. Those crises forced the population to establish different types of inter-group relationships, identities, and forms of agency, and resulted in a broad range of competitive/cooperative behaviors across the varied social and physical landscapes, including warfare, trade, alliance-building, co-residence, and any combinations of these and other practices. That past Indigenous knowledge is manifested in the long-term adaptation to the natural

P. Prządka-Giersz (✉) · M. Giersz · J. M. Chyla
University of Warsaw, Warsaw, Poland
e-mail: p.przadka@uw.edu.pl

© The Author(s) 2024
J. Olko and C. Radding (eds.), *Living with Nature, Cherishing Language*,
https://doi.org/10.1007/978-3-031-38739-5_2

environment, which can be traced by archaeological and ethnohistorical sources (Bethell, 1984a, 1984b; Salomon & Schwartz, 1999; among others).

Access to vital resources (like water, land, plants, marine resources, and wildlife) has always been a major concern, especially in the vast desert areas on the coast of today's Peru, located between the Andean flanks and the Pacific Ocean. Since ancient times, the people who inhabited the coastal region of Peru had to undergo processes of adaptation to ensure optimal living conditions and thus be able to develop their societies. As is observed in our own times, Andean peoples had to face various natural crises, such as earthquakes, tsunamis, or El Niño phenomena (ENSO/El Niño Southern Oscillation) that influenced numerous changes in the settling of the coast, causing floods and intense rains that in turn strained human migration processes. Some researchers consider that the impact of various ecosystems, climate change, and the El Niño phenomenon in the pre-Hispanic communities of the Andes was a key factor in the sedentarization, the rise and fall of polities and cultural continuity in individual regions, particularly in the desert areas of the Pacific coast (Moseley, 1975; Moseley & Feldman, 1982; Nials et al., 1979; Sandweiss & Quilter, 2009; among others). The access to drinking water in sufficient quantity to supply the population with agricultural products was also important, especially in the desertic Pacific coast where water resources are diminishing significantly during the dry season, creating in this region one of the most unreliable water supplies in the world. However, as different studies show, in pre-Hispanic times the Indigenous peoples had various strategies and technologies to address all these difficulties and problems. This situation conditioned many changes, among them the general reorganization of the Andean social space. The population dispersion that characterized the ancient communities formed a system that helped to explore a maximum number of ecological niches to achieve a multi-production of diverse goods (Murra, 1972).

Indigenous knowledge accrued over centuries of developing experiences and practices in the use of land, water, and other natural resources for agricultural needs. Hence, paradoxically, modern scholars demonstrate the need to make use of ancient knowledge and modern science to help prevent natural and anthropogenic disasters on the Peruvian coast (Buytaert & De Bièvre, 2012; Buytaert et al., 2007, 2014; among others). It can be assumed that the profound understanding of fragile environment and skillful dealing with natural crises was grounded in the Andean oral tradition and religious beliefs. The Andean concept of *pachacuti*, deeply rooted in the Inka religion (and most probably pre-Inka religions as well), apprehends the relationship between order and chaos to explain how and why various forms of time, meaning, and power come to be upset and overturned. Whether past, present, or future, the *pachacuti* "moment" (a time of upheaval, end of the world) represents a new epoch and a change in social and political order, whether constructed as a positive or negative outcome (Flores Galindo, 2010, pp. 22–27).

In the coastal valleys, side ravines reinforced by defensive and anti-alluvium (*huaycos*) walls were used to support terraces for housing and other uses (camelid breeding, agriculture, and production areas, among others). Rapid communication between these scattered centers in different regions was made easier by the developed network of roads and trails, and the construction of a dense network of irrigation canals helped to enlarge the agricultural acreage. Astute use of diverse ecological and geographical features, such as riverbanks, side ravines, fossil terraces, and natural springs (*puquios*) was a strategy designed to prevent the recurrence of social crises caused by various natural catastrophes.

Researchers in cultural geography, archaeology, and history have long debated human–environment relationships in the central Andes, engaging a full range of methods and theoretical frameworks. Archaeology has taken an interest in humanly crafted environments almost since its origins. In Peruvian archaeology, natural processes, including long-term climatic shifts and abrupt events, such as the El Niño, prehistoric earthquakes, tsunamis, massive debris flows, and volcanic eruptions has been a focus of many influential publications (e.g., Kosok, 1965; Moseley, 1975; Murra, 1972). Taking into consideration both early advances in archaeology and a recent work that investigates themes of broad archaeological relevance, the central Andean environments have been variously conceived as structuring, modified, and sacred, thus examining both ecological ("environment") and ideological ("landscape") implications of archaeological landscapes (Contreras, 2010). Thus, landscape is understood as a "palimpsest," a constantly overwritten manuscript, with overlapping traces of consecutive human activities (Bailey, 2007). In this sense the modern landscape, what we see today, stores information of interaction of past societies with their natural surroundings in the form of archaeological relics like settlements, public and defensive architecture, roads, or canals. Studying and documenting distribution and function of those relics helps to understand more fully human–environmental interplays throughout the millennia.

Research on such pre-Hispanic human–environment interactions in the ancient central Andes was mainly based on archaeological field prospection. The character of Andean environments has also influenced the development of specific research methods, including an innovative archaeological field prospection workflow (Willey, 1953), where archaeological relics were located through aerial photographs, verified in the field and described by their function and chronology. In the past decade, development of digital archaeology, access to remote sensing and non-invasive methods, and increasing use of Geographic Information Systems (GIS), have transformed field prospection methodologies. Identification of archaeological sites through archival photos to prepare archaeologists for fieldwork in advance started to be supported by other data, like satellite images and their analysis with the use of multispectral indexes, Digital Elevation Models (DEM) which represent modern landscapes. The visibility analysis, created on the base of DEMs, helps in understanding how past societies perceived their surroundings and what elements of the environment were important for them to be more exposed or hidden in the landscape (Gillings, 2015, 2017). Parallel to this process, researchers started to pay attention not only to developing a better understanding of the archaeological landscape, but also to document and counteract the destructive processes, e.g. looting, urbanization, and agricultural development (see Brodie & Renfrew, 2005; Casana, 2015; Chyla, 2017; Contreras, 2010; Lauricella et al., 2017; Tapete et al., 2016).

In this chapter, using above-mentioned spatial, diachronic, and multidisciplinary approaches, and employing fieldwork data from documentation of the archaeological relicts located within modern landscapes of the Culebras Valley, we try to understand more clearly the nature and outcomes of distinct group interactions across the varied social and physical landscapes of the desert coast and highlands. Through the spatial–temporal distribution of different types of archeological sites presented below, we examine the different entanglements that shaped the geo-political landscape of this area throughout the pre-Hispanic and early Hispanic periods, from the beginning of the first millennium AD until the arrival of the Spaniards. In terms of functional, political, and economic perspectives, our research sheds new light on the role of environmental resource management for the emergence and maintenance of different pre-Hispanic complex societies, exploring these topics tackled from both economic and symbolic perspectives.

The Culebras Valley and Its Environmental Setting, Past, and Present

The province of Huarmey is located in the extreme southern part of the north coast of Peru, about 300 km from its capital, Lima. This province reports diversified life zones, which are distributed in two natural regions of the Andes: the *Chala* and the *Yunga Marítima* (Pulgar Vidal, 1996). This part of the coast is characterized by the presence of areas of deep ravines, rocky and sandy deserts, and finally narrow coastal valleys—being the valleys of the Culebras River and the Huarmey River, the two most fertile in the area. The predominant coastal desert landscape of the Huarmey Province is due to the influence of two factors: the Humboldt current of cold waters that prevents the rains and the barrier of the Andes mountain range that prevents the passage of the rains from the eastern slopes of the Cordillera Negra.

The Culebras Valley is one of the smallest on the Peruvian coast, as it barely exceeds 40 kilometers in length. The valley rises near the town of Quián, at the junction of two major ravines, Huanchay and

Cotapuquio (Prządka & Giersz, 2003, pp. 15–16). The Culebras basin has a much larger area, covering some 671 km² and reaching a height of 4483 meters above sea level at its headwaters in the Saccho lagoon (National Institute of Natural Resources, 2007, p. 127). The valley is narrow, with steep and unstable slopes. Only in some sections it widens, making agriculture possible on the riverside lands and on the terraces of the slopes. The most open part and the largest cultivable area is its delta, enclosed by discontinuous hills, pampas, and wind erosion surfaces. The hills and mountains generally correspond to the final western foothills of the Cordillera Negra, and together with the plains, make up the great morphological physiographic ensembles of the coast. In the strip immediately to the coastline, dunes resulting from wind processes that carry materials from the beaches and bays accumulate.

The Culebras River has a markedly irregular stream flow, depending on the rains in the mountains or the events related to the appearance of the El Niño phenomenon. Like other rivers on the north coast of Peru, the Culebras has water only for five or six months of the year (November–April). When it comes to the hydrographic characteristics of the coastal valleys of the province of Huarmey, it should be noted that the current agricultural industry had a significant impact on the natural ecosystem (reed beds, carob trees), requiring a sufficient layer of vegetation to store water. The swamps and lagoons that formerly existed in the lower part of the valley helped to store water, but unfortunately today these places continue to disappear because they have been burned, dried, and converted into crop fields.

The environment is characterized by an extremely arid and semi-warm climate, with an average annual air temperature of 18–19 °C on the coastal coastline, down to levels below 3 °C in the high Andean areas of the Pariacllanque (Alto Culebras). Precipitation on the coast does not exceed 30 mm/m², except in the highest part of the basin, where annual precipitation can reach levels greater than 500 mm/m² (National Institute of Natural Resources, 2007, pp. 31–36). Most of the land suitable for cultivation is irrigated with water from the subsoil and from natural springs (*puquios*).

From the economic point of view, the studied region shows evidence of important non-renewable natural resources, both metallic (silver, lead, zinc, copper, antimony, iron, and tungsten) and non-metallic (limestone, pegmatites with quartz, salt deposits). However, the weight of local economic production, both in pre-Hispanic times and today, falls on agriculture and fishing. The dominant edaphic landscape in the Culebras and Huarmey valleys is closely related to the physiographic characteristics. In general terms, the land suitable for permanent cultivation in the Culebras Valley covers an area of 4000 hectares, distributed mostly in the lower part of the basin (National Institute of Natural Resources, 2007, p. 42). The climate of the Culebras Valley also allows acclimatization and production of most of the vegetables that grow in other climates on earth. Currently, the traditional staple crops, in particular corn (*Zea mays*), manioc (*Manihot esculenta*), sweet potato (*Ipomea batatas*), or gourd (*Cucurbita* sp.) are replaced by asparagus (*Asparagus officinalis*) and by fruit trees for wider regional and international commercial profit (Szpak et al., 2013). As the old inhabitants of the valley affirm, in the times of full prosperity of the old haciendas in the area, the grape vine was also cultivated with the purpose of producing wines and pisco. In pre-Hispanic times, such a warm climate was generally considered favorable for coca cultivation on the coast (Rostworowski, 1989).

The interval coastal sectors, located between the valleys of Culebras and Huarmey, as well as in the southern and northern limits of the Province of Huarmey, between approximately 300 and 700 meters above sea level, remain seasonally covered by clouds forming the *lomas* vegetation in the coastal hills (Ferreyra Huerta, 1953; Oka & Ogawa, 1984; Pulgar Vidal, 1996; Weberbauer, 1945; among others), seasonally concentrating plant and animal communities (tillandsias, mosses and lichens; reptiles, mollusks, arthropods, crustaceans, and myriapods). In ancient times, the *lomas* hills—which can expand during El Niño events (Wells & Noller, 1999)—once offered refuge to herds of guanacos (*Lama guanicoe*) and white-tailed deers (*Odocoileus virginianus*), as well as their predators, such as mountain lion (*Puma concolor*). The hills could have had a primordial importance for pre-Hispanic peoples because it was the only local ecosystem that did not require irrigation to develop agriculture and camelid grazing (Dufour et al., 2014; Thornton et al., 2011).

Territorial Dynamics in a Local Context: The Case of the Culebras Valley

The province of Huarmey was relatively little studied through archaeological methods, especially regarding its pre-Hispanic past. Although the first data on its antiquities come from the time of contact with the Europeans (Cieza de León, 2005 [1553]; Pizarro, 2013 [1571]; Mogrovejo, 2006 [1593–1605]; Calancha, 1976–1981 [1638]), who describe some aspects of the life and customs of the Native people the first information about the archaeological sites in this area was mentioned by buccaneers and adventurers of the seventeenth century, such as Joris van Spilbergen (2014 [1619]) or Lionel Wafer (2004 [1903]), and the seminal archaeological investigations were carried out by Julio C. Tello (1919) who visited the Province of Huarmey during his first expedition to the Huaylas region in 1919. Later, archaeological surveys undertaken by Donald Thompson (1962, 1966), Hans Horkheimer (1965) and Heinrich Ubbelohde-Doering (Prümers, 1990), Duccio Bonavia (1982), and Heiko Prümers (1990) focused on the Huarmey Valley; however, little research has been done since then in the adjoining Culebras Valley. Nevertheless, it is necessary to emphasize that although the part located on the ecological and cultural border between the central and north coast of Peru did not receive much attention from archaeologists—at least compared to other coastal valleys—it was a focus of interest of treasure seekers and tomb raiders. Locally known as *huaqueros*, they affected a large part of the important archaeological sites. Since 2002, Polish and Peruvian archaeologists led by the authors of this chapter, have been conducting a series of investigations in the Huarmey Province. The first stage of this interdisciplinary research program focused on archaeological surveys and excavations in the Culebras Valley. The Culebras Valley Archaeological Research Project (CVARP) consisted of a detailed survey, as well as archaeological excavations of key sites in the area. Thanks to these surveys, 109 archaeological sites have been identified in the lower valley and 43 in the upper Culebras basin, of which 140 were previously unregistered. Those archaeological findings provide new insights into occupational and territorial dynamics, as well as natural resource management strategies over many periods of pre-Columbian chronology. In this text, however, we will focus on the last millennium and a half of Indigenous cultural development before contact with Europeans (100–1532 AD), when the north coast of Peru witnessed the rise and fall of many pre-Columbian complex societies. With the help of digital archaeology we hope to bring more details on human–environmental interactions within the landscape and how past societies saw, understood, and functioned within their surroundings and how it changed with time. Research on the Culebras Valley, presented in this chapter, shows that through such analysis of the distribution of archaeological sites, we can try to interpret changes of pre-Hispanic environment and how past societies interacted with it.

The Early Intermediate Period (100–700 AD)

During the Early Intermediate Period (100–700 AD), throughout the northern Peruvian coast, from the Piura Valley to the Huarmey Valley, regional artistic styles were consolidated, and the first powerful regional state called Moche dominated the region. Moche civilization flourished between approximately AD 100 and 800. The Moche left a vivid artistic record and developed sophisticated ceramics, metallurgy, and weaving. They inhabited a series of valleys surrounded by an arid coastal plain. Despite some uncertainty about the Moche presence south of the Nepeña Valley and its monumental center of Pañamarca, in the Culebras Valley the archaeological evidence corresponding to the Moche presence is very well marked. Thanks to the surveys and excavations carried out by the CVARP, it was possible to register many archaeological remains and excavate key sites, as well as review private collections (Giersz, 2007; Giersz & Prządka-Giersz, 2008; Prządka & Giersz, 2003; Prządka-Giersz, 2009; among others). In the Culebras Valley, during the first part of the Early Intermediate Period (the local Mango Phase [100–400 AD]; Giersz & Prządka-Giersz, 2008), as in other parts of the north coast, along with undecorated utilitarian pottery stylistically related to the examples attributed in other valleys to the Virú-Gallinazo style, the classic vestiges of the Early Moche style appear. In the second part of the Early Intermediate Period (the local Quillapampa Phase [400–700 AD]; Giersz & Prządka-Giersz, 2008), when the Moche elites of the Chicama and Moche Valleys organized themselves to form the first state (Bawden, 1994; Shimada, 1994) and the primary centers, such as the Huacas del Sol y la Luna and Huaca Cao Viejo in the El Brujo complex, acquired their urban and monumental characteristics, Moche III and Moche IV ceramics appear in the Culebras Valley, in the funerary contexts registered in association with elite architecture. In primary centers of the Early Intermediate Period in the lower Culebras Valley, in both the Mango and Quillapampa sites (Fig. 2.1), the first construction phase corresponds to the provisional campsite with enclosures made with *quincha* cane walls. During the second phase it was replaced by monumental stone and *adobe* bricks architecture. The total number of the Early Intermediate Period archaeological sites in the Culebras Valley reaches 33. Among the sites, it is possible to distinguish 2 primary centers with public architecture, 2 secondary centers, 14 village settlements of less than one hectare of surface, 7 cemeteries, 2 hilltop defensive settlements, and 6 watch points (Fig. 2.1). There are no fortifications, except for two hilltop settlements located in the lower part of the valley, which could fulfill the double function of settlements and fortresses. A peculiar, direct visual surveillance system, for military and movement monitoring, was made up of six watch points that guard the accesses to the main nucleus of Moche public centers in the lower-middle valley from the mountains and to the main north–south road that crosses the side ravines. Neither the village settlements nor the hypothetical elite residences have defensive characteristics, they are all located near the valley floor, in open and non-defensible areas. The primary Moche center in the valley was the Quillapampa (Fig. 2.4a), located on the main north–south inter-valley coastal pre-Hispanic road. Our archaeological excavations at Quillapampa I have revealed the presence of Moche palatial building of the so-called *audiencia* type, with stone retaining walls, access ramps, and peculiar roof decoration made with ceramic maces. A Moche burial chamber was hidden beneath the solid clay floor. The palatial residence visually dominates the upper-middle part of the valley where most Moche archaeological sites of different characteristics and functions are located, including rural settlements, pottery workshops, cemeteries, and adobe enclosures (Giersz, 2007, pp. 198–217). In the upper part of the Culebras basin, by way of contrast, no archaeological sites related to the Moche tradition were recorded. Furthermore, the settlement pattern reveals that this part, being a buffer zone between the coast and the highland, has very little human occupation. Only two Early Intermediate Period archaeological sites were recorded in that part of the region.

The settlement pattern from the Early Intermediate Period in the Culebras Valley clearly indicates that human occupation was centered in the lower part of the Culebras basin and is related to the strong Moche presence in the coast. The sites have no defensive features and are generally located near the valley floor, in open, non-defensible areas. The only defensive sites are the hilltop settlements and lookout posts, spread out in very strategic locations. In fact, they dominate the entire

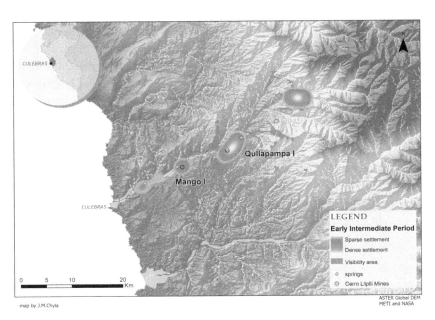

Fig. 2.1 Settlement pattern and visualscape map of the Early Intermediate Period in Culebras Valley. © Julia Chyla

valley floor. The GIS analysis of visualscape suggests a direct relationship between the archaeological sites and the control of intra and inter-valley communication routes (Giersz, 2012). There was no clear integration between the coast and the highlands. The Indigenous people settled in the Culebras Valley during the Early Intermediate Period had a strong agricultural vocation and they occupied the small villages centered around the best farmland and provided with the most abundant water natural springs (*puquios*). Local access to convenient clay outcrops used to produce ceramics and *adobe* mud bricks, as well as important mineral deposits (copper, gold, and silver mines) were also of great importance.

The Middle Horizon (700–1050 AD)

Between the eighth and ninth centuries AD the north coast of Peru saw the decline and final collapse of Moche culture. It did not happen suddenly, nor was it the same throughout the land the Moche occupied. There are different proposals in the scientific literature regarding this issue. One of them considers that the Wari cultural phenomenon was the major dynamic element in the process of reorganization that took place inside the Late Moche groups. The proponents of this interpretation support this proposal with the arrival of highland and Central Coast stylistic elements to the North Coast (Donnan, 1973; Menzel, 1964) as well as the apparent introduction of architectural models supposedly of highland origin (Schaedel, 1951). The second proposal instead views the process of social and political reorganization that the North Coast experienced in the Middle Horizon as a result of the restructuration of the late Moche polities under foreign influences (Bawden, 1994, 1996; Castillo Butters, 2001; Shimada, 1994; among others). Other scholars have attributed the decline of Moche culture to changes in the climate brought about by a paleo-ENSO (Shimada, 1994).

Our recent research at Culebras and Huarmey Valleys, especially those at Castillo de Huarmey archaeological site, clearly showed that this north coast border area had a major role in the attempt to incorporate this region into the Wari Empire. Castillo de Huarmey is the first excavated example of a large Wari mausoleum and site of ancestor worship on the Peruvian North Coast, an area that lies on the borders of the world controlled by the first Andean empire. The burial chamber contained fifty-eight funerary bundles of noble women and over one thousand three hundred exceptionally rich objects that formed the ceremonial offerings and grave goods (Giersz & Pardo, 2014). This discovery provides important data regarding the identity of high-status women and the political structure of the Wari Empire, especially in this part of the coast (Giersz, 2017; Przadka-Giersz, 2019). Evidence of the observed change in the funerary pattern, and the appearance of monumental funerary architecture visible and accessible from the surface, called *chullpa*, is one of the hallmarks of this strong sociopolitical change (Giersz, 2017). Questioning of the preceding religious and funerary practices is also interpreted as somehow related to the change of paleoclimatic regimes on a large scale (Lau, 2016).

In the Culebras Valley, the Wari episode was marked by a sharp increase in cultural activity and population as compared to the previous periods. Moche elite residences are abandoned or turned into cemeteries. There is a notable change in the location of settlements. The densely populated area moves to the lower valley and its center is located near the modern town of Molino, where the new north–south inter-valley road was designated. The new road axis ensures communication with the main Wari provincial center of Castillo de Huarmey, located in the neighboring Huarmey Valley to the south. We found a total of 54 discrete sites belonging to this period (Fig. 2.2). These sites were distributed widely across the valley in both branches of the Culebras River. In the lower Culebras Valley, the sustained growth of the number of registered sites begins: the total number of sites reaches 44, of which we can differentiate 1 primary center, 3 secondary centers (among them, 2 hilltop sites), 16 settlements, 21 cemeteries, 1 defensive hilltop, and 2 watch points. In the upper valley, on the other hand, a sudden change in occupational dynamics is noted compared to the previous period. The settlement pattern is made up of 1 primary center, 1 secondary center, 5 settlements, including one hilltop site with defensive characteristics, in addition to 3 watch points, strategically located on the tops of the highest mountains that visually dominate the entire network of settlements and roads.

Evidence from the Culebras Valley suggests that the Middle Horizon is a period of population and economic growth. The increase in the number of settlements and their location at the bottom of the valley, and in places that are difficult to defend, indicate in turn that this was a time of strong political integration on a macro-regional scale. The change in the settlement pattern and the appearance of exotic Wari goods at the new administrative centers, with orthogonal architecture, such as Cerro León I in the lower valley and Añil Punta in the upper valley, seems to imply that a new authority of foreign origin has managed to impose itself and exercise power directly. Changes in funerary patterns point in this same direction. Construction of new type of necropolis with *chullpas* funerary towers facilitated the celebration of a new cult of the dead and legitimized the authority of

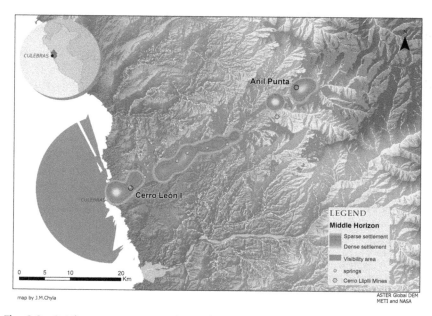

Fig. 2.2 Settlement pattern and visualscape map of the Middle Horizon Period in Culebras Valley. © Julia Chyla

mummified ancestors (most probably of foreign origin) in a symbolically shaped landscape.

However, the construction of fortified sites suggests the existence of possible conflicts with political entities located north of Culebras. The location of the sites at the bottom of the valley and in places that are not very defensible indicate that the defense system based on visual control of the occupied territory was effective. Compared to the Moche domain, the Wari occupation imposed a very different mark on the cultural landscape of the Huarmey and Culebras valleys, which necessarily implies differences in the strategy of power. Our recent findings in Castillo de Huarmey (Giersz, 2017, Prządka-Giersz, 2019; among others) provide arguments in favor of a successful conquest of this territory by the warriors from the south. As a hypothesis, the evidence from Culebras also suggests that the Wari administration converted this valley into a fortified frontier, during an episode of preparation for the conquest of other northern valleys.

The Late Pre-Hispanic Periods (1050–1532 AD)

At the dawn of the eleventh century AD, diverse cultural processes and sociopolitical transformations occurred in the Central Andes. Their consequences lasted until the conquest by the Incas and in some cases, even until the first contact with Europeans in the sixteenth century. This period appears to mark the rise of independent chiefdoms on an unprecedented scale, characterized as much by their artistic richness, funerary customs, and settlement patterns, as by their ability to mobilize and coordinate a large workforce. The cultural transformations were accompanied by the growth of the diversity of specializations and production techniques. Those social and political reconfigurations were largely due to climatic and environmental changes on the coast that occurred around the tenth–eleventh centuries AD, as a consequence of arduous events related to the El Niño Southern Oscillation (ENSO). Experts agree that every pre-Hispanic Andean environmental crisis (Moseley & Feldman, 1982) was followed by political and ideological changes and transformations. It's not surprising that after aforementioned climatic events, characterized among others by higher temperatures and rise in groundwater levels, favorable conditions were created for agricultural expansion and population growth, which is reflected in the size of settlements and in the construction of administrative centers of particular importance, such as Pacatnamú, Chan Chan, or Túcume in the north coast of Peru. According to archaeological evidence, population growth was generally conditioned by the expansion of agricultural fields and by some technological advances, such as the introduction of arsenical bronze as a raw material for making agricultural tools (Hocquenghem & Vetter, 2005). For the first time in Andean prehistory, high-level irrigation systems were developed, such as La Leche-Zaña-Lambayeque, connecting three river basins on the Peruvian north coast. It is very remarkable that these canals are sometimes still in use, and some of them, such as the Taimi or the Reque, can be mistaken for rivers due to their huge dimensions (Kosok, 1965; Kus, 1972).

Our archaeological work in the Culebras Valley revealed that after the collapse of the Wari empire, the coastal area between the Chao and Huarmey valleys also witnessed great sociocultural transformations related to the emergence of a new cultural entity commonly known as the Casma culture. As in the entire north coast, the Culebras Valley experienced population growth that was greater than in any other periods in the prehistory of the region, which is reflected in number, size, and function of human settlements, as well as the growth of agricultural potential (for this period, a minimum of 31 plant species were identified, 17 of which were regularly cultivated and consumed; Prządka-Giersz, 2009). Also, several features of the settlement pattern suggest that the valley reached its maximum sociopolitical complexity during that time (Fig. 2.3).

We found far more sites dated to the local Ten Ten Phase (1000–1450 AD) than to any other time in the entire sequence—with a total of 58 occupations (including 2 primary centers, 25 settlements, 14 cemeteries, 10 hilltop settlements, and 7 hilltop watch points) distributed throughout both branches of the Culebras River. Most of them are in strategic locations: in entrances to dry side ravines intentionally protected against land/mud slides by stone walls, on hill

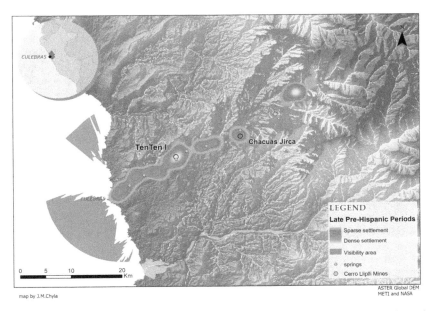

Fig. 2.3 Settlement pattern and visualscape map of the Late Pre-Hispanic Periods in Culebras Valley. © Julia Chyla

slopes, and on natural flattened heights with a commanding view of the valley. An increased density of settlements is also related to the road network created during this period. The nearly continuous distribution of sites appears to support an argument for valley-wide sociopolitical integration. There appears to be a distinct hierarchy of site size and function during the Late Pre-Hispanic Periods—with Ten Ten being the primary site throughout all local chronological phases of the epoch. This center is situated on the northern border of the valley, at the entrance to a small adjacent side ravine, some 16 km directly east of the Pacific Ocean, in the *Yunga Marítima* natural habitat, a short distance from an important *puquio*, a water source active throughout the year, which during ancient times ensured the constant supply of water resources throughout the year. The site extends over 100 hectares (Fig. 2.4b). Architectural compounds at Ten Ten are clearly multifunctional, incorporating space for administrative complexes, temples, fortifications, and residential settlements, as well as public

Fig. 2.4 a Quillapampa archaeological site within the Culebras Valley landscape. b Ten Ten archaeological site within the Culebras Valley landscape. c Chacuas Jirca archaeological site within the Culebras Valley landscape. d a newly constructed neighborhood in the center of a dry riverbed at the outskirts of the Huarmey city; © Miłosz Giersz

warehouses to accumulate and distribute food, as well as other necessary products. Minor hilltop sites surround the main center. The results of GIS analysis of visualscape suggest that all the fortified sites must have served as sentry points for the main north–south coastal route that ran directly through the center of the site, suggesting that Ten Ten could be an important spot on a larger, regional road network, which controlled transports of goods from north to south.

During the Inca influence in the Culebras Valley the Casma culture sites were adapted and transformed. New sites arose mainly in the upper part of the valley, particularly around local copper and silver deposits in the vicinity of Cerro Lliplli. These were mainly administrative centers, with workshops and imposing warehouses built according to new architectural canons typical of Inca constructions. During this last episode before the Spanish Conquest (the local Chacuas Jirca Phase (1450–1532 AD) the Chacuas Jirca center, located in the upper-middle part of the basin, at the foot of Cerro Perolito, takes on importance. The settlement is made up of four sectors of terraced structures with administrative, residential, productive and possibly ceremonial functions that extend over 27 hectares (Fig. 2.4c). As in the case of the center of Ten Ten, Chacuas Jirca is located in a strategic point from which the basin of the Culebras River and the natural path to the highlands were controlled. The adjacent watch points on the high tops of the hills offer generous visual control of the side ravine that leads to the local quartz, gold, and copper mines located in the Cerro Lliplli area. Additionally, near the center of Chacuas Jirca there was a very large lagoon, which still preserves a water source covered by a dense reed bed.

It is evident that during the Late Pre-Hispanic Periods the environment was very important in both cultural and symbolic dimensions. The studies revealed that the main centers had a close relationship with the natural environment, especially with the surrounding highland landscape of the upper valley, being located at the foot of the rocky mountains that are distinguished in the morphology of the valley or stand out for their shape and size. In the local legends and myths, these mountains represent divinities, ancestors, heroes, and heroines. There is archaeological evidence that Indigenous people worshiped them and made offerings of food, vessels, Spondylus shells, anthropomorphic (male and female) clay figurines, and beads of semi-precious stones. Furthermore, those sites were connected by alternative routes that passed through the slopes of the hills and were employed at times when the water level in the river was significantly high. What is interesting, some of these roads and paths in the highlands are still used by rural communities.

Conquest and Aftermath

The spatial reorganization of the Native population carried out a couple of decades after the Spanish conquest on the north coast led to the creation of new settlements or nucleated Indigenous towns built in the European style. This contributed to the transformation of the old settlement patterns and the ways of conceiving the occupation of the territory, as well as the use of natural resources. Very often, as a result of colonial policy, people were moved to different places and ecosystems (generally from the highland area to the desert coast), sometimes at distances of a hundred kilometers, losing ties to their community, beliefs, cultural identity, and traditional habitat. These changes were aggravated by intense demographic transformations related to abandonment of dispersed settlement landscapes in spatial planning, especially on the coast in the vicinity of Spanish towns. The old system that contained and prevented natural disasters was no longer maintained in the new economic policy of the colonial era. Likewise, the *quollqas* public warehouses that served in cases of emergency were no longer maintained, nor were the pre-Hispanic public centers located along the main roads, called *tambos*, that apart from serving as rest points, had guaranteed rapid change of llamas, the pre-Hispanic beasts of burden.

According to archaeological evidence from the Culebras Valley, the most important pre-Hispanic centers such as Ten Ten and Chacuas Jirca were abandoned shortly after the conquest. It seems that the central area of Ten Ten was depopulated in the first place. No remains of the European presence were registered in the archaeological material. The different situation took place at the site of Chacuas Jirca where numerous beads made of Venetian glass (including the Nueva Cádiz type) which were popular in the first phase of contact with Europeans, in the sixteenth century, serving as gifts of luxury for the natives were found during the survey (Feinzig, 2017; Menaker, 2016; Rice, 2013). In addition, small fragments of glazed pottery corresponding to the early colonial period were found. It is interesting to note that according to the material that remained on the surface, it is very likely that the site was suddenly and forcibly abandoned by its inhabitants. This coincides with Viceroy Francisco Toledo's ordinances of 1570, according to which the population to be relocated generally had only ten days to leave their homes and begin moving to a new place. In addition, according to this order, the displaced community had to "demolish and destroy the old houses they once had, but without burning firewood that could be used to build new buildings" ("*demoler y destruir las casas viejas que alguna vez tuvieron, pero sin quemar leña que pudiera servirles para construir nuevos edificios*"; Toledo 1986 [1569–1574], p. 35; English translation by the authors). In this context, it is very likely that the damage to the main domestic structures of Chacuas Jirca was also the result of these abandonment practices, in compliance with the royal order. During the early colonial period, no traces of colonial settlement were recorded within the Huarmey and Culebras valleys.

Several factors explain this situation, especially the tendency to locate houses, *estancias*, or farms in the lower parts of valleys, in the vicinity of agricultural areas, which were more vulnerable to damage caused by floods, *huaicos* (land/mudslides), earthquakes, or other types of natural disasters. In addition, the lack of early colonial architectural remains could be directly related to their remodeling by modern human settlements, as is currently observed in this part of the coast. This last factor was generally caused by changes in riverbeds as a result of landslides or floods. Although we do not have the very early historical sources for the studied region, in a document dated to 1713 we find information about a great landslide in the Huarmey Valley that destroyed most of the lands of the *hacienda* Congon. Its owner, Doña Domissilia Farias, testified that she lost her hacienda and therefore had not visited it for three years "because the river had washed it away and there was hardly any land left on which to plant" (*"por haversela llebado el rio y no aver dejado casi tierras en que poder sembrar"*; "Titulos de propiedad," AGN, Lima, Leg. 44, Cuad. 770, 1713; English translation by the authors). This kind of dynamic relocation of *haciendas* or ranches in this part of the coast was repeated cyclically in historical times as shown by later documents. There is no or little correspondence between the names of the *haciendas* that functioned in the eighteenth century and those that prospered in the nineteenth and twentieth centuries" (Tello, 1919), Libreta de campo 1, Caj.1, fol. 1–28). Returning to the early documents, ethnohistorical sources also reveal that, in the case of the Culebras and Huarmey valleys, the population numbers drastically reduced shortly after the conquest due to the radical environmental changes and introduction of European pathogens. According to the chronicler Pedro Cieza de León (2005 [1553]), p.193), the coastal part of the Huarmey Valley (Guarmey) was formerly heavily populated by Indigenous people, which changed after the arrival of the Spanish settlers.

According to the historical sources, Huarmey Province was appropriate for the agriculture of fruits and vegetables and was also abundant in trees, especially carob trees, which served both for coal and as a raw material sent by ships to the City of the Kings (Lima), the capital of the Viceroyalty (Cieza de León, 2005 [1553], p. 193; Lizarraga, 1987 [1605], p. 85; Stiglich, 1918, p. 220). It is interesting that the majority of the local population was Spanish speaking at that time (so-called *ladinos*), that emphasizes the sway and control that Europeans had over this region. It is also worth mentioning that the province of Huarmey borders the highland province of Huaylas, which in the early phase of contact with Europeans in the sixteenth century was considered one of the most attractive provinces due to its richness in both natural (mineral deposits, water abundance, fertile lands) and human resources. It was in this region that one of the first grants of specified number of Indigenous people from a precise community, including their agricultural land and herds of livestock, to European conquistadors was made within the *encomienda* system (Huaylas and Recuay). The Indigenous people were also subject to the labor system, being assigned to work for the benefit of the Spanish settlers. Francisco

Pizarro himself came into possession of one of the most important *encomiendas* in Huaylas, distributing the neighboring areas among his most devoted comrades. According to the historical documents, the Huarmey *repartimiento* was delivered in 1542 to Don Martin Lengua, one of the most faithful interpreters of Francisco Pizarro during the conquest and the civil wars (Calvete de Estrella, 1964–1965 [1565–1567], p. 389).

As a consequence of the sociopolitical transformations brought about by the Toledo reforms (1569–1581), the dispersed population of this region was reduced to the new Spanish towns, contributing eventually to a strong decrease in the population (Mogrovejo, 2006 [1593–1605]). The pastoral visits made in 1593 by Santo Toribio de Mogrovejo to that part of the coast of present-day Peru, identified the towns founded by representatives of religious orders, such as Huarmey and Santiago de Guamba, both located in the Huarmey Valley, and Santo Domingo de Xanca, located in the Culebras Valley (Mogrovejo 2006 [1593–1605]). Regarding the upper parts of the two valleys, there are also some references to two new important settlements, founded by Spaniards probably at the same time as the previous ones: Huanchay and Pampas, which were key centers for this region, and where numerous visits related to idolatry processes were made at the beginning of the seventeenth century—(visits of idolatry; Arana Bustamante, 2010, pp. 107–111). It is striking that most of the new settlements and towns founded in early colonial times have remained basically unchanged until the present day, as is the case of Huarmey, the capital of the province of the same name.

Conclusions

The present research has shown that our understanding of the pre-Hispanic past of the Huarmey Province, until now considered a marginal zone and of little archaeological interest, is of crucial importance to fully understand the processes that occurred in pre-Hispanic Peru. Huarmey and Culebras Valleys constituted a political and cultural border, a crossroad between great ancient poles of development. The archaeological evidence recorded in the Culebras Valley demonstrates the sociocultural and environmental transformations of pre-Hispanic times, providing interesting data on the character of human adaptation to the extreme climatic conditions of the coastal desert. In our opinion, the prestige and status of the elites were generally based on the degree of ability to manage the flow of goods and natural resources, at a macro-regional and local scale through the control of networks of reciprocity and exchange. In addition, it seems that in the case of the Andes, the cultural borders did not have the character of fortified limes, but rather of a border or buffer zone, with flexible borders and after all permeable territories. An intriguing question remains: why was the Culebras Valley favored by the pre-Hispanic inhabitants as the place selected for establishing a large regional center? We believe that this is due to the advantages offered by the valley's geomorphology. Unlike the neighboring Huarmey valley, in the Culebras Valley settled terraces are located right next to the cultivated fields and natural springs (*puquios*). The valley itself narrows in several sections creating a kind of natural stronghold with easy-to-control access. The extensive fields of the Huarmey Valley delta are nearby, a few hours' walk away.

This knowledge, sensitivity, and respect for nature were mostly lost after the Spanish conquest and the forced reductions of Indigenous inhabitants and their home places. Paradoxically, the specific conditions of the natural environment have not changed much. The colonial, republican, and modern populations living in these areas were and are still exposed to the same random catastrophic events. During the early colonial period different natural phenomena took place, such as torrential rains, droughts, and earthquakes that resulted in population displacements in different regions of the Andes. For example, in 1578, as reported in historical sources, this part of the coast was affected by great floods and torrential rains caused by the ENSO phenomenon that destroyed many towns, fields, and irrigation canals leading to a starvation and housing crisis (Gridilla, 1936). In our own time, new districts have emerged in the Huarmey Province with the tendency of people to migrate from the countryside to the cities. Never inhabited places become new homelands for many desperate people looking for better life prospects. Just a few years after the catastrophic floods associated with recent 2017 ENSO event, which left homeless thousands of families in Huarmey, a new neighborhood in the precise center of a dry riverbed was founded. Soon, with another inevitable ENSO event this area will surely flow into the ocean (Fig. 2.4d). Let us end, then, with a somewhat sad observation. From our contemporary perspective, the pre-Hispanic management of human–environment relations seem to fit into the classical definition of knowledge as the ability to adapt effectively to the environment, to learn from experience, and to overcome obstacles by taking thought (Neisser et al., 1996). Something that we have lost in the unbreakable race for a better tomorrow.

Bibliography

AGN—Archivo General de la Nación, Lima

Arana Bustamante, L. (2010). *Sin malicia ninguna. Transformación indígena colonial y estrategias sociales y culturales en un kuraka ilegítimo (Huaylas 1647–48)*. Asamblea Nacional de Rectores.

Bailey, G. (2007). Time perspectives, palimpsests and the archaeology of time. *Journal of Anthropological Archaeology, 26*(2), 198–223. https://doi.org/10.1016/j.jaa.2006.08.002

Bawden, G. L. (1994). La paradoja estructural: la cultura Moche como ideología política. In S. Uceda & E. Mujica (Eds.), *Moche: propuestas y perspectivas* (pp. 389–412). Universidad Nacional de la Libertad.

Bawden, G. L. (1996). *The Moche*. Blackwell Publishers.

Bethell, L. (1984a). *The Cambridge history of Latin America, Volume 1: Colonial Latin America*. Cambridge University Press. https://doi.org/10.1017/CHOL9780521232234

Bethell, L. (1984b). *The Cambridge history of Latin America, Volume 2: Colonial Latin America*. Cambridge University Press. https://doi.org/10.1017/CHOL9780521245166

Bonavia, D. (1982). *Los Gavilanes. Mar, desierto y oasis en la historia del hombre*. Corporación Financiera de Desarrollo and Instituto Arqueológico Alemán.

Brodie, N., & Renfrew, C. (2005). Looting and the world's archaeological heritage: The inadequate response. *Annual Review of Anthropology, 34*(1), 343–361. https://doi.org/10.1146/annurev.anthro.34.081804.120551

Buytaert, W., & De Bièvre, B. (2012). Water for cities: The impact of climate change and demographic growth in the tropical Andes. *Water Resources Research, 48*(8), 8503. https://doi.org/10.1029/2011WR011755

Buytaert, W., Iñiguez, V., & De Bièvre, B. (2007). The effects of afforestation and cultivation on water yield in the Andean paramo. *Forest Ecology and Management, 251*(1–2), 22–30. https://doi.org/10.1016/j.foreco.2007.06.035

Buytaert, W., Zulkafli, Z., Grainger, S., Acosta, L., Alemie, T. C., Bastiaensen, J., De Bièvre, B., Bhusal, J., Clark, J., Dewulf, A., Foggin, M., Hannah, D. M., Hergarten, C., Isaeva, A., Karpouzoglou, T., Pandeya, B., Paudel, D., Sharma, K., Steenhuis, T., … Zhumanova, M. (2014). Citizen science in hydrology and water resources: Opportunities for knowledge generation, ecosystem service management, and sustainable development. *Frontiers in Earth Science, 2*(26), 1–21.

Calancha, A. (1976–1981 [1638]). Crónica Moralizada del Orden de San Agustín en el Perú con sucesos ejemplares vistos en esta Monarquía. In I. Prado Pastor (Ed.), *Crónicas del Perú* (Vols. 4–9). Universidad Nacional Mayor de San Marcos.

Calvete de Estrella, J. C. (1964–1965 [1565–1567]) Rebelión de Pizarro en el Perú y vida de don Pedro Gasca. In J. Pérez de Tudela y Bueso (Ed.), *Crónicas del Perú* (Vol. IV–V). BAE.

Casana, J. (2015). Satellite imagery-based analysis of archaeological looting in Syria. *Near Eastern Archaeology, 78*(3), 142–152. https://doi.org/10.5615/neareastarch.78.3.0142

Castillo Butters, L. J. (2001). The last of the Mochicas, a view from the Jequetepeque Valley. In J. Pillsbury (Ed.), *Moche art and archaeology in ancient Peru* (pp. 307–332). Studies in the History of Art 63.

Chyla, J. (2017). How can remote sensing help in detecting the threats to archaeological sites in Upper Egypt? *Geosciences, 7*(4), 97. oai: oai:mdpi.com:2309-608X/3/4/54/; setSpec: Article.

Cieza de León, P. (2005 [1553]). *Crónica del Perú el señorío de los Incas.* Biblioteca Ayacucho.

Contreras, D. A. (2010). Landscape and Environment: Insights from the Prehispanic Central Andes. *Journal of Archaeological Research, 18*, 241–288.

Donnan, Ch. B. (1973). *Moche occupation of the Santa Valley, Peru.* University of California Publications in Anthropology (Vol. 8). University of California Press.

Dufour, E., Goepfert, N., Gutiérrez, L. B., Chauchat, C., Franco, J. R., & Vásquez, S. V. (2014). Pastoralism in Northern Peru during pre-Hispanic times: Insights from the Mochica period (100–800 AD) based on stable isotopic analysis of domestic camelids. *PLoS ONE, 9*(1), e87559. https://doi.org/10.1371/journal.pone.0087559

Feinzig, K. M. (2017). *Tracing sixteenth century beads in South America to understand their impact on indigenous ritual practices and material culture at the time of the Spanish conquest.* Master's thesis, Harvard Extension School.

Ferreyra Huerta, R. (1953). Comunidades de vegetales de algunas lomas costaneras del Perú. *Boletín 53. Estación experimental agrícola de La Molina.* Ministerio de Agricultura.

Flores Galindo, A. (2010). *In search of an Inca: Identity and Utopia in the Andes.* Cambridge University Press.

Giersz, M. (2007). *La frontera sur del estado Moche y el problema de la administración wari en la costa norcentral del Perú.* Ph.D. dissertation. Institute of Archeology. University of Warsaw.

Giersz, M. (2012). Los guardianes de la frontera sur: la presencia moche en Culebras y Huarmey. Andes, *Boletín del Centro de Estudios Precolombinos de la Universidad de Varsovia, 8,* 271–310.

Giersz, M. (2017). *Castillo de Huarmey. Un centro del imperio Wari en la costa norte del Perú.* Ediciones del Hipocampo.

Giersz, M., & Pardo, C. (2014). *Castillo de Huarmey. El mausoleo imperial wari.* Museo de Arte de Lima.

Giersz, M., & Prządka-Giersz, P. (2008). Cronología cultural y patrones de asentamiento prehispánico en el valle del Río Culebras, costa norcentral del Perú. In J. K. Kozłowski & J. Źrałka (Ed.), *Polish contributions in new world archaeology, New Series, fasc. 1* (pp. 7–40). Polish Academy of Arts and Sciences. Jagiellonian University and Institute of Archaeology.

Gillings, M. (2015). Mapping invisibility: GIS approaches to the analysis of hiding and seclusion. *Journal of Archaeological Science, 62,* 1–14. https://doi.org/10.1016/j.jas.2015.06.015

Gillings, M. (2017). Mapping liminality: Critical frameworks for the GIS-based modelling of visibility. *Journal of Archaeological Science, 84,* 121–128. https://doi.org/10.1016/j.jas.2017.05.004

Gridilla, A. (1936). *Ancash y sus Antiguos Corregimientos, Tomo I "La Conquista".* Editorial La Colmena S.A.

Hocquenghem, A. M., & Vetter, L. (2005). Las puntas y rajas prehispánicas de metal en los Andes y su continuidad hasta el presente. *Bulletin de l'Institut Franais d 'Études Andines, 34*(2), 141–159.

Horkheimer, H. (1965). Identificación y bibliografía de importantes sitios prehispánicos del Perú. *Arqueológicas, 8*, 1–51.

Kosok, P. (1965). *Water, land and life in ancient Peru.* Long Island University Press.

Kus, J. S. (1972). *Selected aspects of irrigated agriculture in the Chimu heartland, Peru.* Ph.D. dissertation. University of California.

Lau, G. F. (2016). Peligros ambientales y el archivo arqueológico: culturas y vulnerabilidad antigua en la sierra de Ancash, Perú. In N. Goepfert, S. Vasquez, C. Clément, & A. Christol (Eds.), *Las sociedades andinas frente a los cambios pasados y actuales: dinámicas territoriales, crisis, fronteras y movilidades* (pp. 51–87). Actes & Mémoires de l'Institut Français d'Études Andines. Instituto Francés de Estudios Andinos.

Lauricella, A., Cannon, J., Branting, S., & Hammer, E. (2017). Semi-automated detection of looting in Afghanistan using multispectral imagery and principal component analysis. *Antiquity, 91*(359), 1344–1355.

de Lizarraga, R. (1987 [1605]). Descripcion del Perú, Tucumán, Rio de la Plata y Chile. In I. Ballesteros (Ed.), *Historia 16: Crónicas de América 37.* Madrid: Historia 16.

Menaker, A. (2016). Las cuentas durante el colonialismo español en los Andes peruanos. *Boletín De Arqueología PUCP, 21*, 85–97.

Menzel, D. (1964). Style and time in the Middle Horizon. *Ñawpa Pacha, 2*(1), 1–114.

Mogrovejo, S. T. (2006 [1593–1605]). *Libro de visitas de Santo Toribio Mogrovejo, 1593–1605* (J. A. Benito Rodríguez, Ed.). Fondo Editorial de la Pontificia Universidad Católica del Perú.

Moseley, M. E. (1975). *The maritime foundations of Andean civilization.* Cummings Archaeology Series.

Moseley, M., & Feldman, R. (1982). Vivir con crisis: percepción humana de proceso y tiempo. *Revista Del Museo Nacional, 46*, 267–287.

Murra, J. V. (1972). El control de un máximo de pisos ecológicos en la economía de las sociedades andinas. In I. Ortiz de Zúñiga (Ed.), *Visita de la Provincia de León de Huánuco (1562)* (Vol. 2, pp. 429–476). Universidad Hermilio Valdizán.

National Institute of Natural Resources. (2007). *Evaluación de los recursos hídricos en las cuencas de los ríos Casma, Culebras y Huarmey. Inventario de fuentes de agua superficial en la cuenca del río Culebras.* Ministerio de Agricultura, Instituto Nacional de Recursos Naturales—INRENA, Intendencia de Recursos Hídricos, Administración Técnica del Distrito de Riego Casma–Huarmey.

Neisser, U., Boodoo, G., Bouchard, B. T. J., Jr., Boykin, A. W., Brody, N., Ceci, S. J., Halpern, D. F., Loehlin, J. C., Perloff, R., Sternberg, R. J., & Urbina, S. (1996). Intelligence: Knowns and unknowns. *American Psychologist, 51*(2), 77–101. https://doi.org/10.1037/0003-066X.51.2.77

Nials, F., Deeds, E., Moseley, M., Pozorski, S., Pozorski, T., & Feldman, R. (1979). El Niño: The catastrophic flooding of coastal Peru. A complex of oceanographic and meteorologic factors combine in one of earth's most devastating recurrent disasters. Part. II. *Field Museum of Natural History Bulletin, 50*(8), 4–10.

Oka, S., & Ogawa, M. (1984). The distribution of Lomas vegetation and its climatic environments along the Pacific Coast of Peru. *Geographical Reports of Tokyo Metropolitan University, 19*, 113–125.

Pizarro, P. (2013 [1571]). *Relación del descubrimiento y conquista de los reinos del Perú.* Fondo de Cultura Económica.

Prümers, H. (1990). *Der Fundort "El Castillo" im Huarmeytal, Peru. Ein Beitrag zum Problem des Moche-Huari Texstil-Stils.* Mundus Reihe Alt-Amerikanistik 4.

Prządka-Giersz, P. (2009). *Patrones de asentamiento y transformaciones sociopolíticas en la costa norcentral del Perú durante los Períodos Tardíos: el caso del valle de Culebras.* Ph.D. dissertation. Institute of Archeology. University of Warsaw.

Prządka-Giersz, P. (2019). *Mujer, poder y riqueza. La tumba de elite femenina Wari del Castillo de Huarmey.* Ediciones del Hipocampo.

Prządka, P., & Giersz, M. (2003). *Sitios arqueológicos de la zona del valle de Culebras, Vol. I: Valle bajo.* Sociedad Polaca de Estudios Latinoamericanos, Misión Arqueológica Andina.

Pulgar Vidal, J. (1996). *Geografía del Perú. Las ocho regiones naturales, la regionalización transversal, la sabiduría ecológica tradicional.* Peisa.

Raimondi, A. (1873). *El Departamento de Ancachs y sus riquezas minerales.* Imprenta de "El Nacional" por Pedro Lira.

Rice, P. M. (2013). *Space–time perspectives on early colonial Moquegua.* University of Colorado Press.

Roosevelt, A. (1999). The maritime, highland, forest dynamic and the origins of complex culture. In F. Salomon & S. Schwartz (Eds.), *The Cambridge history of the native peoples of the Americas* (pp. 264–349). Cambridge University Press. https://doi.org/10.1017/CHOL9780521630757.006

Rostworowski, M. (1989). *Costa peruana prehispánica.* Instituto de Estudios Peruanos.

Salomon, F., & Schwartz, S. (1999). *The Cambridge history of the native peoples of the Americas.* Cambridge University Press. https://doi.org/10.1017/CHO L9780521630757.006

Sandweiss, D. H., & Quilter, J. (2009). *El Niño, catastrophism, and culture change in ancient America.* Dumbarton Oaks Research Library and Collection.

Sandweiss, D. H., & Richardson III, J. B. (2008). Central Andean environments. In H. Silverman & W. Isbell (Eds.), *The handbook of South American archaeology* (pp. 93–104). Springer. https://doi.org/10.1007/978-0-387-74907-5

Schaedel, R. P. (1951). Mochica murals at Pañamarca (Peru). *Archaeology, 4*(3), 145–154.

Shimada, I. (1994). *Pampa Grande and the Mochica culture.* University of Texas Press.

Spilbergen, J. (2014 [1619]). *The East and West Indian Mirror. Being an account of Joris Van Spilbergen's voyage round the world, 1614–1617 and the Australian Navigations of Jacob Le Maire—Primary source edition.* J. A. J. Villiers. Nabu Press.

Stiglich, G. (1918). *Derrotero de la costa del Perú.* P. Berrio & Company.

Szpak, P., White, Ch. D., Longstaffe, F. J., Millaire, J.-F., & Vásquez Sánchez, V. F. (2013). Carbon and nitrogen isotopic survey of northern Peruvian plants: Baselines for paleodietary and paleoecological studies. *PLoS ONE, 8*(1), e53763. https://doi.org/10.1371/journal.pone.0053763

Tapete, D., Cigna, F., & Donoghue, D. N. M. (2016). Looting marks' in spaceborne SAR imagery: Measuring rates of archaeological looting in Apamea (Syria) with TerraSAR-X staring spotlight. *Remote Sensing of Environment, 178*(June), 42–58. https://doi.org/10.1016/j.rse.2016.02.055

Tello, J. C. (1919). *Huarmey y parte del camino a Huambo. Manuscrito inédito en el Museo de Arqueología y Antropología de la Universidad Nacional Mayor de San Marcos.* Archivo Tello.

Thompson, D. E. (1962). The problem of dating certain stone-faced stepped pyramids on the north coast of Peru. *Southwestern Journal of Anthropology, 18*(4), 291–301.

Thompson, D. E. (1966) Archeological investigations in the Huarmey Valley, Peru. In *Actas y memorias del XXXVI Congreso International de Americanistas* (Vol. I, pp. 541–548). España 1964.

Thornton, E. K., Defrance, S. D., Krigbaum, J., & Williams, P. R. (2011). Isotopic evidence for middle horizon to 16th century Camelid Herding in the Osmore Valley, Peru. *International Journal of Osteoarchaeology, 21*(5), 544–567.

Toledo, F. (1986 [1569–1574]). *Disposiciones gubernativas para el Virreinato del Perú (1569–1574). Tomo I.* Publicaciones de la escuela de estudios hispanoamericanos (G. Lohmann Villena & M. J. S. Viejo, Eds.). Escuela de Estudios Hispano-Americanos.

Wafer, L. (2004 [1903]). *A new voyage and description of the Isthmus of America* (G. P. Winship, Ed.). Kessinger Publishing.

Willey, G. R. (1953). Prehistoric settlement patterns in the Virú Valley, Peru. *Bureau of American Ethnology Bulletin, 155*, 1–453. Smithsonian Institution.

Weberbauer, A. (1945). *El mundo vegetal de los Andes peruanos. Estudio fitogeográfico.* Ministerio de Agricultura, Estación experimental agrícola de La Molina.

Wells, L. E., & Noller, J. S. (1999). Holocene coevolution of the physical landscape and human settlement in Northern Coastal Peru. *Geoarchaeology, 14*(8), 755–789.

3

Ihuan Yehhuan Tlacuauh Tlamauhtiah in Ichcapixqueh. "And the Shepherds Are Inspiring Great Fear". Environment, Control of Resources and Collective Agency in Colonial and Modern Tlaxcala

Justyna Olko

Introduction

During my research stays in Tlaxcala, I had the honor and pleasure to meet Doña Rosaria, who is a healer in the town of San Miguel Xaltipan, part of the municipality of Contla de Juan Cuamatzi, known in colonial times as San Bernardino Contla.[1] In 2017, during our language revitalization field school organized in collaboration with this community, she generously agreed to share her knowledge of plants and healing with the participants at our event, including a young Nahua healer from a community in the region of Texcoco. I continued my

[1] Research reported in this paper was developed within the project no. 2018/29/B/HS3/02782 funded by the National Science Center in Poland. It also benefited from the activities carried out within the Coling project (European Union's Horizon 2020 research and innovation program under the Marie Skłodowska-Curie grant agreement No. 778384), including the Ethnohistory conference panel organized in 2018 as well as other workshops and discussions.

J. Olko (✉)
University of Warsaw, Warsaw, Poland
e-mail: jolko@al.uw.edu.pl

© The Author(s) 2024
J. Olko and C. Radding (eds.), *Living with Nature, Cherishing Language*,
https://doi.org/10.1007/978-3-031-38739-5_3

conversation with her in the summer of 2019 and at that time she revealed her sense of loss regarding the changing local environment and the key resources necessary to continue with traditional healing practices:

> *Ce quitemoa ....nochi in medicina ye ocuparohuaya para ce mopahtiz, non (nimi)tzonilhuia in tlamacaz, itztohyatl, mirto, topoya tlen onon occe totzoniltac (tzotzoniztatl) axxan can? Cah yocmo cah, ye opoliuh. Nin zurcoz nochi oyeya yenoz de xihuitl pero ahorita para ce mopahtiz ce quitemoti ihtec in atlauhtl pero yocmo cah quimin den oyeya nican, yocmo cah.*
>
> One looks for it …. [but] all of the medicine that was being used in order to cure people, that which I tell you would be given to people, *itztoyatl*, myrtle, *topoya* (*tlalchichinolli*; *Tournefortia mutabilis*), this other [herb] *tzotzoniztatl*, where are they now? They are gone, they have already disappeared. These furrows were all full of herbs but now if you want to cure yourself you have to go and look in the ravine. But it is not as it used to be, [the plants] aren't here anymore.

The sense of loss permeates the history of Indigenous communities right from the first encounters with the Europeans. As the Native people of the Americas confronted the outsiders, they suffered a broad array of injuries. They both resisted and fell victim to violence, oppression and forced assimilation; they acted to counter the loss of their autonomy, land and economic base; they engaged in unending and ultimately not always successful struggles to preserve their beliefs, social order, land, resources, language, local knowledge, ways of healing and intimate connections to nature and the environment. The case of Tlaxcala in Mexico shows that no group was exempt from this long-term struggle and loss, no matter what their arrangements and relationships with the colonizers were. Moreover, the present situation of the Native peoples of the Americas cannot be disentangled from their colonial history as it, in fact, transcends the conventional "end" of the colonial period, with numerous forms of colonialism continuing until the present day.

Scholars often share the view that the position of Tlaxcallan[2] under Spanish rule was privileged in the sociopolitical stage of early New Spain (e.g., Baber, 2005, 2011; Lockhart et al., 1986, p. 2; Navarrete Linares, 2019; Sullivan, 1999, p. 55), and in some ways, it was indeed. Unlike many other pre-Hispanic political organizations, the native state of Tlaxcallan was granted special royal privileges and considerable political autonomy in recognition of their martial and political alliance with Spaniards during their war with the Aztec Empire, and then of their participation in numerous conquests in southern and northern areas of Mesoamerica. Importantly, the territorial and to a certain degree sociopolitical and economic integrity of their preconquest state—complex *altepetl* (see Lockhart, 1992)—was preserved. Its land could not be given in *encomienda* to individual Spaniards as the inhabitants enjoyed the status of being "free vassals" of the king (e.g., Baber, 2005, pp. 103–105).[3] The privilege of March 13, 1535, placed the Tlaxcaltecah perpetually under royal authority, giving them important legal leverage to protect their land claims and territorial integrity (Baber, 2011, p. 50). Many of these early privileges and legal acts were elevated to the status of royal laws through their incorporation in the *Recopilación de leyes de los reinos de las Indias* of 1681 (Martínez Baracs, 2008, p. 414). These laws recognized the Tlaxcaltecah as "the first who in New Spain received the holy Catholic faith and declared obedience to us" and whose request to "keep their ancient customs" should be secured by means of the royal provision.[4] In addition, the law not only guaranteed that their governors must be Indigenous, as in other *altepetl* in New Spain, but also that this position was closed to external functionaries coming from outside of the province; as royal subjects, the Tlaxcaltecah were also granted the option of writing directly to the king, without any obstruction from the authorities of

---

[2] I use the term "Tlaxcala" and the adjective "Tlaxcalan" to refer to the territory roughly delimited by the modern state of Tlaxcala; on the other hand, I use the terms "Tlaxcallan." in reference to the pre-Hispanic and colonial complex altepetl and province, and "Tlaxcaltecah", to designate the Indigenous people inhabiting this territory, both in historical and present times.

[3] Recopilación de leyes de Indias, Libro VI, titulo I, Leyes 39 & 40.

[4] Recopilación, vol. II, Libro VI, titulo 1, Leyes 38 & 40.

New Spain.[5] Moreover, in 1585 the king issued a tribute exemption, and although in reality, the province had paid various taxes since the sixteenth century, this original immunity was invoked by local elites as an important argument in legal struggles and in political rhetoric (Gibson, 1967, pp. 160, 170–181; Martínez Baracs, 2008, p. 418; Skopyk, 2020, p. 105). One of the key components of the early colonial discourse of the Tlaxcaltecah was that of claiming the status of allies and winners during the Spanish conquest (Navarrete Linares, 2019).

These and many other legal acts and arrangements with the Crown notwithstanding, Tlaxcallan was in many ways as precarious as other *repúblicas de indios* in New Spain. The highly dynamic and complex colonial reality exposed them to a number of external pressures and threats—political, economic, legal and religious—from right after the Spanish conquest. As recently argued by Bradley Skopyk, "This society coped with the most severe climatic shift of the Holocene: the Little Ice Age. The impacts of disease, famine, cold, drought, or even excessively wet weather triggered social and biological responses that transformed colonial society with little regard for class, ethnicity, or subregion. [...] When the ground that fed and clothed society moved, humans had little choice but to respond" (Skopyk, 2020, p. 209).

Indeed, the ruling elites, community leaders and individual community members did not remain passive during any single part of the colonial period and beyond. The form of Indigenous government, both at the central level of the Tlaxcallan *cabildo* and in the organization of numerous towns subject to it, tirelessly engaged in internal and external politics, maintaining, readjusting and expanding spaces of collective responsibilities and collective action. Facing these challenges, they forged diverse strategies, which are reflected in the rich corpus of Nahuatl and Spanish documents. These embrace municipal records and an abundance of legal documentation that reflects the everyday concerns and affairs of Indigenous inhabitants of this state. Their language, Nahuatl, was an inherent part of the struggle to maintain their autonomy and protect their province, flourishing as the medium of negotiation, complaint, call to action, law-making,

---

[5] Recopilación, vol. II, Libro VI, titulo 1, Leyes 32 & 45.

economic and busInéss activities and in multi-faceted administrative and official documentation. They built on a long and successful tradition of political, military, economic and cultural resistance to the expansion of the Aztec empire before the Europeans appeared on the Mesoamerican scene and have continued to resist external pressures under Spanish rule and beyond.

The long-term trajectory just mentioned becomes even more clear if we acknowledge that this history is not only about the past. And this is not just because, as historians, we take part in an "unending dialogue between the present and the past" (Carr, 1987[1961], p. 30) and we "always ask questions from the point of view of the present" (Carr & Lipscomb, 2021, p. 13); it is also because the aftermath of apparently remote historical processes continues unfolding to this day. Therefore, striving to offer a *longue durée* view of Tlaxcalan history from the specific point of view of strategies aimed at the protecting of land and associated resources, I attempt to make the connection between past and present facets of the same longer narrative. In order to do so, I complement this study with selected contemporaneous anthropological and sociolinguistic data, including insights gathered during several years of field research in Tlaxcala. It is through the voices of today's Tlaxcaltecah that we can better understand, on the one hand, why local knowledge and language are at risk and being lost, but, on the other, how access to such vital resources relates to the well-being of community members. While attempting to grasp a long-term temporal perspective, this study is not a systematic overview of the processes in question; rather, it builds on the key Nahuatl sources from the sixteenth century and selected microhistorical records from the end of the seventeenth and beginning of the eighteenth century. The emphasis on human agency has been a particular focus of microhistory that has brought into sharp focus the lives, decisions, motivations and actions of past individuals seen "not merely as puppets on the hands of great underlying forces of history, but they are regarded as active individuals, conscious actors" (Magnússon & Szijártó, 2013, p. 34). The voices and actions of the Indigenous people today complement those of their ancestors living and acting in the same communities in the past.

Protecting the Territory

Directly after the first phase of the Spanish conquest of Mesoamerica, or more precisely after the war waged by Spaniards, Tlaxcaltecah and other local allies of Cortés against the Mexihcah and their supporters, the *altepetl* of Tlaxcallan retained much its autonomy and was initially little affected by political control and Christianization (Fig. 3.1). The situation started to change after the arrival of the first Franciscans and the establishment of a temporary convent in a palace belonging to one of the rulers. Soon the instruction of the local youth from noble families (often against the will of their parents) began, leading thereafter to a violent upheaval and persecutions of "idolaters" and culminating in the execution of several Tlaxcaltecah leaders (Gibson, 1967, pp. 29–37; Lopes Don, 2010; Martínez Baracs, 2008, pp. 113–120; Olko & Brylak, 2018). Even though religious, political and economic forms of control over the province were strengthened in the decades that followed, it was also during this time that the structure and prerogatives of local government were consolidated, taking the unique form of a municipal council and a huge electing body composed of over 200 nobles (Lockhart et al., 1986, pp. 1–3).

Right from its first preserved records onward, the Indigenous *cabildo* maintained a very clear position that foreigners, except for the Franciscans and the Spanish *corregidor* with his retinue, were unwelcome in the province. Accordingly, the policy pursued was to exclude them from living in the Tlaxcallan territory entirely. In the council minutes for August 8, 1550, we read that *tiquixpantizque yn visorey ynic techmocneliliz ayac totlah yezqui español vel quiçazque* ("we should propose to the viceroy that he grant us the favor that no Spaniard be among us and they leave entirely"; fol. 65v–66; Lockhart et al., 1986, p. 76). Regarding Indigenous settlers from other *altepetl* in Central Mexico, it was made clear in 1547 that they should perform the same public duties as local commoners (Lockhart et al., 1986, p. 30).

This policy of the Indigenous government of excluding Spanish intrusions was, in fact, supported by the Crown, officially at least, as reflected in a number of royal provisions issued between 1535 and 1586, making grants of land in the Tlaxcallan territory void. Yet

San Bernardino
Contlan

Tlaxcala · San Miguel Xaltipan

Huamantla

Santa Ana
Chiyauhtempan

Cuauhtzincola/
San Felipe Cuauhtenco

Nativitas

Matlalcueye/Malintzin

Santa Ines Zacatelco

San Toribio Xicohtzinco

Fig. 3.1 Map of Tlaxcala showing the localities mentioned in this chapter. © Joanna Maryniak

Spanish officials and colonists repeatedly violated this rule as many grants of land followed from the 1530s onward. Pursuing their own economic benefits, individual colonists also managed to defy the Crown's policy, as well as the efforts of the Indigenous government to block land sales to foreigners (Gibson, 1967, pp. 79–85). This happened despite the fact that *cabildo* did have some success in abolishing Spanish estancias (e.g., *Actas de cabildo*, fol. 98; Lockhart et al., 1986, p. 53), but this was a short-term success. By 1585, the king and viceroy recognized the validity of grants to Spanish land owners in Tlaxcallan (Gibson, 1967, p. 84).

The main legal means for increasing the Spanish presence was the purchase of Indigenous land. The royal order of 1535 allowed the Spaniards to acquire lands from local owners if the latter made the sale of their own free will (Gibson, 1967, pp. 85–86). Again, quite early on, the Native *cabildo* tried to stop this process, first in May 1553 by prohibiting nobles from selling their lands under the threat of dispossession and eviction (fol. 93v–94v; Lockhart et al., 1986, pp. 85–86; Martínez Baracs, 2008, p. 183), justifying these drastic measures with reference to the growing impoverishment of the nobility and the rise of wealth among commoners. This is an early sign of the growth of internal-community conflicts and tensions between the nobility and commoners as the colonial economy developed. Then in April 1562, the municipal council made it clear in its decree that no resident of the province, regardless of social status, should sell land to the Spaniards, for they should live in their own cities and not in Tlaxcallan at all: *yn cabildo quinnauati yn ixquichime Tlaxcalan tlatoque pipiltin yuan nochi tlacatl ipanpa ayac quinamacaz ysolar yn nican ypan ciudad amo quinnamaquiltiz españolme yehica amo yz totzalan nemizque oncah ymaltepeuh ciudad de los Angeles yuan Mexico* (*Actas de cabildo*, fol. 149v; Celestino Solís et al., 1984, p. 197) ("The cabildo orders that all the Tlaxcallan rulers, nobility and all the commoners are not to sell their house lots here in the city, they are not to be sold to Spaniards because they are not to live among us, they have their city in Puebla de los Angeles and Mexico"). Since purchases made from Indigenous people by Spaniards were often below the market value, the Crown ordered that sales were to be announced at auction for thirty days before the transaction, but this, too, proved to have a limited impact on the social reality (Gibson, 1967, pp. 86–87). This is clear through microhistorical insights into early colonial processes of land sale and the ongoing dispossession of local land owners (e.g., Acocal, 2020, pp. 211–223).

The efficacy of counteractions undertaken by the Native government was circumscribed not only by the scale of disobedience among Natives and Spaniards but also due to the huge population decline as a result of devastating epidemics, the growing debts of Indigenous land owners and the sizeable amount of abandoned land that became available as a consequence. After initial incursions, Spaniards rapidly increased in numbers in the last quarter of the sixteenth century, while the Native population constantly declined. As in other regions of Mexico, deadly epidemics took their toll in Tlaxcallan in 1520, 1532, 1545, 1575–1579, 1585–1588, 1595 and later on, accompanied by other disasters, especially famine (Gibson, 1967, pp. 141–142, 188; Martínez Baracs, 2008). The 1520 epidemic reduced the Indigenous population from some 350,000 down to 250,000 or less; then in 1545, roughly 50% of people died, reducing the number of inhabitants to ca.

150,000. The 1576 epidemic was the last year with high-magnitude mortality (ca. 48,000 dead), to be repeated only in the 1690s (ca. 40,000) and in the 1730s (ca. 50,000), reducing the total population to levels as low as just over 50,000 immediately after the periods with the heaviest death toll, to ca. 100,000 in the periods of temporary population increase; at the end of the colonial era the population was only 75,000 (Skopyk, 2020, pp. 77–78; Fig. 8). In a parallel way, the initially marginal but bustling Spanish community grew steadily, oriented toward the expanding city of Puebla and specializing in cattle and sheep herding, textile production and transportation, and gradually becoming more and more significant for the province's internal economy, social life and politics (Lockhart et al., 1986, p. 31). It is also impossible to overestimate the impact of extreme pluvial anomalies, including the excessive rainfall, floods, frosts and drought that occurred from 1541 to the mid-1580s and that were also behind "an unparalleled and sustained eighty-year socioecological crisis" (Skopyk, 2020, p. 32), perhaps even directly contributing to the scale and severity of the epidemics (Skopyk, 2020, p. 34).

Collective Threats and Losses

The considerable efforts of the Tlaxcallan *cabildo* were invested into stopping the spread of Spanish *estancias* or livestock enterprises and summer pasturing; the employees of Spaniards, especially their shepherds, were perceived as violent, destroying crops and stealing from and terrorizing people. Another essential threat was the forced resettlement or *congregaciones* of the Indigenous population into more compact settlements, more closely following European models, a policy pursued by the Spanish Crown as a response to depopulation and the need for control over the Native subjects and their Christian instruction. Such resettlements of people implied not only the abandonment of households and ancestral cultivated land but also, paradoxically, a greater exposure to epidemics as well as a threat to existing social structures and tribute obligations (Sullivan, 1999, pp. 38–39). The Tlaxcaltecah strongly and successfully opposed these relocations, even if some concentrations of more dispersed settlements indeed occurred in different parts of the province as early as the sixteenth century (Acocal, 2020, pp. 110–111). Formal congregations were greatly delayed in the region in comparison to other parts of New Spain, even into the first decade of the seventeenth century. One of the most expressive testimonies comes from a 1560 *cabildo* session (*Actas de cabildo*, fol. 136r–136v) as it conveys "the feared effect of uprooting on the ordinary individual" (Lockhart et al., 1986, p. 104) and envisions a combined disastrous effect of the two parallel threats:

> A great deal of suffering and affliction will result from it (*vel miec netoliniztli netequipacholly ye mochivaz*). [...] They will leave behind their houses and all they take care of: their edible cactus fruit, their cochineal-bearing cactus, their American cherry trees, their maguey, and their fruits, sweet potatoes, sapotes, chayotes, and quinces, peaches, etc., and then also the household fields which they clear and cultivate, and their dogs and turkeys [...] Where will he cultivate if they are entirely lost to him? And if he thus leaves behind his maize, chia, or cactus fruit, and his burial grounds, who will guard them for him? And the shepherds are inspiring great fear; (they wander about the grasslands) all over Tlaxcala. And even though commoners are there now, sometimes (the shepherds) beat them and take their children from them; sometimes they snatch their daughters away, and they take their turkeys, mats, etc., from them. (Lockhart et al., 1986, pp. 104–105)

The members of the Indigenous government expressed their deep concern about the loss of cultivated fields with all the local plants and fruit that would remain behind, while resettled commoners would have to establish their homes elsewhere. The harm was aggravated by the abandonment of burial grounds, which implied not only leaving these places with no protection or care but also the loss of vital links with the ancestors that would be disastrous for individual families and whole communities. Notably, Indigenous officials linked this threat to another peril: the various kinds of violence committed by Spanish shepherds invading Indigenous countryside and cultivated fields. After the first successful attempts at the expulsion of Spanish-owned cattle and sheep, in the second half of the sixteenth-century herders kept coming, spreading all over Tlaxcallan through grants, concessions, land seizures and transactions. According to the legal framework applied in New Spain, stock-raising depended not on exclusive property rights, but on access for grazing to the so-called *baldías*, or unused wasteland, as well as community or privately owned land, provided no crops were currently growing there (Greer, 2018, p. 254; Radding, 1997, pp. 175–176).

The 1550s saw the adoption of the policy of seasonal restrictions regarding the entering of cattle into the province of Tlaxcala (only to be waived at the end of the century) as well as areal limitations excluding cattle from within half a league of Indigenous fields and from trespassing on public lands before harvest (Gibson, 1967, p. 85). Despite this legal protection of Indigenous land and resources, Spanish cattle, especially before the harvest season, caused constant turbulence and severe damage to the local people, their establishments and resources. The environmental conquest and colonization have been extensively documented in the Valley of Mezquital with the expansion of pastoralism, which contributed to domination over the Indigenous populations and vast areas of rural land (Melville, 1999). In the words of the Tlaxcallan *cabildo* members recorded in 1562, *y nican agustar yc quichiua espanolme cenca yc motolinia maceualli* ("the pasturing that the Spaniards do here causes great harm to the Native people"; fols. 155r–155v). As vividly described by Gibson, "houses were forcibly entered and ruined; public works were damaged; whole pueblos were destroyed; and boundaries that had been made were overrun. In 1594 one of the largest towns of the province, Hueyotlipan, was temporarily abandoned by the Indians after the destruction of its nopal and fruit crops by roving cattle" (Gibson, 1967, pp. 152–153). Also in other regions of Mexico, the relationships between local people and herdsmen became increasingly violent during this time (Melville, 1999, p. 122, 134).

Another profound change in the economy of Tlaxcallan, driven by Spanish trade and replete with far-reaching consequences, was cochineal production. Though known in preconquest times, it massively increased in response to the demands of external markets opening through the city of Puebla (Lockhart et al., 1986, p. 79; Skopyk, 2020, p. 83). By the mid-sixteenth century, the cochineal industry was booming to the extent that in 1552 the *cabildo* attempted to forbid the cultivation of the cochineal cactus for fear of the "idleness" it causes: *cenca tetlatzocuitiya ipanpa aocmo tlayznequih maceualtin* ("it makes commoners very lazy because they do not want to till the land anymore") and of impending famine (*yntla quizaqui mayanaliztly tlen cualoz* (fols. 85r–85v; Celestino Solís et al., 1984, p. 128) ("if the famine comes, what will be eaten?"). This catastrophic vision—no doubt enhanced by scary memories of the devastating epidemic and its aftermath, famine, that had decimated the area of Tlaxcala in the previous decade—was further elaborated in 1553 (fols. 91r–93v). In its reaction to growing social tensions between nobles and commoners driven by new economic opportunities, the *cabildo* set limits on cochineal production. Indigenous officials expressively affirmed that the local social, economic and moral order was in jeopardy due to the accumulation of wealth of commoners active in the cochineal trade and acquiring luxuries not appropriate for their social standing, as well as feasting, consuming alcohol, succumbing to moral decline and failing to provide their service as providers of food staples. Not only was their "idleness" and dedication to cochineal causing many fields to turn into a grassy wasteland, wreaking havoc and threatening an impending famine but it also resulted in many sins committed against God through their non-attendance at mass and their pervasive drunkenness.

uh niman vel xohxocomiqui vel yvinti yn cepan ynçiçiuauan yn canpa mohololoua oncan vehvetzi vel yhivinti miec yn oncan mochiva tlatlacolly/ çan ya yeh ytech quiça yn nochiztly / / no ·yehuan nochiznecuiloque aocmo tlayznequi yn manel cequin cuemeque aocmo tlayznequi ça yeh quixcaviya yn nochiztli quitemoua. (fol. 92r–92v).

And then they become entirely inebriated and senseless, together with their wives; they fall down one at a time where they are congregated, entirely drunk. Many sins are committed there, and it all comes from cochineal. Also these cochineal dealers no longer want to cultivate the soil; though some of them own fields, they no longer want to cultivate; they do nothing but look for cochineal". (Lockhart et al., 1986, p. 83).

While it is true that cochineal production caused a considerable contraction of land use by Indigenous farms, it should also be seen as an adaptation to the changing climate and economic demand. And it was also climate—not the *cabildo's* efforts—that turned out to be the dominant factor in the decline of cochineal production in Tlaxcallan. It began to fall after the 1576 epidemics, in the coldest decades of the Colonial Mexican Pluvial (Skopyk, 2020, p. 67, 85). In part, the threats associated by the *cabildo* members with cochineal production were seen as derived from establishing improper relationships with Spanish merchants and consuming their product: Castilian wine. Thus, in addition to the demoralizing impact of cochineal trade, the contact with Spaniards posed serious moral and social threats, including public drunkenness, which was associated with both local and foreign beverages: *vino hano nican octly miec tlacatl ya neci teixpan yn uel iuintiua* (fol. 164v; Celestino Solís et al., 1984, p. 217), "It seems that many people do indeed get drunk publicly with wine and local agave drink", complained Indigenous officials in 1566. On that occasion, the *cabildo* forbade the drinking of wine in public, making flnés mandatory and putting in jail drunkards of any social standing, with the help of public officials, including watchmen and constables.

Individuals of Spanish and mixed origin could also be seen as a potential source of moral decline for the Indigenous inhabitants of the province. A document composed in 1557 focuses on a "woman of bad living" (*mujer de malvivir*):[6] Juana Ramirez, *mestiza*, was expelled by the *alcalde mayor* Francisco Verdugo and moved to the city of Mexico because her way of life presented a very bad example for the Native people. According to a similar order from 1581,[7] the Native people of Tlaxcallan requested the *alcalde mayor* to implement the viceregal order of removing from the city four mestizos currently held in prison for being "men of bad living and bad examples for the Indigenous people". They were to stay at least four leagues from the city and, if the order was violated, be punished with 200 lashes each. Thus, while, despite many efforts of the *cabildo*, the Tlaxcaltecah were not able to stop

---

[6] AHET, Fondo Colonial, Sección Judicial, Serie Criminal, Caja 1, Exp. 13.
[7] AHET, Fondo Colonial, Sección Judicial, Serie Criminal, Caja 9, exp. 7.

Spanish civilian intrusions, they did not cease in their legal actions against what they perceived as threats to the integrity and well-being of their society.

It has been claimed that the legal arguments of the Tlaxcaltecah "were creative stories that were influenced by the rhetorical norms and political culture of Castile" and while they contained "elements of truth and recognizable descriptions of reality" they should be viewed as "artifacts of human creativity, first and foremost", reflecting "native people's ingenuity and agency" (Baber, 2011, p. 57). However, the complaints, alarming diagnoses and even denunciations expressed by the Tlaxcaltecah need to be cast in the direct spotlight of the catastrophic and recurring crises affecting the inhabitants of the province from the onset of the colonial period and begetting profound social and economic upheavals: extreme depopulation, waves of famine and "cataclysmic" climatic changes whose status as "historical facts" historians simply must recognize (Martínez Baracas, 2008, pp. 257–259; Skopyk, 2020). In fact, royal privileges improved local affairs only to a very minor degree: orders prohibiting Tlaxcaltecah labor and personal service in the city of Puebla, banning Spanish settlements and *estancias* and controlling monopolies were constantly violated (Gibson, 1967, p. 169). Emerging *obrajes* (manufacturing workshops) became the places of exploitation of the local population despite royal legislation aimed at counteracting the abuse. The Spanish-owned *obrajes* in the province included closed workshops (*obrajes cerrados*) where Tlaxcaltecah workers were practically imprisoned. In addition, both the forced and voluntary involvement of local people in the *obrajes* reduced the number of people engaged in agriculture, whereas the development of the city of Puebla itself contributed to the depopulation of Tlaxcallan (Sánchez Verín, 2002, pp. 43–54, 79).

Moreover, the content of the official petitions and legal suits takes on a different perspective when illuminated by the proceedings of the *cabildo*, recorded in Nahuatl for the internal needs of the ruling class of the *altepetl*. As much as they are replete with traditional rhetorical tools relating to perceived threats and the needs of the community, this layer of historical records underpins and shores up the concrete measures and actions undertaken by the Indigenous government. Their goal was to address, contain and counteract threats and forge favorable solutions in the new reality by means of the available resources and structures, very much in accordance with the contemporary understanding of agency (Giddens, 1984; Sztompka, 1991). And indeed, the policies and fears of the Indigenous *cabildo* were no doubt perspicacious and far-sighted; the multi-faceted effects of Spanish infiltration into the province proved disastrous for local land-holding, control over resources and linguistic-cultural assimilation, both in the mid- and long-term perspectives.

Other Early Responses

The responses of Indigenous authorities and residents of the province to both threats and opportunities emerging from the contact with Spaniards were by no means limited to legal actions and legal rhetoric. While the influx of Spanish cattle herders and their animals eventually slipped out of the control of both local and Crown authorities, an important strategy was aimed at protecting the environment and preventing damage to Indigenous settlements and fields, especially in key moments of the agricultural production cycle. In February 1548, the *cabildo* decided to expel the sheep from the wetlands where Indigenous people would sow amaranth and chili: *chiyauhtently ca quizazque yn ichcame ypanpa yn zan cuel toca maceoaltzintly zan cuel quitoca auhtly yuan chilli quitepeua yuan quequeza tonalchilli* (*Actas de cabildo*, fol. 10v; Celestino Solís et al., 1984, p. 51) ("The sheep are to leave the wetlands because the commoners will soon be planting, they will soon sow amaranth, they will scatter chili seeds and plant dry-land chili").

Struggles over containing the negative impact of the presence of Spanish cattle spiraled into conflicts between the colonists and the local people, who resorted to their own retributive responses. Although in 1548 the Spanish *corregidor* issued an order obliging the local government to build a corral for cattle and horses caught consuming or destroying Indigenous crops so that the animals could be retained until released to their owners (*Actas de cabildo*, fol. 12v), the Tlaxcaltecah would slaughter roving cattle and sell their meat. In response, in 1568, the viceroy Martín Enriquez banned all slaughter houses in Native communities, only excluding pork under appeal, and then allowing more exceptions for towns which also housed friars and other Spaniards (Gibson, 1967, p. 153). However, daily interactions followed their own course as the rigidity of legal regulations gave way to the more practical retributive responses of the Indigenous people. An archival testimony of 1582[8] reveals the incident of three oxen that were caught devastating Indigenous fields in Apizaco. A man named Francisco Martín put them in safekeeping so that the Indigenous people would not kill the roving animals; he only let the owner, Francisco López, retrieve them when the latter had recompensed both the losses to the Indigenous people and the rescuer's work of saving the animals' lives.

While striving to limit the harm done by Spanish cattle, the *altepetl* also owned significant numbers of sheep. In December 1547, the municipal government ordered the sale of 580 sheep and to inquire about how many sheep were currently held in the wetlands of Amalinalco (*Actas de cabildo*, fol. 3v; Celestino Solís et al., 1984, p. 42). This evidence attests to the careful management of resources in economically profitable ways. Although foreign intrusions were recognized as a source of threat, the economic exchange and imports provided essential opportunities that the Indigenous people could not ignore. An important factor in play could also have been the need to solve the problem of the huge amount of unused land, which increased after the drastic demographic declInés in 1545 and in 1576–1578. Rearing livestock made it possible to avoid falling under the dreaded *tierra baldías* category that threatened ownership titles (Skopyk, 2020, p. 47). Thus, quite early on, the *cabildo* began to explore and exploit such new resources, engaging in novel economic enterprises, including buying sheep, in a conscious effort to secure sufficient food supplies in these times of climatic anomalies and epidemics that brought about depopulation, famine and growing areas of wasteland.

---

[8] AHET, Fondo Colonial, Sección Judicial, Serie Criminal, Caja 10, exp. 1,

Accordingly, three years after the traumatic epidemic of 1545, the *cabildo* decided that planting wheat on the fields belonging to the *altepetl* was an important opportunity to "multiply" its property (*Actas de cabildo*, fol. 12v: *yn atltepetl ycuentla motocaz Castilla cently yuan yn nican cently ynic momiequiliz atltepetl ytlatqui*; "Wheat [lit. Spanish corn] will be sown along with local corn on the fields of the *altepetl* so that its property will be multiplied"). This strategy required entering into arrangements with Spaniards to manage herding and to provide expert support in new agricultural investments involving European plants (Lockhart et al., 1986, p. 31). This happened in 1549 (fols. 30v–31r), when a Spanish expert was hired for two years in order to take care of the *cabildo*'s sheep and teach Tlaxcallans about breeding, as well as wool and cheese production. In 1550, the Indigenous government obliged the Spanish *corregidor* to find a knowledgeable Spaniard to herd the city's sheep, pastured over the plains of Amalinalco (fols. 55v–56r), while another foreigner was recruited to supervise agricultural works and the sowing of wheat in Chicueimalinalco, in the same region (fols. 95r–95v). This was not just a short-term strategy of assimilating and making a profit from newly available resources: evidence from the end of the seventeenth century reveals that rates of Indigenous draught animal usage were equal to or higher than those found on the majority of Spanish estates (Skopyk, 2020, p. 114).

The Tlaxcaltecah successfully employed intimate knowledge of their natural environment for agroecological adaptations, developing new niches of agricultural production and farming. As convincingly argued by Bradley Skopyk, based on local environmental knowledge they were able to develop and maintain an active and creative response to the changing social ecologies and the adverse climate change associated with the Colonial Mexican Pluvial, with its excessive cold and humidity: "native communities in central Mexico acted as the primary sites of ecological creativity, assembling knowledge and skills for cultivating native and exotic plants and animals to form new, and successful, agrosystems that defy the binary native and non-native" (Skopyk, 2020, p. 19).

Communities' Stories of Resilience: Legal Battles in the Latter Part of the Colonial Period

Along with the troubling climatic and demographic change, in the seventeenth- and eighteenth-centuries power shifted from the central government of Tlaxcallan to elected town functionaries at local levels. At the same time, the significant number of Spaniards living in the area resulted in Spanish pressure on the Indigenous authorities and the emergence of a separate Spanish jurisdiction that was not envisioned in the original sixteenth-century organization of the province. The duality of power spurred conflicts and debilitated the Indigenous *cabildo* (Martínez Baracs, 2008, pp. 352–502). Between 1590 and 1610, "caciques sold vast quantities of land to Spaniards and thereby created a new rural geography that interspersed Spanish haciendas with indigenous lands" (Skopyk, 2020, p. 116). This new socioeconomic and ethnic landscape resulted in growing pressure on Native land and other resources, despite the fact that the Indigenous population was still vulnerable to devastating epidemics and very much reduced when compared to the pre-Hispanic and early colonial periods.

I began this paper with a quote from the interview with Doña Rosaria, a healer from the municipality of Contla, formerly San Bernardino Contlan. This place is well-attested in the archival record, which offers glimpses into the collective actions of local people in the past. Thus, for example, the year 1688 witnessed a major legal struggle by the representatives of San Bernardino Contlan against Francisco and Diego González Gallardo, his son. These Spaniards and hacienda owners from Santa Ana Chiauhtempan allegedly sowed barley and cleared trees on community land located on the lower slopes of the sacred mountain, the volcano Matlalcueyeh (known today as Malinche), and bordering with the *barranca* of Quauhtzincola.[9] The barrio of Quauhtzincola was located on the wooded slopes of the mountain, in

---

[9] Los naturales y común del pueblo de San Bernardino contra Francisco y Diego González, AHET, Fondo Colonial.
Sección Administrativa, Caja 141, Expediente 13, Año 1688.

the area corresponding today with San Felipe Cuauhtenco. This forested mountainous area at the southeastern edge of the town of Contlan is clearly seen in the late colonial map of Tlaxcala (Figs. 3.2 and 3.3).

With the support of the town attorney (*procurador por el comun y naturales del pueblo*), the Natives presented their petition in the Audiencia court in Mexico, arguing that land recently restored to Francisco and Diego González Gallardo in another legal dispute with Contlan had been illegally extended to what was the property of the town. The Spaniards invaded the area and began to clear the trees, thus

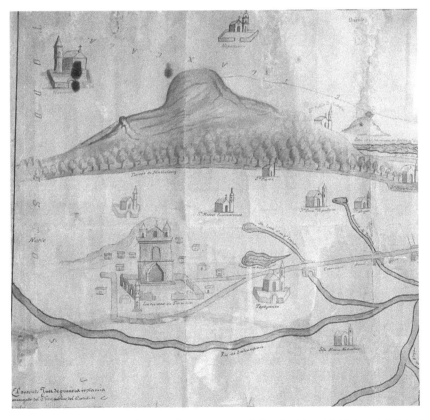

Fig. 3.2 Colonial map of Tlaxcala, courtesy of the Archivo Histórico del Estado de Tlaxcala

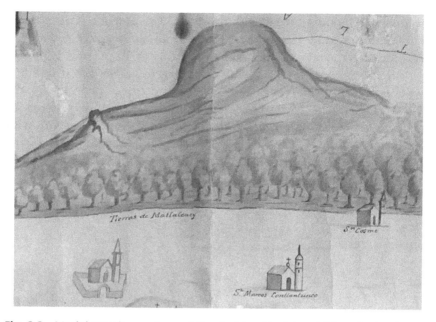

Fig. 3.3   Matlalcueyeh, courtesy of the Archivo Histórico del Estado de Tlaxcala

damaging the community's livelihood and limiting the local people's access to the *monte*, the wooded slopes of Matlalcueyeh, which provided them with firewood and other natural resources.[10] As a result, the people of Contlan secured a royal provision prohibiting the two Spaniards' usage of this land, including cultivation and the extraction of trees, thus leaving it under Indigenous control and restoring unrestricted access to the mountain to the community.

The hacienda owners rejected the accusation in January of 1689, contesting the Indigenous ownership of the land in question. In turn, the community of Contlan responded in May 1689 that the foreigners "have entered this mountainous piece of land, plowing it and clearing trees to extract what belongs to our community, having been inherited

---

[10] Pasando a desmontarlas y ararlas en grave perjuizio de d < ic > ho pueblo Y sus naturales Como tambien a ympedirles y embarazarle las entradas y salidas del monte donde no podian escusarse a yr y cortar la leña y madera para sus preçiosos menesteres.

from our ancestors".[11] Moreover, they convinced the Tlaxcallan authorities to imprison the Spaniards for crimes and violence against the town. Several Indigenous witnesses presented their testimony through the translator, attesting to having seen Francisco and Diego Gonzalez entering their land, clearing and sowing it, thus causing the local people a "violent dispossession" (*violento despojo*). In response, Don Diego de Arcos y Rivera, the *teniente general de Tlaxcala*, proceeded with the imprisonment of the accused: Francisco Gonzales was indeed arrested but his son could not be found. However, the *hacienda* owner argued that the land in question was part of another dispute with the community and hence not in their legal possession. He also asked to be released from prison because of his health issues and advanced age.

At that point the *teniente general* resumed the case, admitting more witnesses and agreed to release the Spaniard with bail. In the course of the litigation, Francisco Gonzales accused the people of Contlan of being "trouble-makers and instigators who go around pursuing lawsuits without legal basis or right"[12] and claimed to have worked on the land for two preceding years without any objections from the local people. Indeed, some Spaniards testified in favor of the *hacienda* owner, stating that they had seen him working this parcel in 1688 and 1689 without complaints from the community and that it formed part of the disputed land in the Real Audiencia court. Nonetheless, a number of Indigenous witnesses confirmed that this piece of land was separate from that under litigation. In the testimonies they referred explicitly to their intimate knowledge of local territory ("por el conosim<ien>to que tiene de ellas"), an important detail that the notary decided to record each time. The authorities of Tlaxcallan, headed by the Spanish *teniente*, recognized the validity of the Indigenous claims, refusing the *hacienda* owner the possibility of appeal as they considered his claims groundless. The documentation of the case in the archive of Tlaxcala

---

[11] "nos querellamos criminal mente de fran<cis>co gonsales gallardo Díego gonsales gallardo su hijo en racon de que Los suso d<ic>hos dem autoridad se han entrado en vn pedaso de tierra grande montuoso harandolo í descascarando los arboles p<ar>a derívarlos lo qual es proprio de n<uest>ra comunidad. havído y heredado de n<uest>ros antepasados"

[12] "inquietadores y perturbadores y andar moviendo pleytos sin accion ni derecho"

ends with the request of Gonzalez Gallardo to the Real Audiencia in Mexico City to revoke the royal provision conceded to Contlan and to transfer the documents to the court of appeal.

The legal regulations for New Spain maintained the principle of equal access for both Indigenous people and Spaniards to unoccupied forests (*monte*) and grasslands for foraging, firewood collection and hunting (Greer, 2018, p. 254). In the case of the disputed land, the court case implies that its possession was strategic to the community, probably not only because of the access to the forest, as claimed in the legal arguments, but also for cultivation purposes on the lower slopes of the volcano. There were already many Spanish *ranchos* located on the slopes of Matlalcueyeh at that time, including in the area around Santa Ana Chiyauhtempan (Skopyk, 2020, p. 122), so the pressure on Indigenous land was growing. In a responsive adaptation to adverse climate conditions, farmers in Tlaxcallan developed large-scale cultivation of maguey (agave) plants on hilltop terraces for the purposes of the pulque production that was booming for most of the seventeenth century. But this form of land management turned out to be destructive to the environment: in the long term this cultivation strategy, combined with violent climatic anomalies, caused serious erosion problems. The Crown's withdrawal of tax exemption for *pulque* production additionally contributed to the decline of maguey cultivation, field abandonment and mass landslides, as well as both desiccation and flooding (Skopyk, 2020, pp. 123–129).

Maguey cultivation was also vital in the area of San Bernardino Contlan and this community was responsible for violent *pulque* riots in 1680, opposing the abolition of the tax exemption (Martínez Baracs, 2008, p. 389). The legal struggle for land on the slopes of Matlalcueyeh that unfolded in the same decade confirms that hilltop cultivation areas and forest resources were vital for the community. Its residents showed particular determination in their collective action against Spanish intruders: we can only speculate as to whether the crisis of *pulque* production was among the factors that contributed to the community efforts undertaken to secure the vital resources of the town. What we do know, however, is that Contlan and other towns in Tlaxcallan were entering into turbulent times that made their economic and demographic situation very precarious, especially in the subsequent two decades.

The importance of *monte*, or wooded land adopted for local farming and settlement, is also salient in the history of San Toribio Xicohtzinco, located in the lower area of Río Zahuapan where it joins the upper basin of the Atoyac River. In 1720, the Indigenous people of San Toribio Xicohtzinco brought a charge against a mulatto, Francisco Antonio, who was the slave and *mayordomo* of the infantry captain, Juan de Torres.[13] They accused him of closing the town's points of entry, and of beating and threatening the Indigenous community members who tried to pass. As for physical abuse, the slave knocked a local constable (*topileh*), Joseph Marcos, to the ground with a horse and beat him up. On a Sunday morning, he then whipped another *topileh*, Juan Thomas, when the constable was on his way to attend mass. The *mayordomo* also tried to shoot him with a musket, but others intervened. The Natives managed to put the mulatto in prison for the duration of the legal proceedings, while the judge made sure that Don Juan de Torres did not leave the town during the course of the trial.

Juan Thomas, the *topileh* who had suffered violence at the hands of the slave of the *hacienda* owner, stated in his testimony that the community possessed a map and writing proving that from time immemorial they had had "the usage and custom" (*el uso y costumbre*) of carrying out commercial activities on the roads that the mulatto and his owner had closed. This, according to this official, violated the laws of the Indies, which guaranteed free access to public roads. Indeed, the map preserved as part of the legal documentation features not only the land of the community, but also all the roads, where they lead to and even who is in charge of maintaining them (Fig. 3.4). The dimensions of the town, which is identified with the church, are specified, as is the area with individual houses called *quauhtla*, or "monte", probably a wooded area adapted for residence, and the official responsible for it, the *quauhtlatlalpixqui*. It also contains a short genealogy going back to Don Luis Xicotencatl (Fig. 3.5). Below him is a person called

---

13 AHET, Fondo Colonia, Siglo XVIII, Serie Administrativa, año 1720-1722, Caja 56 Exp. 41; fragments of this legal case were published by Meade de Angulo (1985).

Tecpaxochitzin Maxixcatzin, also mentioned in documents from the local archive in Xicohtzinco as one of the women who donated land to the town (Macuil Martínez, 2017, p. 100).

We learn more about the meaning of this imagery from the Nahuatl text on the reverse side of the page. According to this late colonial rendering of local sixteenth-century history, when the viceroy Don Diego de Mendoza[14] arrived, the rulership began in the town and was established precisely by Don Luis Xicotencatl. He also divided the land and assigned it to local rulers along with the *tlahuiztli*—the insignia in form of a heron device. The establishment of the *cabildo* followed and its officials were elected on a yearly basis. The rulers and officials were

Fig. 3.4 Map of San Toribio Xicohtzinco, AHET, Fondo Colonia, Siglo XVIII, Serie Administrativa, año 1720–1722, Caja 56 Exp. 41. Courtesy of the Archivo Histórico del Estado de Tlaxcala

---

[14] No doubt reference to Don Antonio de Mendoza, the first viceroy of New Spain who held office from 1535 to 1550.

Fig. 3.5 Genealogy accompanying the lawsuit between San Toribio Xicohtzinco and hacienda owners, AHET, Fondo Colonia, Siglo XVIII, Serie Administrativa, año 1720–1722, Caja 56 Exp. 41. Courtesy of the Archivo Histórico del Estado de Tlaxcala

to keep *yn tlahuistli motocayotia astatetl* ("the insignia called heron" [literally: "heron stone"]) along with the rights to the land referred to as both *tlalli* and *quauhtla* ("land and forest"), also depicted in the document. This corresponds with *aztatototl xicotzinco axcatl,* a strange non-possessive construction apparently meaning "the heron bird, the possession of Xicohtzinco", found in the genealogy section. It was indeed the pre-Hispanic insignia most closely identified with the Tlaxcallan *altepetl* and also became its main heraldic device in the colonial period. By highlighting this association, the people of San Toribio Xicohtzinco not only argued for the legitimacy of their town and titles to land but also attempted to link it to the very core of the Tlaxcallan *altepetl* as a source of legitimacy, including on the symbolic and visual level. The document was supposedly made 25 years after the arrival of the Viceroy Mendoza (1535), who confirmed the rights of the town so that their borders would never be questioned. Orthographic details, however, leave no doubt about the late colonial date of this manuscript, which is roughly contemporary with the legal suit.

By now it is clear that the whole fight was not about roads: the roads were only a means to open access to community's land. In response to the presentation of the pictorial and textual documents and to substantiate his own claims, the *hacienda* owner brought in much earlier legal documentation that he had received from his *causantes*, the persons from whom he derived his own rights to the *hacienda*. According to this evidence, on June 22, 1682, in the presence of the Indigenous authorities of Tlaxcallan, including the governor, the land was measured 500 *varas*[15] from the church in San Toribio toward each of the cardinal directions. The borders were marked with crosses and maguey plants. During the process, two neighbors intervened. They pointed out a contradiction based on the alleged circumstance that the land around the church belonged to Doña Ana de Anzures and her *hacienda*. However, the authorities assured that the measurements were correct from the legal point of view. To confirm this, all of the community members threw stones in an act of taking possession. The legal basis for the ceremony was provided by the royal decree of 1567, which determined the legal status of the so-called land *por razón de pueblo* ("by right of township"; called *fundo legal*, or "legal basis", in the nineteenth century), specifying its perimeters at a minimum distance of one thousand *varas* that should be kept between an Indigenous town and the nearest private estate (the center being measured from a church or *hermita*) and setting the town center with five hundred *varas* in each of the four cardinal directions (Wood, 1990, p. 118). A significant change came with the 1684 *ordenanza* of the Real Audiencia that gave actual possession to the Indigenous people of the land contained within a radius of 500 *varas*, measured from the most peripheral construction in the town (Castro Gutiérrez, 2015, p. 77). This was only two years after the process started and a year before it officially ended.

---

[15] The *vara* ranged from ca. 0.83 to 0.85 meters depending on time and place (Chardon, 1980, p. 295).

The *hacienda* owners appealed this decision. They brought the following argument: there was no church, but only *una hermita* constructed by an Indigenous woman on a little piece of land that belonged to her. Along with other people, she then obtained permission to celebrate holy mass there. So, according to the Spaniards, the Indigenous people of the larger neighboring town of Santa Inés Zacatelco in fact tried to establish a separate town based on this hermit shrine. The *hacienda* owners also claimed that San Toribio Xicohtzinco did not have a governor, *alcaldes* or other officials, nor a sufficient number of houses; in fact, there were only five houses and that it was a subject settlement of Santa Inés Zacatelco. However, Xicohtzinco's existence is confirmed in the census of 1556–1557 (Gibson, 1967, p. 131). Around 1554, it had an *iglesia de visita* dependent on the convent of Tepeyanco, which in 1646 shifted under the jurisdiction of the *doctrina* of Santa Inés Zacatelco (Gonzáles Jácome, 2008, p. 266). In response to the allegations of the *hacienda* owners, the community made a legally valid argument that there is a general *altepetl* of Tlaxcallan and its governor is also the governor of all towns and *sujetos* in the province, including San Toribio Xicohtzinco. However, the Audiencia decided to restore the land to the *hacienda* owners in 1685, although they would not be able to execute the decision for years. The community opted to ignore the decisions coming from Mexico City. Ironically, the decree of 1687 by king Charles II extended the lands surrounding and pertaining to any Indigenous center by another one hundred *varas* in each of the four cardinal directions from the last houses and called for eleven hundred *varas* to the nearest estate; in other words, all settlements became entitled to the land *por razón de pueblo*, not only head towns with a governor or a church but also all *sujetos* (Goyas Mejía, 2020, p. 79). This meant that the case made by the people of Xicohtzinco was perfectly justified even if they did not modify the original claim of taking the location of their church as the point of reference. Later on, the decree of 1695 revoked the "last house" rule, ordering that the six hundred *varas* be measured from the principal church in each town (Wood, 1990, p. 118), but local claims were made in accordance with this general rule.

We do not have information regarding what happened after that, until the next trial in 1720, when the community brought the case against the slave of the *hacienda* owner. We know, however, that across New Spain communities' interest in securing the legal endowment *por razón de pueblo* gained new strength after the legal revisions of 1687, although not without opposition from land owners. Their efforts continued even after the less favorable ruling of 1695, also invoking the legal acts as a means to restore the possession of land that had earlier been lost (Wood, 1990, pp. 118–122). When the people of Xicohtzinco restored the case, the owner at that time, Juan de Torres repeated the same argument that had been successful forty years before, entirely ignoring the fact that the legal circumstances had changed. He stated that "It was not for any other reason that they were expelled and their houses demolished, than that they were illegally built on the land of the *hacienda* that today is mine". In the same way, he says, they do not have any rights to roads "because they have nowhere to enter and from nowhere to leave, so for what purpose do they need them?".

The community responded in 1721 that access to the shrine was fundamental because "this is the way to the church of this parish where the mass is celebrated on Sundays [...] and it is necessary for the divine worship". Moreover, "today the parish is composed of eighty families, which is sufficient to form a town". What is more, according to their argumentation it was in the best interests of the Crown that "the native people have land to sustain their families and contribute to the royal tribute and responsibilities toward the church". Accordingly, the people of Xicohtzinco challenged whether defending the Spanish *hacienda* owner was indeed in the best interests of the Crown: "What is the utility of an individual who with his rancho causes harm to the poor *indios,* mistreating them as we have documented?" As had happened with the previous litigation with Doña Ana de Anzures, Don Juan de Torres also died during the process. The documentation ends with his son Don Joseph de Torres asking for an appeal. The individual *hacienda* owners died and changed during the course of the litigation, while the community persisted with its claims, actions and occupation of the land they considered their own.

We can speculate about how much the progressing climatic, environmental, socioeconomic and demographic changes could have contributed to the Indigenous determination to secure the *monte* area for their settlement and probably also for cultivation. The time period during which the long-term battle between the people of San Toribio Xicohtzinco and Spanish *hacienda* owners took place was harsh: in 1691 a huge climatic crisis began with disastrous frosts, while epidemics, hunger and cold weather persisted until 1697, only to be followed by hard years of drought, forcing many people to emigrate to Mexico City (Skopyk, 2020, pp. 123–125). While depopulation deepened due to repeated waves of epidemics, there were also other factors affecting the socioeconomic strength and structure of Indigenous settlements. The long-term colonial project of *congregaciones* gained some success not earlier than in the first decade of the seventeenth century, but it wasn't until 1749 that the Tlaxcallan *cabildo* announced that the resettlement has been completed because people had abandoned their lands and homes in the "hills and ravInés", a process that reinforced the division between center/periphery and infield/outfield (Skopyk, 2020, p. 87). However, as the concentration of centralized Indigenous towns increased, they suffered from a growing encroachment of Spanish *estancias* and *haciendas* that made it difficult for Native settlements to expand in moments of demographic rise, as, for example, in the case of San Juan Ixtenco, which was forced to establish the satellite community of San Pedro Cuauhtla in 1681 (Martínez Baracs, 2008, p. 460). This seems to be a paradox considering that throughout the seventeenth and eighteenth centuries, a huge percentage of land in Tlaxcala remained unused or underused (Skopyk, 2020, p. 103), but what Indigenous people fought for in the eighteenth century was quality, well-watered land or areas safe for settlement.

In the case of Xicohtzinco, the litigation makes it clear that this wooded land was crucial for the community, probably because the valley land was insufficient, too greatly transformed by geomorphological processes or too prone to inundation, as it was located in the area of confluence of the Zahuapan and Atoyac Rivers. The flood frequency of the Zahuapan in Tlaxcala increased from 1650 and especially in the period 1700–1774, while sand sedimentation affected low-lying agricultural fields; the situation was especially alarming downstream of the city of Tlaxcallan. The town of Nativitas, located between the city of Tlaxcallan and Xicohtzinco, was at the mercy of the unpredictable waters of the Zahuapan, which destroyed bridges and buildings and flooded land and roads, including the *camino real* between Tlaxcallan and Puebla (Skopyk, 2020, pp. 136–148). Encroached on by Spanish *hacienda* owners and affected by climatic anomalies and diseases, the people of Xicohtzinco struggled for their rights and for the integrity and well-being of their community through persistent collective action. Eventually, in 1741, they managed to purchase the *hacienda*, but were forced to sell it two years later because they lacked a sufficient workforce due to the epidemics of 1737 (Gonzáles Jácome, 2008, pp. 266–267).

The microhistorical insights into the struggles undertaken by Tlaxcaltecah communities show that the premonitions of the Indigenous *cabildo* in the sixteenth century regarding the adverse impact of Spanish settlers on the Native population, land and natural resources were indeed right. In this late colonial era, replete with disasters and socioeconomic challenges, the struggle over vital resources was often linked to the activities and abuses of Spanish land and cattle owners. At much the same time as the two cases just discussed, on the other side of the volcano Malintzin, where it descends toward the town of Huamantla, housing not only an Indigenous but also a growing Spanish and mestizo population, a water crisis developed in 1705.[16] According to the denouncement, Cristóbal and Gabriel Báez, mestizos from the *hacienda* of Andrés Báez, caused serious damage to the pipes that brought water from the mountain's springs to the town, causing it to "suffer from many calamities because of the lack of water".[17] The mestizos filled the aqueduct with plants so that water would not descend to Huamantla, but redirected its flow to feed their own cattle. As a result of the investigation, the people of the *hacienda* were imprisoned, fined 25 pesos for the aqueduct repair costs and obliged to

---

[16] AHET, Fondo Colonial, Sección Judicial, Serie Criminal, Caja 6, Exp. 17, Año 1705.

[17] "a benído el pueblo a padeser munchos Calamidades por la falta del agua"

make sure that neither they nor other *hacienda* workers would cause any damage to the water pipes in the future but that they would respect the use of this important resource by the residents of the town. Indeed, climatic anomalies and provincial authorities' inadequate political and administrative crisis management, as well as the economic and demographic challenges of that turbulent period, affected Indigenous, mestizo and Spanish residents of the province, sometimes creating space for cross-ethnic alliances directed against abusive landowners.

Traditional Knowledge, Healing, Language and Well-Being

But stories of resilience and resistance are not limited to struggles for land. They also extend to defending key traditional practices and ways of healing as components of local well-being. For example, Spaniards were astounded by the Native custom of regular sweat bathing and scandalized by the fact that women and men bathed together. In the oldest known dated document in Nahuatl, a 1543 act against idolatry from Tlaxcallan, "those of us men who bathe themselves together with women, those who bathe themselves in public" are listed among the abominable sinners subject to persecution (Olko & Brylak, 2018). Such baths, performed in steam houses (*temazcalli*), had crucial ritual, purifying and curative functions. It was essential to keep the ritual equilibrium of female and male participants in this liminal place. Spaniards had a hard time eradicating this custom. One example of Indigenous resistance to the Spanish attempts to eliminate this fundamental practice comes from 1706. It is a criminal case that followed the reissued decree, announced in public plazas in the communities,[18] prohibiting joint female–male sweat baths under severe penalties. The Natives rebelled against this prohibition. Indigenous laborers of the *hacienda* of Masegualpa, after receiving this warning from a Spanish official who found them cleaning a *temazcalli*, became extremely angry (*se alteraron*), to the extent that they started a

---

[18] AHET, Fondo Colonial, Sección Judicial, Serie Criminal, Caja 6, Expediente 51, Año 1706.

commotion (se *yvan juntando con alboroto*), chasing away the intruder, who later filed a complaint against them.

As this eighteenth-century archival document shows, Spaniards met with persistent local resistance from the Native people with regard to their obedience to this prohibition. Today, the *temazcales* persist in numerous Nahua communities and are often to be found in Tlaxcalan households. However, according to local women from San Miguel Xaltipan in the municipality of Contla, keepers of both the Nahuatl language and intimate knowledge of traditional healing, "before all people would bathe in the sweat bath" (*achto nochi omotemaya de temazcal*), including small children, who "would take sweat baths, but not anymore" (*ometemaya ahorita yocmo*), but in the present day just a few of them are taken to *temazcales* (*motema, pero zan cequi, occequi yocmo*, "they take sweat baths, but only some, others not anymore"). The lack of continuity of traditional practices—expressed by the oldest community members—is accompanied by the sense of loss of natural resources. As signaled at the beginning of this text, many herbal plants used in healing and grown in and around the town have disappeared, while community members are more and more reluctant to rely on traditional knowledge transferred across generations: *Yocmo catqui in xihuitl de non achto, den achto oyeya pero miec xihuitl itech in tlalli* ("The herbs of the past are gone, before there were many plants [growing] on the land"); *yocmo cah quimi in achto, achto oyeya miec xihuitl tlen timopahtiayah* ("there are not [plants] any more like as in the past, earlier there were many herbs that we used for curing"); *yopolihqueh, yopoliuh ihuan non tanto cualimeh, cualli in pahtli* ("they have already disappeared, it is gone and they were so good, good medicine").

The loss of crucial healing rituals and natural resources assuring individual and collective health and well-being is accompanied by the shift to Spanish and reluctance to learn the language of the grandparents: "They do not want to learn" (*Amo quinequih zalozqueh*), the elders repeatedly say about their grandchildren. And yet, our most recent research shows that speaking Nahuatl is related to speakers' well-being in Contla; surprisingly, when compared across four different Mexican regions where this Indigenous language is still spoken, the effect is the strongest in the most urbanized and assimilated region of

Contla, despite the fact that it suffers from the deepest loss of the heritage language and of the traditional knowledge conveyed in it (Olko et al., 2022).

Summary and Conclusions

It has been argued that the rhetoric of the Tlaxcaltecah employed in their colonial-period petitions and court documents reflects their high-level ability to forge skillful strategies and formulate effective legal arguments, but does not necessarily reflect their "lived experience" or "a native reality" (Baber, 2011, p. 42). However, when confronted with the internal sixteenth-century records of the Native *cabildo*, along with the disastrous demographic, climatic, environmental and socioeconomic data, the voices of the inhabitants of the province struggling to protect their land, resources and well-being can be viewed in a much sharper and more dramatic light. The mid- and long-term consequences of processes set into motion in the early colonial period, such as the increasing role of Spanish landholders, also proved that the concerns and sinister prognoses of the Native government were not exaggerated. However, both the Tlaxcallan elites and the commoners were by no means passive observers of the calamities and challenges that became an inadvertent part of their daily reality. As aptly summarized by Skopyk, "even in the context of sickness and death, the pluvial fueled a creative and productive response by local communities as they sought to protect resources and find ways to profit from the changing ecology" (Skopyk, 2020, p. 34). The pressures and threats that initially manifested themselves in the first decades after the arrival of Spaniards continued across the colonial era, but the Tlaxcaltecah were able to use a number of strategies to protect their land and their well-being.

Colonial legal battles show that collective threats generated collective responses. Local resilience was grounded in the ability to maintain and mobilize social networks for action. While in the sixteenth century, the response was at the level of the *cabildo*, of Tlaxcallan, later on the struggle became decentralized, with local communities taking on a number of initiatives and strategies to safeguard their own resources.

Both leaders and community members actively and resourcefully protected essential assets and components of their well-being, responding not only to abuse and violence but also to climatic and epidemiological challenges. And despite all the complexity and magnitude of challenges, they have been able to keep much of their traditional economy and industries, as well as their language, cultural practices and forms of organization, until the second half of the twentieth century, and in some aspects until the present day. Social approaches to agency developed in the context of contemporary societies emphasize the dualism of the "individual" and "society" or "structure" that is both "enabling and constraining" (Giddens, 1984, p. 162, 169), while social actors are seen as reproducing or transforming social systems through praxis (Giddens, 1984, p. 171; see also Sztompka, 1991). The history of Indigenous people in the colonial period reveals a demonstrable collective and social level of agency and resilience based on the ability to mobilize social networks for support and action, grounded in traditional power structures, even if channeled through specific groups of actors. The latter could be the *cabildo*, composed of local nobility or town representatives acting in the name of whole communities, as seen in the cases of San Bernardino Contlan and San Toribio Xicohtzinco.

Many legal steps turned out to be of little effect, and some adverse processes were ultimately unstoppable. However, historical texts reveal complex battlegrounds upon which the Tlaxcaltecah strove to maintain control over land, environment and various kinds of resources. It is through these texts that we can see how the Indigenous inhabitants of this particular territory protected essential assets and components of their well-being that were embedded in traditional concepts of power and local knowledge. When seen separately and in their own contexts, particular historical documents do attest to Indigenous agency but often do not immediately make us aware of a longer historical process in which some of these battles were won and some were eventually lost. Yet this can be seen when one ties together the common threads of microhistories—some of which are discussed in this paper—across longer periods of time, different places and available documentary genres; and it is even more salient if we acknowledge that the aftermath of these apparently remote historical processes continues to unfold to this day. Thus, there is no contradiction between the *longue durée* approach to history and microhistory (Frankopan, 2021, p. 25), as only together they provide the necessary bridge between past and present. In the case of Tlaxcala, the erosion of the traditional economic base, accompanied by accelerating assimilation to the dominant Spanish/Mexican culture, can be seen as socioeconomic and sociocultural processes set into motion in the colonial period and related to land ownership, control of natural resources and acculturation processes.

Modern communities in Tlaxcala have suffered intense and prolonged discrimination, forced assimilation and an accelerated language shift to Spanish (e.g., Hill & Hill, 1986; Nutini & Isaac, 2009; Robichaux, 2005). The loss of language and profound changes in ethnic identification are accompanied by the loss of traditional knowledge and of environmental resources such as medicinal plants. The ongoing urbanization processes and changes in local economy, including farming methods, are often perceived today, especially by elder community members, as threats to their well-being and as direct causes of health problems affecting the Indigenous population. As shown in this paper, the sense of loss can be traced back to the sixteenth century, even if the partial nature of our evidence often leaves us with a history of silences, gaps and erasures. This reminds us that "the remembrance of absence, of loss that endures, a void from which we can still learn" (Hicks, 2021, p. 114) should be recognized as an inherent part of Indigenous history.

Tlaxcalan history is yet another example of the deep intertwining of human action and environmental change, with nature playing an active role in historical processes. It showcases the complexity of human–environment relations that have been affected both by climate change and by colonization and postcolonial domination. Intimate links with the environment, and especially with sacred mountains such as Matlalcueyeh/Malintzin, have marked crucial relationships that continue to be perpetuated in local communities through healing practices and story-telling (e.g., Nava Nava & Cuahutle Bautista, 2015). And not unlike in colonial times, the Tlaxcaltecah today continue to face ongoing threats to the integrity of their communities, the control of resources, a healthy environment, cultural survival and a local, relational sense of well-being.

References

Acocal, S. (2020). *Nobleza india de Tlaxcala en el siglo XVI. Gobierno de San Pablo Quauhtotoatlan* (Doctoral dissertation). Escuela Nacional de Antropología e Historia.

Baber, J. R. (2005). *The construction of empire: Politics, law and community in Tlaxcala, New Spain, 1521–1640* (Doctoral dissertation). University of Chicago.

Baber, J. R. (2011). Law, land, and legal rhetoric in colonial new Spain a look at the changing rhetoric of indigenous Americans in the sixteenth century. In S. Belmessous (Ed.), *Native claims: Indigenous law against empire, 1500–1920* (pp. 41–62). Oxford University Press.

Carr, E. H. (1987[1961]). *What is history? Second edition*. Penguin Books.

Carr, H., & Lipscomb, S. (2021). Prologue: Ways in. In H. Carr & S. Lipscomb (Eds.), *What is history, now? How the past and present speak to each other* (pp. 3–16). Weidenfeld & Nicolson.

Castro Gutiérrez, F. (2015). Los ires y devenires del fundo legal de los pueblos indios. In M. del Pilar Martínez López-Cano (Ed.), *De la historia económica a la historia social y cultural. Homenaje a Gisela von Wobeser* (pp. 69–104). Universidad Nacional Autónoma de México, Instituto de Investigaciones Históricas.

Celestino Solís, E., Valencia R., & Medina Lima, C. (1984). *Actas de cabildo de Tlaxcala, 1547–1567*. Archivo General de la Nación, Instituto Tlaxcalteca de la Cultura.

Chardon, R. (1980). The Elusive Spanish League: A problem of measurement in sixteenth-century New Spain. *The Hispanic American Historical Review, 60*(2), 294–302. https://doi.org/10.2307/2513219

Frankopan, P. (2021). Why global history matters. In H. Carr & S. Lipscomb (Eds.), *What is history, now? How the past and present speak to each other* (pp. 17–32). Weidenfeld & Nicolson.

Gibson, C. (1967). *Tlaxcala in the sixteenth century*. Stanford University Press.

Giddens, A. (1984). *The constitution of the society. Outline of the theory of structuration*. Polity Press.

González Jácome, A. (2008). *Humedales en el suroeste de Tlaxcala: Agua y agricultura en el siglo XX*. Universidad Iberoamericana.

Goyas Mejía, R. (2020). Tierras por razón de pueblo ejidos y fundos legales de los pueblos de indios durante la época colonial. *Estudios De Historia Novohispana, 63*, 67–102.

Greer, A. (2018). *Property and dispossession: Natives, empires and land in early modern North America*. Cambridge University Press.

Hicks, D. (2021). Glorious memory. In H. Carr & S. Lipscomb (Eds.), *What is history, now? How the past and present speak to each other* (pp. 101–115). Weidenfeld & Nicolson.

Hill, J. H., & Hill, K. C. (1986). *Speaking Mexicano. Dynamics of syncretic language in Central Mexico*. University of Arizona Press.

Lockhart, J. (1992). *The Nahuas after the conquest: A social and cultural history of the Indians of Central Mexico, sixteenth through eighteenth centuries*. Stanford University Press.

Lockhart, J., Berdan, F., & Anderson, A. J. O. (Eds.). (1986). *The Tlaxcalan Actas. A compendium of the records of the Cabildo of Tlaxcala (1545–1627)*. University of Utah Press.

Lopes Don, P. (2010). *Bonfires of culture. Franciscans, indigenous leaders, and the inquisition in early Mexico, 1524–1540*. University of Oklahoma Press.

Macuil Martínez, R. (2017). *Los tlamatque, guardianes del patrimonio: dinámicas interculturales en la Sociedad Naua (México)* (Doctoral dissertation). Leiden University.

Magnússon, S. G., & Szijártó, I. M. (2013). *What is Microhistory?* Routledge.

Martínez Baracs, A. (2008). *Un gobierno de indios: Tlaxcala, 1519–1750*. Fondo de Cultura Económica, Colegio de Historia de Tlaxcala and CIESAS.

Meade de Angulo, M. (1985). *Dos códices del pueblo de San Toribio Xicohtzinco, Tlaxcala. Facsimilares y transcripcin del documento*. Colección de Amatlatocayotl. Instituto Tlaxcalteca de Cultura.

Melville, E. G. K. (1999). *A plague of sheep. Environmental consequences of the conquest of Mexico*. Cambridge University Press.

Nava Nava, R., & Cuahutle Bautista, B. (2015). *Tlahtolcozcatl. In tlapohual tlen mocaqui nican Tlaxcallan*. Faculty of "Artes Liberales", University of Warsaw.

Navarrete Linares, F. (2019). "Las historias tlaxcaltecas de la conquista y la construcción de una memoria cultural." *Iberoamericana, XIX* (71), 35–50. https://doi.org/10.18441/ibam.19.2019.71.35-50

Nutini, H. H., & Isaac, B. L. (2009). *Social stratification in Central Mexico*. University of Texas Press.

Olko, J., & Brylak, A. (2018). Defending local autonomy and facing cultural trauma. A Nahua order against idolatry, Tlaxcala 1543. *Hispanic American Historical Review, 98*(4), 573–560. https://doi.org/10.1215/00182168-716 0325

Olko, J., Lubiewska, K., Maryniak, J., Haimovich, G. de la Cruz, E. E., Cuahutle Bautista, E. Dexter-Sobkowiak, B., & Iglesias Tepec, H. (2022). The positive relationship between Indigenous language use and community-based well-being in four Nahua ethnic groups in Mexico. *Cultural Diversity and Ethnic Minority Psychology, 28*(1), 132–143. https://doi.org/10.1037/cdp0000479

Radding, C. (1997). *Wandering peoples: colonialism, ethnic spaces, and ecological frontiers in Northwestern Mexico, 1700–1850*. Duke University Press.

Robichaux, D. (2005). Identidades cambiantes: "Indios" y "mestizos" en el suroeste de Tlaxcala. *Relaciones, 104*(26), 58–104.

Sánchez Verín, C. A. G. (2002). *Obrajes y economía en Tlaxcala a principios del siglo XVII, 1600–1630*. Archivo General de la Nación.

Skopyk, B. (2020). *Colonial cataclysms. Climate, landscape, and memory in Mexico's little Ica Age*. University of Arizona Press.

Sullivan, J. (1999). Un diálogo sobre la congregación en Tlaxcala. *Colonial Latin American Review, 8*(1), 35–59.

Sztompka, P. (1991). *Society in action. The theory of social becoming*. University of Chicago Press.

Wood, S. (1990). The fundo legal or lands Por Razon de Pueblo: New evidence from Central New Spain. In A. Ouweneel & S. Miller (Eds.), *The Indian community of colonial Mexico fifteen essays on land tenure, corporate organizations, ideology and village politics* (pp. 117–129). Centrum voor Studie en Documentatie van Latijns Amerika, CEDLA.

4

Ñudzahui Custom, Contracts, and Common Lands in Eighteenth-Century Oaxaca

Yanna Yannakakis

Land represented a nexus for Indigenous customary claims in Mexico's colonial courts across three centuries of Spanish rule. The Crown recognized the lands of the Native nobility, who were considered Mexico's "natural lords," as entailed estates under the designation of *cacicazgo*. The question of which noble lineages controlled what lands provoked much controversy, as did the use-rights of commoners. From the 1530s through the 1570s, in response to challenges mounted by Native rivals or Spanish colonists, Native people appeared in front of the Real Audiencia to claim and dispute ownership or use-rights based on pre-Hispanic custom. They often substantiated their claims with recourse to painted histories and maps. Due to the moral authority that Spanish judges accorded to pre-Hispanic institutions and the practicality of maintaining aspects of Indigenous land tenure and labor, they often affirmed Native claims based on custom (Kellogg, 1995, pp. 45–51; Menegus Bornemann, 1994a, 1994b; Ruiz Medrano, 2010,

Y. Yannakakis (✉)
Emory University, Atlanta, GA, USA
e-mail: yanna.yannakakis@emory.edu

J. Olko and C. Radding (eds.), *Living with Nature, Cherishing Language*,
https://doi.org/10.1007/978-3-031-38739-5_4

pp. 31, 38–39; Villella, 2016). In doing so, though, they did not simply affirm established Native rights; they produced new ones by incorporating Native customs into a Spanish normative order (Herzog, 2013).

Native customary claims to land fell off considerably by the end of the sixteenth century, in part because of Spanish officials' changing attitudes toward the pre-Hispanic past and Indigenous forms of knowledge, and also because of the exploitative economic demands of Spanish officials and colonists, and the ravages of epidemic disease. In response to a precipitous decline in the Indigenous population, and with the ambition of nucleating the survivors into concentrated settlements to facilitate evangelization, organization of corvée labor, and tribute collection, the missionaries spearheaded a program of forced resettlement, known as *congregación* or *reducción*, depending on the region. This took place in two waves, the first in the mid-sixteenth century from 1550 to 1564, and the second from the 1590s through the first decade of the seventeenth century (Borah & Cook, 1963; Cline, 1949; Cook & Borah, 1971). Congregación and population decline displaced Native people from their ancestral lands and left much of it vacated, opening local communities to expropriation by Spanish colonists (García Martínez, 1987). In order to stem the chaotic expansion of Spanish property, and harness land distribution and titling to royal power, Philip II issued a royal *cédula* (decree) in 1591 requiring Spanish colonists to present their land claims and titles so that they could be validated by Spanish law. If they possessed land but had no title, they could acquire one for a fee. Any vacant lands without proper title—known as *tierras baldías*, or simply, *baldíos*—would escheat to the Crown so that they could be used, sold, or distributed at royal discretion (Menegus Bornemann, 1994b; Solano, 1984, pp. 273–274). The fees would go into the royal treasury to ameliorate the Crown's ailing finances. The royal land titling program, known as the *composiciones de tierras*, persisted through the seventeenth century, allowing for a massive transfer of lands from Indigenous to Spanish control (Menegus Bornemann, 1994a; Ots Capdequí, 1946).

Native communities went to court to defend their lands, and they presented varied kinds of evidence to prove possession and make customary claims to land use and tenure, including Spanish legal instruments that recognized Indigenous possession. Native people also resorted to local knowledge and forms of representation to support their claims, despite Spanish disinclination to take such evidence seriously. These included maps that blended European and Indigenous styles, and a genre of painted histories and genealogies known as the Techialoyan codices (Wood, 2007, 2012). Native towns also generated a written genre of Native-language documents known as primordial titles that recounted the migration of the community's founding ancestors, the marking of territorial boundaries, the consecration of the community's Church, and the establishment of the Native town council (*cabildo*) (Cortés Márquez & Reyes García, 2004; Haskett, 2005; Oudijk, 2003; Menegus Bornemann, 1994b, p. 208; Oudijk & Romero Frizzi, 2003; Romero Frizzi & Vásquez Vásquez, 2003, 2011; Sousa & Terraciano, 2003). Across these encapsulations of Indigenous memory, Native writers and painters reimagined antiquity, the anchor of custom, by braiding together pre-Hispanic and colonial symbols, histories, and chronologies (Megged & Wood, 2012). Despite Native efforts to stem the tide of dispossession, however, the *composiciones de tierras* continued to transform Indigenous lands into Spanish haciendas and ranches in many regions of New Spain and across Spanish America (Goyas Mejía, 2015; Borchart de Moreno, 1980; Florescano, 1971, p. 44; Glave, 2008; Rivera Marín de Iturbe, 1983; Torales Pacheco, 1990).

The recovery of the Indigenous population at the end of the seventeenth century combined with the Crown's desperate need to generate revenue to pay for its expensive imperial wars created a shift in land policy, which in turn reinvigorated Native claims to land.[1] Until 1691, the policy of composición applied to Spanish and mestizo colonists, but from 1692 forward, the Crown widened its net to require Native towns to title their lands. In order to do so, they needed to

---

[1] Based upon the Montemayor census of 1661, Borah and Cook's classic study estimates the population of the Mixteca Alta to have been 22,841 during the period 1660–1670. Based upon censuses of colonial jurisdictions, parishes, and towns, their population estimates for the eighteenth century are as follows: 1715–1720: 41, 687; 1742: between 49, 133, and 54, 200; 1750–1755: 61,272; 1777: between 68, 554, and 73, 911; 1803–1804: between 75,990 and 80,835. See Cook and Borah, 1968, pp.75, 89.

prove possession since time immemorial and pay a "voluntary donation" to the Crown to have their lands surveyed and their boundaries marked. Royal regulations required that communal lands, known in Spanish as *fondo legal* or *bienes de comunidad*, should measure at minimum 600 *varas* in diameter, from the village church or the center of the pueblo outward. Any territory beyond the limits of officially designated communal land lacking documentation of ownership had to be titled for a fee. If not, it could be declared royal land, and subject to confiscation and public auction (Wood, 1990). Native participation in the *composiciones* of the 1692–1696, 1707–1709, 1717–1718, and the composiciones that followed in the rest of the eighteenth century, produced a range of results, including Native land titles, reduction of Native communal lands, and official recognition of customary and "irregular" forms of land tenure that sat at the margins of the law (Carrera Quezada, 2015; López Castillo, 2010, 2014; Mendoza, 2011, pp. 54–59; Menegus Bornemann, 2017; Pastor, 1987; Radding, 1997, pp. 171–207, 2005, pp. 89–116; Torre Ruiz, 2012; see also Radding's chapter in this volume).

Prior to the eighteenth century, most Native communities possessed land, rather than owned it, since securing or producing a title of ownership was more difficult than claiming possession in Spanish courts (Owensby, 2008, pp. 90–129). Possession, in fact, constituted a primary means by which individuals and communities held land in the Spanish empire, and the Mediterranean-Atlantic World more broadly (Bastias Saavedra, 2020; Greer, 2018; Seed, 1995). The *Siete Partidas*—a statutory code from medieval Castile that strongly influenced legal practice in Spanish America-defined possession as lawfully entering, occupying, and holding a piece of land, a concept that was distinct from ownership, which required legal title. Central to this definition was the absence of force or coercion, which made a claim to possession unlawful and unjust. Evidence of possession included longstanding use and cultivation, manifested by crops or structures.[2] The community's judgment often determined the legitimacy of claims to possession in courts of law, anchored in the expression "since time immemorial."

---

[2] Burns, ed., *Las Siete Partidas*, Partida III, Title XXX, Laws I, II, III, VI, 850–851.

Whereas titles carried heavy weight as evidence of ownership, so too did custom—local practice as accepted and recognized by the community.

This chapter analyzes Native claims to customary land tenure and possession since time immemorial in response to the *composiciones de tierras* and other challenges to communal territory in the Mixteca region of Oaxaca. The land titling program dovetailed with the expansion of the livestock economy, population growth, and an increase in tribute and taxes during the late seventeenth and eighteenth centuries. In a context of increasing scarcity and pressure to normalize landholding, many Native communities went to court with competing claims to land. But conflict and litigation were not the only strategies deployed by Native authorities to address the need for subsistence and income. Indigenous pueblos also came together to create plural ownership that allowed them to pool resources and share territorial jurisdiction. Partnership contracts—the form in which plural ownership was legally instantiated—were more legible to Spanish authorities than codices, maps, and primordial titles, and had practical benefits since they were much less costly than litigation. Through partnership contracts, Native authorities preserved or extended the territorial expanses of their communities, challenged or whittled away at the property of powerful caciques, and transformed customary claims into new legal rights with an eye to securing the territorial integrity of their communities for the future.

Ñudzahui Territory, Land Tenure, and a 1690-Partnership Contract

Oaxaca provides a counter-narrative to Native dispossession during the early rounds of *composiciones de tierras* of the sixteenth and seventeenth centuries. In fact, most Indigenous nobles and communities in Oaxaca maintained their landholdings from the conquest until the end of the colonial period. Oaxaca's economy, which was dominated by commerce and fueled by Indigenous production rather than mining and Spanish-controlled haciendas, provides a central explanation for this trend (Chance, 1989; Menegus Bornemann, 2009; Pastor, 1987; Taylor,

1972; Arrioja Díaz Viruell, 2011).[3] In the case of the Mixteca region, the few Spaniards who settled there rented land from Native nobles and communities and focused their energies on trading in the valuable products of native labor, most notably, cochineal dyestuff, wheat, cattle, leather, and cloth (Romero Frizzi, 1990).

The persistence of Native landholding meant that community land tenure remained deeply informed by Indigenous notions of territory, which did not conform to clearly delineated boundaries. In the Mixteca, this fungible relationship was expressed by the Ñudzahui institution of the *yuhuitayu,* sometimes shortened to *tayu,* a political entity made up of two communities (*ñuu*) joined through the marital alliance of lords from each. The yuhuitayu was not a geographical designation, as was a European kingdom or señorial estate, but rather a shifting mosaic of constituent communities and sub-units, known as *ñuu,* that periodically realigned depending on élite inter-marriage and the tributary claims of their lords. Often, they were not contiguous territories nor evenly distributed geographically. A small settlement in close proximity to the palace complex of one yuhuitayu might have been subject to or affiliated with another yuhuitayu. Autonomy often defined the relations of the ñuu within the yuhuitayu and between individual ñuu and the seat of the yuhuitayu. Although lordly marriage served to combine the resources of the constituent ñuu, it did not compromise the autonomy of either. Sometimes ñuu seceded from yuhuitayu and shifted allegiances to others (Dahlgren, 1954; Spores, 1974).

The *yuhuitayu* as a form of territorial and political organization endured well into the eighteenth century, though it had been modified somewhat by the late sixteenth and early seventeenth-century process of congregación, spearheaded by Dominican missionaries who were the dominant religious order in Oaxaca. The effort to impose the Spanish administrative arrangement of the *cabecera* (administrative and parish

---

[3] In the districts of Teposcolula and Villa Alta, with a few exceptions, Native caciques and communities held their lands throughout the colonial period. In Oaxaca's central valleys, there was a rise of Spanish owned haciendas and ranches during the late sixteenth and seventeenth centuries, but by the late eighteenth century, Zapotec nobles and communities began to expand back onto the lands.

seat, literally "head town") and its *sujetos* ("subjects"), which were geographically proximate and politically subordinate villages or dependencies, and whose residents owed tribute and labor to the authorities of the cabecera met with only modest success. Dispersed settlement patterns persisted, and many ñuu maintained their identities, locations, and lands. The territorial imprint of the yuhuitayu persisted, as did its political meaning, as evidenced by the pervasive use of the term in colonial-era Ñudzahui-language documentation (Martín Gabaldón, 2018a, 2018b; Spores, 1967; Terraciano, 2001, pp. 119–120).

Although it did not affect the territorial organization as much as Spanish officials had intended, the imposition of the cabecera-sujeto model had important political implications for inter-community relationships. Spanish officials designated some yuhuitayu, and not others, as cabeceras, and recognized some lords (yya) as caciques by granting them señorial title to land (cacicazgos), while discounting the claims of others. The yuhuitayu that were assigned the lesser status of subject towns resisted the imposition of new hierarchies strongly, bringing legal cases to the Audiencia of Mexico from the 1550s forward in which they argued for the right to "secede" from their cabeceras. In their petitions, they railed against paying tribute and performing services for cabeceras and caciques, and they were aggrieved that their neighbors, as parish seats, had become the centers for sacred rituals (Martín Gabaldón, 2018a, 2018b; Pastor, 1987, pp. 175–178; Terraciano, 2001, p. 124). For their part, caciques and cabeceras took advantage of their status, and competition between yuhuitayu took new forms, including legal disputes over the boundary lands that separated communities, which according to Ñudzahui *lienzos* (Indigenous cartographic histories painted on cloth), included sacred sites (Aguilar Sánchez, 2015/2016, 2020; Van Doesburg, 2001). These conflicts heated up at the end of the seventeenth century as the Indigenous population recovered from its devastating decline during the previous century, and as the livestock economy expanded, putting new pressures on land (Spores, 1984, pp. 210–225).

Royal legislation at the end of seventeenth century contributed to the tensions. In 1687, the Crown issued a cédula that granted subject communities the same expanse of lands that had previously been reserved for cabeceras: 600 *varas* radiating outward from the town church. The idea was to provide expanding settlements with the territorial foundation necessary for subsistence agriculture and pastureland. Consequently, many subject communities built churches, had their land surveyed, and declared themselves independent cabeceras in their own right, to the chagrin of the authorities of their former cabeceras (Menegus Bornemann, 2009, 2017, pp. 62–72).

Litigation in Spanish courts provided one answer to these conflicts, though not a desirable one because it was expensive and time-consuming. Native authorities turned to other means to address land disputes, pivoting away from Spanish courts and attempting to resolve conflict within the ambit of Native jurisdiction. A 1690 Ñudzahui-language notarial record documenting a land-use agreement between the communities of San Juan Sayultepec and San Andrés Sinaxtla provides an example.As occurred with many Native-language legal records, the 1690 agreement was sewn into a voluminous four-hundred and thirty-eight-page land dispute between the two communities, adjudicated in a Spanish court across three and a half decades, between 1713 and 1749.[4]

The 1690 Ñudzahui-language record was a genre of contract, whose origins can be traced to the concept of partnership (*societas*) in Roman Law. In contrast to commercial contracts, which were reciprocal in nature (one party does something to receive something else from another party), and presumed an opposition of interests, the purpose of partnership contracts was to pool resources, such as property or labor, for a common purpose, and sometimes against the interests of a third party. Partners in *societas* were friends and allies rather than antagonists (Zimmerman, 1996, pp. 451, 454–455).

Through the Ñudzahui language contract, the Native authorities of San Juan Sayultepec and San Andrés Sinaxtla aligned the relationship among yuhuitayu into the Spanish relation of partnership. In the text of the contract, they referred to their communities as "chayu" (a variation

---

[4] In my analysis, I use the Ñudzahui-language petition and its Spanish translation transcribed and published in Jansen and Pérez, 2009, pp.338-340. They cite case as AGN Tierras 1717, vol.308, fs. 140-142.

of "tayu," short for *yuhuitayu*), instead of using the Spanish designations of cabecera and sujeto. For place names, they used Christian-Ñudzahui hybrids, rather than the Christian-Nahuatl names imposed on their communities by Spanish and Mexica conquerors: San Juan Sayultepec was written as "Sa Juan tiyuqh" and San Andres Sinaxtla, as "San Andres atata." (Jansen & Pérez, 2009, p. 339). The Native authorities stated that the purpose of the contract was to protect the agricultural lands of the two communities against the territorial predations of a third community, Santa María Asunción Nochixtlan, an important pre-Hispanic and colonial-era commercial and political center. In the 1680s, Nochixtlan, which had been part of the Spanish province of Teposcolula-Yanhuitlan, became a Spanish administrative seat, with jurisdiction over the pueblos of Tilantongo, Chachoapan, Etlatongo, Huaclilla, Tejutepec, Tiltepec, and Jaltepec (Spores, 1984, p. 98). The concentration of Spanish and Indigenous political power in Nochixtlan produced tensions with Sinaxtla and Sayultepec, powerful native communities in their own right. Perhaps emboldened by newfound status, the local authorities of Nochixtlan saw an opportunity for territorial expansion. In the text of the contract, the officials of Sayultepec and Sinaxtla expressed their common outrage that the natives of Nochixtlan sought to expropriate valuable irrigated land where corn fields cultivated by each community came together at the boundary among all three communities.

Through their partnership, the officials of Sayultepec and Sinaxtla joined together in common cause against another mutual antagonist: the cacique don Domingo de San Pablo. The narrative of the agreement devoted significant space to a shared past in which the communities united in "friendship" against the cacique and the authorities of the town of San Mateo Yucucuihi who made heavy demands of them, presumably in labor and tribute. The authorities of Sayultepec and Sinaxtla stated that they would no longer recognize don Domingo or any other Native lord as their cacique, and that "only the lord God and lord King are our lords."[5] Their refusal to recognize the cacique's

---

[5] See Jansen and Pérez, 2009, p. 339 for the Mixtec original, and p. 340 for the Spanish translation.

customary authority and prerogatives represented a growing trend across the Mixteca. Economic and cultural transformations spurred by colonialism stoked these conflicts. Spanish entailment of cacicazgos in the early colonial period transformed don Domingo de San Pablo and other native lords in the region into a powerful rentier class who earned significant income from the lease of their lands to Spaniards, mestizos, and other Natives. Don Domingo and other Ñudzahui caciques like him were often wealthier than the region's Spanish merchants, and had the goods to show it: luxurious European clothing, horses, high-quality wooden and silver home furnishings, and elaborate Christian art. Their easy assimilation to Christianity and migration to urban centers where they could live comfortably off of their earnings exaggerated the social distance between themselves and Ñudzahui commoners. Caciques' detachment from their pueblos loosened the reciprocal obligations that bound them to their communities, inflamed resentments, and led to what one historian has called the eighteenth-century "revolt against the caciques." The revolt was primarily a legal one, in which pueblos took caciques to court to protest unjust demands and abuses of authority (Pastor, 1987, pp. 166–175).

Disavowal of don Domingo in the contract may have had something to do with the land under dispute, in that it could have pertained to his cacicazgo. By claiming that they did not recognize don Domingo as their cacique, the authorities of Sinaxtla and Sayultepec cleared the way to claim possession of the land for their communities, which if it had belonged to don Domingo, they might have worked through usufruct in the past. The remainder of the contract recounted legal procedures typical for recognizing possession, including a boundary survey and placement of boundary markers in order to preclude controversy in the future. The fact that the native authorities of the two communities conducted a land survey and produced a legal agreement on their own, without the presence of a Spanish court functionary, points to an autonomous Native forum and set of procedures for addressing conflicts over boundary lands.

The agreement closed with a reassertion of friendship and partnership. Behind the aspiration of social harmony, however, lingered some doubt. The contract closed with the stipulation that if any member of either community were to disturb the peace, the officials of either pueblo could appeal to the Real Audiencia (the highest Spanish court in the land), which would insure the maintenance of the agreement. With this clause, the signatories concurred that only the King's justice could enforce the partnership between the communities.[6] This legal instrument was written, then, with an eye to preventing conflict in the future, not only with Nochixtlan but also between its authors, the communities of Sinaxtla and Sayultepec. Although it was written in the Ñudzahui language and archived in the town halls of the signatories, the Native authorities produced it with the possibility in mind of presenting it to a Spanish judge as evidence of possession in the boundary lands, the most liminal and vulnerable part of a community's territorial grant, and where friends of the moment could become enemies in the future. This indeed came to pass. As discussed above, the officials of San Juan Sayultepec submitted the agreement as evidence in the case, arguing that the natives of San Andrés Sinaxtla had broken its terms by claiming the land as their own.

The 1690 partnership contract written and signed by the Native authorities of Sayultepec and Sinaxtla bore the imprint of Indigenous territorial and political organization. At the same time, it reveals how the European legal category of possession shaped inter-communal relations in the boundary lands between Ñudzahui communities, and how customary claims to land could be used to instantiate new rights through written agreements. The ephemerality of the agreement, and its incorporation into a future land dispute points to an important dynamic in Oaxaca's agrarian history during the late seventeenth and eighteenth centuries. Partnership contracts were but one component in *longue-durée* struggles over land. Although they were supposed to endure, they often did not. Rather than etching the contours and conditions of communal land in stone, they represented a reprieve from open conflict and a space for the re-negotiation of political and territorial relationships. This process was not unique to colonial Mexico but rather occurred throughout the Atlantic World where partnership

---

6 See Jansen and Pérez, 2009, p. 339 for the Mixtec original, and p. 340 for the Spanish translation.

contracts constituted a strategy used by rural communities to negotiate land rights (Blaufarb, 2010; Herzog, 2015). In this regard, the authorities of Sayultepec and Sinaxtla were actively contributing to an Atlantic World legal culture as they forged their Ñudzahui-language agreement in their rural town hall in the Mixteca.

Partnership and Plural Ownership in the Eighteenth-Century Composiciones de Tierras

The eighteenth-century composiciones de tierras provided an opportunity for Ñudzahui communities to reaffirm customary landholding patterns in their boundary lands, while creating new legal norms. Whereas the communities of Sayultepec and Sinaxtla achieved this temporarily within the ambit of Indian jurisdiction, the special court of land titling constituted a higher-order legal forum in which Native officials could negotiate customary access to boundary lands. In July of 1717, the Native authorities of the Ñudzahui communities of Tecomatlan and Magdalena Zahuatlan appeared before don Félix Chacón, Spanish magistrate of Teposcolula-Yanhuitlan with such a petition. The land bureau tasked with overseeing the composiciones de tierras—the Superintendencia del Beneficio y Composición de Tierras—had appointed Chacón as a judge to the royal commission of claims, titling, and sale of land and water in the district of Teposcolula-Yanhuitlan. In this role, he was tasked with overseeing agrarian matters, especially payment for the composiciones de tierras (Carrera Quezada, 2015). The court that he administered for this purpose was known as the Juzgado Privativo de Tierras (Torre Ruiz, 2012).

The Native authorities' petition requested a license to form a partnership contract regarding possession of some land that lay in between their communities, much like the written agreements produced by Sayultepec and Sinaxtla. In the recent past, the Natives of each town had claimed the land as their own. With forensic detail, the officials related the Ñudzahui place names that the land encompassed, citing the river that contoured it, and the location of three crosses that served as boundary markers. The problem was that although both towns had asserted possession, in actuality, farmers from both communities planted corn on it, such that their crops were interspersed. According to Spanish law, occupation manifested by cultivation proved legal possession, so barring the existence of legal title to the land, this dispute would be difficult to resolve in court, a point that the Native authorities understood well. In order to preclude competing legal claims to the land and costly litigation in the future, the two towns had arrived at an agreement to share the land and preserve the custom of interwoven cultivation. Apparently, though, the agreement between the two communities was not enough, which is why they petitioned Judge Chacón to intervene and authorize their agreement. Through the flourish of the Spanish judge's pen, the agreement would produce a relationship of joint-possession over the land, valid for all time, equal in force to a decision rendered by a civil judge. In short, the contract would fix the towns' customary use of the boundary lands unless a third party produced a title to the land.[7]

By 1717, the year that the officials of Tecomotlan and Zahuatlan petitioned Judge Felix Chacón to form the partnership, the population of the core region of the Mixteca Alta had jumped from 28,000 in 1660 to 42,000 in 1720, on an upward trajectory that increased to 76,000 by 1803 (Spores, 1984, p. 223). In the meantime, two rounds of composiciones de tierras had taken place, from 1692 to 1696, and 1707 to 1709, with a third underway from 1717 to 1718, sending Native caciques and communities to Spanish courts to obtain titles to land and firm up their territorial boundaries. The trends that had pushed the communities of Sayultepec and Sinaxtla to form their partnership contract in 1690 had intensified: population growth, commercialization of agriculture, expansion of the livestock industry, increased pressure on land and resources, and an explosion of litigation over land.

As spelled out in their petition, the Native officials of Tecomotlan and Zahuatlan did not seek to clarify their boundaries in order to

---

[7] T Civil Leg.21 Exp. 16.093 1717 "Petición de licencia, carta de compromiso," f.201–210.

produce land titles for their pueblos. Rather, they hoped to maintain the custom of interwoven cultivation in their borderlands. As evident in their petition to Chacón, the Native officials surveyed the boundary lands and consulted with one another to hammer out some of the fundamental terms of the agreement. The composiciones de tierras provided them with a unique opportunity to give agrarian custom the force of law and protect landholdings from the predations of third parties, like caciques or larger pueblos.

Judge Felix Chacón was persuaded by the petition and granted the towns of Tecomotlan and Zahuatlan the license they sought to draw up the partnership contract.[8] From where he stood, maintaining peaceful relations between pueblos was always preferable to rancor and the threat of violence over boundary lands. Furthermore, whereas a primary goal of the composiciones was to make boundaries and titles, another goal, far more pragmatic—and opportunistic since it implied a fee—was to codify customary uses of land that sat on the margins of the law.

Partnership contracts regarding land tenure were as much an agreement about the nature of the partnership—the ties that bound Native communities to one another (and to the Spanish courts)—as about the relationship of the communities to the land. This was evident in the 1690 Ñudzahui-language contract between Sayultepec and Sinaxtla, in which the political purpose of the partnership was intertwined with the integrity of each community's landed possessions at the boundaries. The partnership proposed by Tecomotlan and Zahuatlan was different in that the two communities had asked for recognition of an arrangement that did not align with the Spanish ideal of a territorially bounded community. Since the communities would be farming the lands together, the question of who would be responsible for material losses if one party did not uphold the agreement—in short, questions of harm, injury, liability, and enforcement—had to be taken into account. This required the intervention of a Spanish judge.

The partnership contract signed by the authorities of Tecomotlan and Zahuatlan entailed a promise to one another to uphold special rules and mutual obligations regarding land use. Preservation of customary

---

[8] T Civil Leg.21 Exp. 16.093 1717 "Petición de licencia, carta de compromiso," f.201–210.

agrarian practice, harmonious relations between the two communities, and collective possession of the land provided the contract's stated purpose. The first of its five clauses indicated that the lands that the Natives of each pueblo cultivated would be the lands that they continued to cultivate without alteration. The language used to express this—"sin ynnovar en cosa alguna"—was also part of the medieval Spanish discourse of custom, which was double-edged: it preserved longstanding, continuous practice, but at the same time could be altered and established anew. Custom's flexibility allowed for change and adaptation, but in this case, the Native authorities adopted a staunchly conservative posture toward it in their attempt to preclude any innovation. The fifth clause underscored the imperative to preserve custom by stipulating that if the Natives of one pueblo or the other had more or fewer crops in the commonly held lands, they should not try to sow them equally; instead, each pueblo should sow what they presently sowed even if some of the lands remained *baldíos* (uncultivated lands). The importance of maintaining the status quo "in order to avoid disputes" was accentuated by the risk implied in leaving untitled lands uncultivated. The policy of the composiciones program held that untitled and uncultivated land could be confiscated by the Crown via invocation of eminent domain, to be redistributed according to royal discretion. Despite the risk, the contract held that the land would remain untitled: the second clause stated that neither pueblo nor Natives therein could claim legal title to the lands. The Indigenous authorities appear to have counted on the contract as a form of insurance against expropriation, a process that could be triggered by an *amparo*, or judicial stay based upon legal documents that provided evidence of possession. The objective of maintaining good relations between the pueblos and avoiding litigation was telegraphed clearly through the contract's fourth clause, which maintained that the pueblos "must conserve and continue always the peace, union, and law-abiding manner in which they have lived without disputes." The fifth clause sought to preserve the peace by precluding the land invasions that were increasingly common in the region: each pueblo would possess and continue to possess its parcels without entering into one another's lands.

At the same time that it aimed to shore up horizontal and equitable relations between the pueblos, the contract reinforced the hierarchies of status and power that structured Ñudzahui communities. The fifth and final clause prescribed punishment for any member of the pueblo who broke the agreement. The punishments varied according to social status. If the perpetrator was a town notable (*principal*), he would be fined 100 pesos in common gold (*oro comun*), half destined for the judge of the Real Camara and half for the compliant party. If the perpetrator was a commoner (*macehualli*), the punishment would be 200 lashes. This was serious business. Two-hundred lashes constituted common punishment for highway robbery, murder, and sedition, and could easily lead to the death of the person to whom it was applied. The disparity in punishment was in keeping with Spanish criminal law, which applied punishment unequally according to social rank, and advised harsher punishment for commoners.[9]

Perhaps most importantly, the contract required each pueblo to renounce its own jurisdiction over the land and transfer jurisdiction to one other in order to produce joint-jurisdiction. Additionally, each pueblo had to renounce legal claim to the lands and transfer possession to one another to produce joint-ownership and remain "equal pueblos." Property and jurisdiction were separate but related legal categories as applied to Indigenous communities. Property pertained to the community land base, and jurisdiction to authority over the people and territory of the community. Joint-ownership meant that both communities possessed the land. Joint-jurisdiction implied authority over the land, the capacity to determine its use, and the power to punish those who transgressed the laws that applied to it. The clause about joint-jurisdiction was in keeping with transatlantic developments in legal agreements regarding collective land tenure in which the emphasis was as much on the right to administer common lands and resources as a means of conserving and defending them from encroachment by third parties as it was on possession and ownership (Ingold, 2018).

---

[9] Burns., ed., *Las Siete Partidas*, v.5, Partida VII, Title XXXI "Concerning Punishments" Headnote, Law VIII, 1463, 1466–1467.

The inter-pueblo partnership contract between Zamotlan and Tecomotlan created during the composición program was one among several produced during the early eighteenth century.[10] Taken together, the contracts represented attempts by Native authorities to short-circuit conflict and litigation over land, as well as expropriation by the Crown and third parties, like caciques or more powerful pueblos. They also served to maintain aspects of customary agrarian practice rooted in the yuhuitayu, while accommodating the expansion of the livestock economy. At the same time, the contracts instantiated a logic of debt—particularly individual responsibility for joint-liability—into local relations of land tenure. Joint liability meant that every member of the pueblo had to comply fully with the contracts' stipulations, even though the document was signed only by the Native officers. And any member of the pueblo who broke the agreement had to pay for their transgression, but with distinct penalties based on social status. In this light, partnerships of co-ownership did not represent an assertion of egalitarian communalism against powerful local landholders and state actors. Collective land tenure had many faces, depending on the laws that framed them, and the underlying agrarian relations that gave them shape.

Partnership, Plural Ownership, and Cacicazgos

Native authorities also used partnership contracts to transform customary use-rights of cacique lands into joint-possession, thereby expanding and securing their land base at the expense of cacicazgos. As Margarita Mengus Bornemann has shown, almost all communal land in the Mixteca fell under the designation of *propio*, lands used to sustain

---

[10] A T Civil Leg.21 Exp. 16.099 1717 "Petición de licencia, carta de compromiso," fs. 215r-217v; T Civil 1717 Leg. 21 16.101 "Petición de licensia, escritura de compromiso," fs. 219v.-222r; T Civil 1717 Leg. 21 Exp 16.108 "Petición de licensia, escritura de compromiso," fs. 232r-234v; T Civil 1718 Leg. 21 Exp. 16.132 "Petición de Licensia. Escritura de Compromiso," fs. 276r.-279v; T Proto "Escritura de convenio," 19 de junio de 1733,Leg. 6 Exp.4.17, fs. 25v-30v.

the cabildo, and the tax and tribute obligations of the community. The pueblos of the region do not appear to have held *tierras del común repartimiento*, which in other parts of Oaxaca and New Spain were distributed to individual families for their subsistence. Instead, Native commoners often enjoyed usufruct rights on cacique lands, which they cultivated for their own use in exchange for rent paid in specie or labor to the cacique. These relationships were not written down or contractual, but regulated according to custom. During the composiciones de tierras, some of these pueblos claimed that they possessed the land since time immemorial and should therefore be given title. They insisted that they did not recognize the authority of any caciques, only that of the Spanish king. Claiming land in this way was strategic since usufruct rights were distinct from possession. According to Spanish law, usufruct encapsulated the right granted by a proprietor to someone to work his or her land usually in exchange for labor or fees. Even if dependent laborers (*terrazgueros*) had worked the land for decades or even centuries, they could not claim it through possession because it belonged to someone else. Nevertheless, communities of terrazgueros seized upon the opportunity of the composiciones to claim the land through immemorial possession. By disavowing their caciques, they rejected the basis of the cacique's right to their labor and the land. Through this legal sleight of hand, they transformed customary usufruct rights into ownership, making the cacique's land their own (Menegus Bornemann, 2017).

Two partnership contracts from the Mixteca Alta district of Tlaxiaco show how communities and caciques disputed and resolved their competing claims to land outside the confines of the composición program, and in a region in which land rights were especially layered and complex. During the pre-Hispanic period, Tlaxiaco was one of the largest and most powerful yuhuitayu of the Mixteca. Its territory encompassed lands in different ecological niches, including cold, temperate, and tropical, allowing for agricultural complementarity and the production of diverse trade goods. It was also a multi-ethnic polity, comprised of a majority Mixtec population, with Triqui and Nahua minorities. Compound lordship defined its political organization, with many *yya* (lords) controlling specific territories through shared or confederated authority (Martín Gabaldón, 2018a, 2018b, pp. 44–46).

After the congregaciones of the late sixteenth century, Tlaxiaco became a cabecera with multiple subject communities, some of which were cabeceras in their own right, making its jurisdiction layered and at times, conflictive. Due to its strategic location, it became a center for trade, and its fertile lands, especially a territory known as the cañada of Yosotiche, made it a center for sugar production, livestock grazing, and agricultural production more broadly. During the eighteenth century, the cabecera of Tlaxiaco rented the rich agricultural lands of Yosotiche to Spanish sugar producers, positioning it as one of the most important proprietors in the Mixteca region (Martín Gabaldón, 2018a, 2018b; Pastor, 1987, pp. 181–188). Tlaxiaco's commercial success can be explained not only by its location and the quality of its lands but also by its relationship with its subject communities, which was more cooperative and complementary and less hierarchical than that of other yuhuitayu of the Mixteca region. Each community managed grazing and agricultural lands to the benefit of the whole, contributing to a generalized prosperity (Martín Gabaldón, 2018a, 2018b, p. 63).

Tlaxiaco's strength was counterbalanced by the multiple cacicazgos of the region, which concentrated power and wealth, and created additional layers of authority. The compound lordship of the pre-Hispanic period may explain this phenomenon of plural cacicazgos. In some cases, subject pueblos of Tlaxiaco and other cabeceras were embedded within the territorial limits of cacicazgos. Notably, San Pedro Mártir Yucuxaco, San Juan Ñumi, San Antonino, San Sebastián Almoloya, and Santo Domingo, all subject towns of Tlaxiaco, were located within the cacicazgo of don Pedro de Chávez y Guzmán, making them subject to the authority of both the cabildo of Tlaxiaco and the Chávez family. During the sixteenth and seventeenth centuries, various members of the Chávez family served in the cabildo of Tlaxiaco, making the body an instrument of lordly power (Martín Gabaldón, 2018a, 2018b, pp. 43, 59).

In the eighteenth century, the Chávez family's hold on regional power began to wane, as the cabecera of Tlaxiaco and many of its subject communities came to define their interests against those of the caciques. This local trend tracked with a broader regional pattern in which cacicazgos across the Mixteca found themselves in crisis by mid-century due to challenges from Spaniards and Native commoners. As the livestock economy expanded, caciques leased their land to other caciques, Native commoners, mestizos, and even religious orders, some for cultivation, but most for grazing. The rent of cacique lands increased fourfold from 1671 to 1730, with a marked increase from 1700 to 1730. Many of the most significant renters were Spanish ranchers, who often sought to turn the land they leased from caciques into their own property by making dubious claims in court. For the most part, the caciques successfully defended their cacicazgos, but at the high price of endless legal fees. By 1740, almost all of the Mixtecan cacicazgos were embroiled in some form of litigation over land, a process that seriously undermined their economic solvency (Pastor, 1987, pp. 172–173). Not only did they face challenge from Spaniards but they also faced challenges from communities of *terrazgueros* and subject communities located on their lands. Sometimes caciques rented the boundary lands of their subject communities, sending Native authorities to the courts to cry foul.

The caciques of Tlaxiaco played their part in this process. Between 1714 and 1742, don Pedro Martín Chávez de Guzmán entered into seven rental agreements.[11] One of them, signed in 1723, authorized the rent of lands named Yosoñama to Leonor de Aguirre, a wealthy resident of Tlaxiaco, so that she could use it to establish a ranch for cattle, sheep, and goats. Yosoñama sat within the limits of the pueblo of San Juan Ñumi, whose territory was embedded in Chávez' cacicazgo. In addition to an annual fee of twenty pesos, she was expected to pay for all of the improvements to the land required for the founding of the

---

[11] T Proto, "Arrendamiento de tierras" 1714, Leg. 4 Exp 2.163, fs. 376r-377r; T Proto, "Arrendamiento de tierras," 1714, Leg.4 Exp. 2.164, fs. 378r-379r; T Civil, "Arrendamiento de tierras," 1715, Leg.21 Exp. 16.008, fs. 27v-30v (Note, this is a collective agreement in which Chávez rents along with the cabildo of Tlaxiaco); T Proto, "Arrendamiento de Tierras," 30 de octubre de 1722, Leg. 5 Exp. 1.122, fs. 241r-243v; T, Proto, "Arrendamiento de Tierras," 1723, Leg. 5 Exp. 2.43, f. 91r-94v; T Proto, "Arrendamiento de Tierras," 1728, Leg. 6 Exp.2.36, f. 63r-65r; Ta, Proto, "Arrendamiento de tierras," 1733, Leg. 6 Exp. 4.04, 4ry, fs. 7r-10v. ; T Proto, "Escritura de Arrendamiento," 1741, Leg.7 Exp. 5, fs. 41r-45v; T Civil, "Arrendamiento de tierras," 1742, Leg. 30, Exp. 29.7, fs. 14r-17r. ; T Civil, "Licencia, Arrendamiento de tierras," 1742, Leg. 30 Exp 29.7, fs. 14r-17r.

ranch. Like most rental agreements, it was valid for a period of nine years, in this case, until 1732.[12]

Doña Leonor's ranch disappeared from notarial records after the 1730s, and its fate remains unclear. It does appear, though, that ownership of the lands called Yosoñama, nested within the concentric circles of San Juan Ñumi, Chávez' cacicazgo, and the cabecera of Tlaxiaco, was muddy enough that the municipal authorities of Tlaxiaco attempted to rent the lands in 1741 to don Juan Antonio de Ladesa, the lieutenant of the *alcalde mayor* (Spanish magistrate) of Teozacoalco, for nine years at the rate of seventeen pesos per year.[13] This set off a land dispute among the municipal authorities of Tlaxiaco, San Juan Ñumi, and don Pedro Chávez. In 1742, the dispute gave rise to a partnership contract in which the municipal authorities of Tlaxiaco and San Juan Ñumi claimed joint possession of the land for cooperative use, against current and future claims of the Chávez family.[14]

Tlaxiaco and Ñumi had not always enjoyed a cooperative relationship. In their petition for a license to form the contract, the authorities of both towns recounted how since August of 1742, the two pueblos had been engaged in litigation over Yosoñama. The officials of Ñumi asserted that under pressure from don Pedro Chávez, they entered a claim to Yosoñama in court, implying that Chávez was angling for a land grab from Tlaxiaco. For their part, the cabildo of Tlaxiaco defended what they claimed was their right to Yosoñama. To put an end to the dispute, Tlaxiaco agreed to give Yosoñama to San Juan Ñumi, but only under the condition that villagers from Tlaxiaco could continue to enjoy the fruits of the land through usufruct, not rent. In fact, they insisted that the land could not be rented at all. In short, the partnership contract would transform lands that the cacique don Pedro Chávez had formerly claimed as his own to lease out for ranching to agricultural land jointly held by Tlaxiaco and Ñumi.

The problem was that the authorities of Tlaxiaco had already tried to claim the land as their own and rent it to don Juan Antonio de Ladesa.

---

[12] T Protoc "Arrendamiento de Tierras," 1723, Leg. 5 Exp. 2.43, fs. 91r-94v.

[13] T Proto, "Escritura de Arrendamiento", 1741, Leg.7 Exp. 5, fs. 41r-45v.

[14] T Civil, "Contrato de mancomunidad de tierras," 1742, Leg.30 Exp. 29.18.

To address this problem, the contract stipulated that Ladesa would have to remove his livestock by June of 1743. By ceding the land to Ñumi, a community that did not have a rental contract with the Spanish rancher Ladesa, the cabildo of Tlaxiaco did not have to break its contract, and Ladesa and his animals could be sent packing. Once the land was free of livestock, the two pueblos could strengthen their claim to possession by putting the land under cultivation for the necessities of each community, one of the stated purposes of the partnership, and a surefire way to protect the land from expropriation by the Crown.

The other crucial part of the agreement was that neither pueblo would recognize any of the Chávez family as their caciques. As discussed earlier, disavowal of cacique authority represented a strategy used by communities to claim cacical lands through the composiciones. At the same time, relations among Tlaxiaco and its subject communities and the Chávez family had been souring for decades. From 1715 to 1734, Tlaxiaco and some of its subject communities, including San Juan Ñumi, formed a partnership to rent pasture land to a religious order, the Compañia de Jesús del Colegio de la Nueva Veracruz. In the 1715 contract, don Pedro de Chávez de Guzmán and his brother Miguel Chávez formed part of the partnership. Afterward, they did not.[15] It appears that Tlaxiaco and Ñumi had either cut the Chávez brothers out of the agreement or claimed the land as their own. And in 1723, the pueblo of San Martín, which like San Pedro Ñumi was embedded within Chávez' cacicazgo, sought an *amparo* (writ of protection, or judicial stay) to protect communal lands in their possession from Chávez' claims.[16] One year later, Chávez agreed to "donate" the lands under dispute to San Martín, and allowed the pueblo to maintain its writ of amparo, insuring that its possession of the lands would be respected in the future.[17] The resolution of the

---

[15] T Civil, "Arrendamiento de Tierras," 1715, Leg.21 Exp. 16.008, fs. 27r-30v; T Civil, "Arrendamiento de Tierras," 1727, Leg. 25 Exp 2.15, fs. 36v-39v; T Proto, "Obligación de Pago," 1734, Leg. 6 Exp. 4.26, fs. 51r-55r; T Proto, "Arrendamiento de tierras," 1733, Leg. 6 Exp. 4.08, fs. 12v-17r; T Proto, "Arrendamiento de tierras," 1733, Leg. 6 Exp.4.09, 17r-18v; T Proto, "Arrendamiento de tierras," 1734, Leg. 6 Exp. 4.46, fs. 85v-96v.

[16] T Prot "Poder especial," 1723, Leg. 5 Exp. 2.48, fs. 104r-106r.

[17] T Proto, "Escritura de convenio y donación," 1724, Leg. 5, Exp. 2.89, fs. 188r-193v.

dispute raises questions. If the pueblo of San Martín possessed the lands, why did Chávez need to donate them? Perhaps Chávez preferred this extrajudicial arrangement to the costly litigation that a full-blown land dispute entailed. Indeed, Chávez appeared to be facing pressures from many sides. During the same year, the cabildo of Tlaxiaco gave power of attorney to an *alcalde* (magistrate) and principal of their community to represent them in the Real Audiencia in a land dispute against Chávez.[18]

The 1742 partnership contract between Tlaxiaco and Ñumi represents a moment, then, in this long conflict between pueblos and caciques. Perhaps the power-sharing among the multiple cabeceras of the region and the more cooperative relations among subject communities facilitated a collective stance against the cacique. This spirit of cooperation for mutual benefit came through in the conclusion of the contract, which summed up its purpose: to end their legal dispute over Yosoñama, farm the land together for their common necessities, never rent the land, nor recognize *any* cacique.[19] Unlike the partnership contracts drawn up during the 1717–1718 composiciones, the contract between Tlaxiaco and Ñumi did not invoke joint or partial liability, or liability of any sort, and contained no punitive measures with which to enforce its provisions.[20]

Conclusion

In late seventeenth- and eighteenth-century Oaxaca, Native landholding and forms of territorial organization endured to a much greater extent than in other regions of colonial Mexico. For example, in northwestern

---

[18] T Proto, "Poder Especial," 1724, Leg. 5 Exp. 2.77, fs. 167r-170r.

[19] Ibid.

[20] The cabildo of Tlaxiaco formed another partnership in 1742, this time for joint-possession of boundary lands with Chilapa, a cabecera that fell under the jurisdiction of Tlaxiaco, and four of its subject communities. The contract put an end to ongoing litigation, and allowed all of the signatories to enjoy the right to water, palm, maguey, and timber, all customary uses of boundary lands. Each of the signatories had equal claim to the land and would give up any prior or future legal claims to it. T Civil, "Escritura de Convenio," 1742, Leg. 30 Exp. 29.11, fs. 26v.-31r.

Mexico, Hispanic settlement, population growth, expansion of cattle ranching, and the legal regime of private property and fixed boundaries undermined Indigenous communal land tenure and variable use-rights over the course of the eighteenth century (López Castillo, 2010, 2014; Radding, 1997, pp. 171–207, 2005, pp. 89–116). By contrast, in Oaxaca, Indigenous legal claims to customary land tenure, use, and possession produced a broad range of agrarian relationships that preserved old forms of collective ownership in new guises or created new forms altogether.

Although many Native communities in Oaxaca acquired title to their lands during this period through the composiciones de tierras, others resisted drawing firm boundaries around their communities, opting instead to share land through cooperative agreements. In some instances, they resorted to custom to legitimize claims to joint-possession and joint-jurisdiction, and in others, they claimed possession in order to transform customary usufruct into a contractual relationship. Partnership contracts provided an alternative to the bitter, expensive, and lengthy processes of litigation and land titling, at least temporarily. They also represented a strategy to pool land and natural resources, and to join forces against powerful outsiders, whether Spanish ranchers, officials, or caciques. Crucially, although claims to custom pointed to the preservation of traditional agrarian order, when incorporated into the partnership contract they became a potent mechanism for challenging that order and generating new rights, while preserving the privileges of the Native authorities who were their signatories.

Partnership contracts expand our view of the legal repertoire available to and developed by Native people to make customary claims and not only preserve but also produce common land. Partnership contracts were distinct from other forms of Mesoamerican claim-making—like maps, primordial titles, and codices—in significant ways. The contracts framed Native territoriality and social order in forms that appealed to Spanish norms of ownership, and collective and individual responsibility. In addition to providing evidence of customary practice and possession, the contracts also produced new legal effects, like liability, through which burdens for transgression were shared unequally. By petitioning Spanish judges to put into law customary arrangements that sat at the margins of legality, a rising class of legally literate Native officials strengthened their authority over people and land, while ceding some of their limited sovereignty to Spanish judges. The laws of medieval Spain allowed them to move strategically between agreement and conflict, and social harmony and exploitation. The fruits of their efforts demonstrate that Native custom and communal land in Mexico were not primordial or static, but rather works in progress, conditioned by laws, Native legal agency, and political and economic transformations. Joint-ownership and joint-jurisdiction had a lasting legacy in Mexico that stretched beyond the composiciones de tierras. During the nineteenth century and up until the Mexican Revolution, co-ownership (*condueñazgo*) provided a strategy for Mexico's peasant communities—some Indigenous, and some of mixed ethnicity—to respond to state-led efforts to fiscalize, privatize, and legalize communal landholding.[21]

Bibliography

Archives consulted and referenced

Archivo Histórico Judicial de Oaxaca (AHJO) Teposcolula Civil, Teposcolula Protocolos (T Civil, T Proto)
Archivo General de la Nación, México (AGN)

Printed Primary Sources

King of Castile & Leon, A. (2001). *Las siete partidas* (R. I. Burns Ed. & S. P. Scott Trans.). University of Pennsylvania Press.

---

[21] There is a robust literature on nineteenth century *condueñazgo*. See the seminal work of Antonio Escobar Ohmstede, 1993.

Secondary Sources

Aguilar Sánchez, O. (2015/2016). La construcción de espacios sagrados como límites territoriales en los pueblos de la Nueva España. *Thule* 38/39–40/41, 689–707.

Aguilar Sánchez, O. (2020). Ñuu Savi: Pasado, presente y futuro. Descolonización, continuidad cultural y reapropiación de los codices mixtecos en el Pueblo de la Lluvia (PhD dissertation). Leiden University.

Arrioja Díaz Viruell, L. A. (2011). *Pueblos de indios y tierras comunales. Villa Alta, Oaxaca: 1742–1856.* El Colegio de Michoacán.

Bastias Saavedra, M. (2020). The normativity of possession. Rethinking land relations in early-modern Spanish America, ca. 1500–1800. *Colonial Latin American Review, 29*(2), 223–238.

Blaufarb, R. (2010, September). Conflict and compromise: Communauté and seigneurie in early modern provence. *The Journal of Modern History, 82*, 519–545.

Borah, W., & Cook, S. F. (1963). *The aboriginal population of Central Mexico on the eve of the Spanish conquest.* University of California Press.

Borchart de Moreno, C. (1980). La transferencia de la propiedad agraria indígena en el corregimiento de Quito hasta finales del siglo XVII. *Cahiers De Monde Hispanique El Luso-Brésilien, 34*, 5–19.

Chance, J. K. (1989). *Conquest of the Sierra: Spaniards and Indians in Colonial Oaxaca.* University of Oklahoma Press.

Cline, H. F. (1949). Civil congregations of the Indians in New Spain, 1598–1606. *Hispanic American Historical Review, 29*(3), 349–369.

Cook, S. F., & Borah, W. W. (1971). *Essays in population history: Mexico and the Caribbean* (Vol. 1). University of California Press.

Cook, S. F., & Borah, W. (1968). *The population of the Mixteca Alta 1520–1960.* University of California Press.

Cortés Márquez, M. M., & Reyes García, L. (2004). Manuscritos coloniales de Santa María Tiltepec, Mixe, Oaxaca. *Cuadernos Del Sur, 10*(20), 121–136.

Dahlgren de Jordán, B. (1954). *La mixteca, su cultura e historia prehispánica.* Imprenta Universitaria.

Escobar Ohmstede, A. (1993). Los condueñazgos indígenas en las Huastecas hidalguense y veracruzana:¿ defensa del espacio comunal? In A. E. Ohmstede & P. Lagos Preisser (Eds.), *Indio, nación y comunidad en el México del siglo XIX* (pp. 171–188). Centro de Estudios Mexicanos y Centroamericanos.

Florescano, E. (1971). *Estructuras y problemas agrarios de México* (1500–1821). SEP.

García Martínez, B. (1987). *Los pueblos de la sierra: el poder y el espacio entre los indios del norte de Puebla hasta 1700*. Colegio de México.

Glave, L. M. (2008). Gestiones transatlánticas. Los indios ante la trama del poder virreinal y las composiciones de tierras (1646). *Revista Complutense De Historia De América, 34*, 85–106.

Goyas Mejía, R. (2015). Las Composiciones de Tierras de 1643 en la Nueva España. *Revista De Historia Iberoamericana, 8*(2), 4–75.

Greer, A. (2018). *Property and dispossession: Natives, empires and land in early modern North America*. Cambridge University Press.

Haskett, R. S. (2005). *Visions of Paradise: Primordial titles and Mesoamerican history in Cuernavaca*. University of Oklahoma Press.

Herzog, T. (2013). Colonial law and 'native customs': Indigenous land rights in colonial Spanish America. *The Americas, 69*(3), 303–321.

Herzog, T. (2015). *Frontiers of possession: Spain and Portugal in Europe and the Americas*. Harvard University Press.

Ingold, A. (2018). Commons and environmental regulation in history: The water commons beyond property and sovereignty. *Theoretical Inquiries in Law, 19*, 425–456.

Jansen, M. E. R. G. N., & Pérez Jiménez, G. A. (2009). *La lengua señorial de Ñuu Dzaui: cultura literaria de los antiguos reinos y transformación colonial*. Colegio Superior para la Educacion Integral Intercultural de Oaxaca.

Kellogg, S. (1995). *Law and the transformation of Aztec Society*. University of Oklahoma Press.

López Castillo, G. (2010). *El poblamiento en tierra de indios cahitas*. Siglo XXI, El Colegio de Sinaloa.

López Castillo, G. (2014). *Composición de tierras y tendencias de poblamiento hispano en la franja costera: Culiacán y Chiametla, siglos XVII y XVIII*. Instituto Nacional de Antropología e Historia-Centro INAH Sinaloa, H. Ayuntamiento de Culiacán-Instituto Municipal de Cultura.

Martín Gabaldón, M. (2018a). New crops, new landscapes and new socio-political relationships in the *cañada* de Yosotiche (Mixteca region, Oaxaca, Mexico), 16th-18th centuries. *Historia Agraria, 75*, 33–68.

Martín Gabaldón, M. (2018b). Territorialidad y paisaje a partir de los traslados y congregaciones de pueblos en la Mixteca, siglo XVI y comienzos del siglo XVII: Tlaxiaco y sus sujetos (PhD dissertation). CIESAS.

Megged, A., & Wood, S. G. (2012). *Mesoamerican memory: Enduring systems of remembrance.* University of Oklahoma Press.

Méndez y Martínez, E., & Méndez Torres, E. (Eds.). (1999). *Límites, mapas y títulos primordiales del estado de Oaxaca: Índice del Ramo de Tierras.* Archivo General de la Nación.

Mendoza, J. E. (2011). *Municipios, cofradías y tierras comunales: los pueblos chocholtecos de Oaxaca en el siglo XIX.* Universidad Autónoma Benito Juárez de Oaxaca.

Menegus Bornemann, M. (1994a). *Del señorío indígena a la república de indios: El caso de Toluca, 1500–1600.* Consejo Nacional para la Cultura y las Artes.

Menegus Bornemann, M. (1994b). Los títulos primordiales de los pueblos de indios. *Estudis: Revista de historia moderna, 20,* 207–230. http://hdl.handle.net/10550/34249

Menegus Bornemann, M. (2009). *La Mixteca Baja. Entre la revolución y la reforma. Cacicazgo, territorialidad y gobierno. Siglos XVIII-XIX.* Universidad Autónoma Benito Juárez de Oaxaca.

Menegus Bornemann, M. (2017). Del usufructo, de la posesión y de la propiedad: Las composiciones de tierras en la Mixteca, Oaxaca. *Itinerarios. Revista De Estudios Lingüísticos, Literarios, Históricos y Antropológicos, 25,* 193–208.

Ots Capdequí, J. M. (1946). *El régimen de la tierra en la América española durante el periodo colonial.* Universidad de Santo Domingo.

Oudijk, M. R. (2003). Escritura y Espacio. El Lienzo de Tabaá I. In M. de los A. Romero Frizzi (Ed.), *Escritura zapoteca: 2.500 años de historia* (pp. 371–376). CIESAS.

Oudijk, M. R., & Romero Frizzi, M. de los A. (2003). Los títulos primordiales: un género de tradición mesoamericana. Del mundo prehispánico al siglo XXI. *Relaciones, 95*(24), 19–48.

Owensby, B. P. (2008). *Empire of law and Indian justice in colonial Mexico.* Stanford University Press.

Pastor, R. (1987). *Campesinos y reformas: la mixteca, 1700–1856.* Centro de Estudios Históricos, Colegio de México.

Quezada, S. E. C. (2015). Las composiciones de tierras en los pueblos de indios en dos jurisdicciones coloniales de la Huasteca, 1692–1720. *Estudios De Historia Novohispana, 52,* 29–50.

Radding, C. (2005). *Landscapes of power and identity: Comparative histories in the Sonoran Desert and the forests of Amazonia from colony to republic.* Duke University Press.

Radding, C. (1997). *Wandering peoples: Colonialism, ethnic spaces, and ecological frontiers in Northwestern Mexico, 1700–1850.* Duke University Press.

Rivera Marín de Iturbe, G. (1983). *La Propiedad territorial en México: 1301–1810.* Siglo XXI.

Romero Frizzi, M. de los A. (1990). *Economía y vida de los españoles en la Mixteca Alta, 1519–1720.* INAH.

Romero Frizzi, M. de los A., & Vásquez Vásquez, J. (2011). Un título primordial de San Francisco Yatee, Oaxaca. *Tlalocan, 17,* 85–120.

Romero Frizzi, M. de los A., & Vásquez Vásquez, J. (2003). Memoria y escritura. La memoria de Juquila. In M. de los A. Romero Frizzi (Ed.), *Escritura zapoteca: 2.500 años de historia* (pp. 393–448). CIESAS.

Ruiz Medrano, E. (2010). *Mexico's indigenous communities: Their lands and histories, 1500–2010.* University Press of Colorado.

Seed, P. (1995). *Ceremonies of possession in Europe's conquest of the new world, 1492–1640.* Cambridge University Press.

Solano, F. (Ed.). (1984). *Cedulario de Tierras, Compilacion de legislacion agraria colonial (1497–1820).* UNAM.

Sousa, L., & Terraciano, K. (2003). The 'original conquest' of Oaxaca: Late colonial Nahuatl and Mixtec accounts of the Spanish conquest. *Ethnohistory, 50*(2), 349–400.

Spores, R. (1967). *The Mixtec kings and their people.* University of Oklahoma Press.

Spores, R. (1974). Marital alliance in the political integration of Mixtec kingdoms. *American Anthropologist, 76*(2), 297–311.

Spores, R. (1984). *The Mixtecs in ancient and colonial times.* University of Oklahoma Press.

Taylor, W. (1972). *Landlord and peasant in colonial Oaxaca.* Stanford University Press.

Terraciano, K. (2001). *The Mixtecs of colonial Oaxaca: Ñudzahui history.* Stanford University Press.

Torales Pacheco, M. C. (1990). A note on the Composiciones de Tierras in the Jurisdiciton of Cholula, Puebla (1591–1757). In A. Ouweneel & S. Miller (Eds.), *The Indian community of colonial México: Fifteen essays on Land tenure, Corporate organizations, ideology and village politics* (pp. 87–102). Centro de Estudios y Documentación Latinoamericanos.

Torre Ruiz, R. A. de la. (2012). Composiciones de tierras en la alcaldía mayor de Sayula, 1692–1754: un estudio de caso sobre el funcionamiento del Juzgado Privativo de Tierras. *Letras Históricas, 6,* 45–69.

Van Doesburg, S. (2001). De Linderos y Lugares: Territorio y Asentamiento en El Lienzo de Santa María Nativitas. *Relaciones, 86*, 17–83.

Villella, P. B. (2016). 'For so long the memories of men cannot contradict it': Nahua patrimonial restorationism and the law in early New Spain. *Ethnohistory, 63*(4), 697–720.

Wood, S. (1990). The fundo legal or lands por razón de pueblo: New evidence from Central New Spain. In A. Ouweneel, & S. Miller (Eds.), *The Indian community of colonial Mexico. Fifteen essays on land tenure, corporate organizations, ideology and village politics* (pp. 117–129). Centro de Estudios y Documentación Latinoamericanos.

Wood, S. (2007). The Techialoyan codices. In J. Lockhart, L. Sousa, & S. G. Wood (Eds.), *Sources and methods for the study of postconquest Mesoamerican ethnohistory* (pp. 1–22).

Wood, S. (2012). *Transcending conquest: Nahua views of Spanish colonial Mexico.* University of Oklahoma Press.

Zimmermann, R. (1996). *The law of obligations: Roman foundations of the civilian tradition.* Oxford University Press.

5

The *Yoreme* Creation of *Itom Ania* in Northwestern Mexico: Histories of Cultural Landscapes

Cynthia Radding

We gathered in the patio of a village elder in the small community of Los Nachuquis ("place of the *huachapori* thorns"). Under the shade of mesquite and *piocha* trees we discussed the vital link between strengthening the *Yoremnokki* language among contemporary *Yoreme* communities and the defense of *lutu'uria*, their rightful claim to the forested spaces of the monte, *huya ania*, and the resources it provides. Several of the men and women present recalled their ancestors' legacy in the stories they had told about their struggles to defend their land, affirming that *territory—itom ania* or "our world"—signifies not only the land but also the culture and wisdom of the people.[1]

C. Radding (✉)
University of North Carolina, Chapel Hill, NC, USA
e-mail: radding@email.unc.edu

[1] Piocha, in the regional vernacular of Sonora, Mexico, is the *Melia azedarach* tree known in temperate climates as the white cedar. Possibly introduced from Asia in the past, it has acclimated to subtropical environments in the Americas. The word *piocha* is derived from

© The Author(s) 2024
J. Olko and C. Radding (eds.), *Living with Nature, Cherishing Language*,
https://doi.org/10.1007/978-3-031-38739-5_5

Radding, "Jornada de trabajo en torno al Pueblo de Cohuirimpo," field notes, 23 October 2021

Language plays a fundamental role in our interpretation of the rich archival sources that allow us to comprehend the deeply rooted knowledge base that Indigenous peoples developed from the material and spiritual worlds through which they moved in seasonal patterns of migration and dwelling. Language is also an essential part of the living histories we construct in collaboration with the Indigenous peoples that maintain their traditions in a radically transformed ecological region through their ritual cycles and collective memories that derive their meaning from the natural world. Indigenous perceptions of their environment were rooted in the landscapes they had created in the course of pursuing subsistence strategies for centuries before European contact. The arid environments of the coastal plains and highland valleys bordering the Sonora and Chihuahua Deserts required access to a diverse range of resources and ecological spaces. Indigenous peoples adapted the traditions they had developed over centuries to secure a livelihood to the demands and the opportunities of the colonial economy. Following the initial violent encounters with early Spanish *entradas*, Native villages coalesced into mission towns, whose internal economies combined Indigenous cropping and irrigation methods with European cultigens and livestock. Their settlement patterns were altered even further as they intersected with the colonial enterprises of mining, livestock raising, and commercial agriculture that developed in northwestern New Spain (Deeds, 2000).

Indians were not explicitly "environmentalists" in their worldview. Nevertheless, their territorial defense arose from longstanding holistic practices of land use that included arable land for cultivation, coastal estuaries for fishing, and the forests and grasslands for foraging. The

---

Nahuatl *piochtli*, meaning "full of grace or magnificent" (RAE Diccionario https://dle.rae.es/piocha). Mesquite (*Prosopis velutina*) is Native to northwestern Mexico.

Los Nachuquis, like Punta de la Laguna, Bachoco el Alto, Buaysiacobe, and other small communities in the middle portion of the Mayo river basin, are linked together by the Traditional Government of the Pueblo of Cohuirimpo, *Batwe Cowiktipo*, a historic *Yoreme* town and one of the eight Jesuit missions founded in the early seventeenth century. *Lutu'uria yo'owe*, the ancient or greater truth (Lerma Rodríguez, 2011, p. 214).

term *monte* occurs throughout the documentary record and in common speech to refer to uncultivated spaces in forests, grasslands, and wetlands near rivers or coastal estuaries. The meanings that Native peoples ascribed to places and the intensity of conflicts that erupted over basic resources stemmed from practices that linked labor to water and land, including floodplain cultivation and the renewable vitality of the *monte*.

This chapter offers new readings of land titles for the colonial provinces of Ostimuri and Sinaloa, focusing on the Mayo river valley. The analysis of changes in land tenure and use that are documented in these archival sources foregrounds ecological conditions and cultural meanings through the dual lenses of environmental history and ethnohistorical perspectives. It privileges Indigenous knowledge of landforms, biological species, and the cultural values that the communities of this region ascribed to the physical features and the territorial extension of the spaces they inhabited and defended. Its objectives seek to highlight the parallel production of oral and written sources and, thus, to suggest points of intersection in the languages and modes of communication that are inferred from both colonial documents and ethnographic registers. Finally, its purpose is to contribute an historical analysis that is useful for the *Yoreme* communities in their present-day defense of their territory and its resources.

The research that supports this chapter is rooted in archived land titles and judicial cases involving Indigenous pueblos and Hispanic settlers during the Spanish colonial administration of northwestern Mexico. These archival materials are supplemented with botanical information and place names collected by the author using hand-written notes during conversations with elders in the Pueblo of Cohuirimpo in 2017–2018 and 2021, as shown in Tables 5.1 and 5.2. In turn, the author has given transcriptions of the archival documents to the community in both printed and digital formats in an effort to collaborate with their vital interests for demonstrating their persistence as corporate communities in the region over centuries and their cultural practices for dwelling in the land. The research methodology is basically historical: to analyze a corpus of selected land titles that was compiled

Table 5.1 Placenames and names of plants derived from Yoremnokki

Name	Meaning
Bacabachi	Local variant of maize or carrizo seeds [*Arundo donax*]
Bachoco	Brackish water, associated with coastal marshes
Bachomo	A plant with medicinal properties that grows in the wetlands
Bachomotahüeca	Place where the bachomo plant stands
Caamoa	Pueblo: where maize or carrizo did not sprout
Capetamaya	Place where the fishing hook is thrown
Cohuirimpo	Pueblo: where the river turns back on itself, forming a tight meander
Conicari	Nest or home of the crow, associated with the gift of maize
Echomocha	Place where the *echos*, a fruit of the Pachycereus cactus is piled high
Etchojoa	Pueblo: the house of *echos*
Hona	[Óna] Salt
Huatabampo	Pueblo: a willow that stands in water
Jito	Distinctive species of trees in the monte: *Forchhammeria watsonii*
Júpare	Place of mesquite trees
Masiaca	[Masiacahui] Centipede hill
Móhua	Carrizo reed, basic material for building houses and fences
Mo'olco	Wild grass
Molcovaso	Grass that has turned yellow or brown, after the summer rains
Navojoa	Pueblo: house of tunas, the fruit of the nopal cactus
Tepagüi	Fattened deer
Tesia	Pueblo: the place of the *teso* tree, *Acacia occidentalis*
Yócuribampo	Rainwater

Sources Crumrine (1977), Camacho Ibarra (2017), Conversation with Elders of Cohuirimpo (2017)

from different repositories in order to create a narrative that is sensitive to change over time and informed by relevant insights from the archaeological and ethnographic literature.

Table 5.2 Selected Títulos Primordiales in Ostimuri and Sinaloa

Archive	Lands Registered	Year	District	Solicitant	Adjacent Properties
AGN IV 5907, 77	Yoricarichi Los Camotes	1715	Ostimuri	D Mateo Gil Samaniego	Pueblo de Macoyagüi, de la misión de Conicari
AGES TP LV,800	Cerro Colorado y Taimuco	1765	Álamos	D Juan José Amarillas	Pueblo de Macoyagüi, de la misión de Conicari
AGES TP XXVIII, 387	Husibampo	1788	Ostimuri	D Manl Ign Vlenzuela	Pueblos de Tesia, Navojoa, And Camoa
AGES TP XXVI,359	Real Viejo de Guadalupe, Mezcales, La Cabeza, Tres Marías, Osobampo	1793	Álamos	D Bartolomé Salido	Rl de S Joseph de Gpe, Rl de Santa Bárbara, D Juan de Sayas (difunto), Pueblo de Camoa
AGES TP XXIV,331	Lo de Ramírez	1790–1794	Álamos	D. Bartolomé Salido de Exsodar	Pueblos de Tesia, Camoa, Osobampo, de los herederos de D. Juan de Zayas, Usibampo de D Manl Ign Valenzuela
AGES TP XXIII,305	Jupsibampo Bachaca El Retiro Soledad	1793–1807	Álamos	D. Manl Ign de Valenzuela	D Bartolomé Salido D Patricio Gómez de Cossío Manl Cayetano Espinoza Pueblos de Camoa, Tesia, Navojoa, Cohuirimpo

(continued)

Table 5.2 (continued)

Archive	Lands Registered	Year	District	Solicitant	Adjacent Properties
AGES TP XIX,244	Echomocha	1796	Álamos	D Marcos de Valenzuela	D José Manl de Campos D Blas Antonio Muñoz Naturales de Echojoa Indio Julián Ontiberos Indio Pablo Ant Escalante
AGES TP V,58	Bacajaquito	1796–1797	Álamos	José María Lucenilla	Pueblo de Macoyagüi, Manuel Anguis, Lucas Ruiloba
AGN Tierras 1421, 9	Los Pilares	1800–1820	Ostimuri	D Francisco García	D José Ramón de Soto, vecino de Río Chico, Rancho de la Dispensa del propio García
AGN Tierras 1422, 2	Capetamaya Tierras agregadas	1806–1819	Álamos	D Juan Tomás González	Pueblos de Navojoa, Cohuirimpo, Chinobampo de D. Manuel de Espinoza, Soledad de D Manuel de Jesús de Valenzuela
AGES TP VI,69	Bacusa	1791–1820	Baroyeca Ostimuri	D Gabriel Felix	Naturales de Tepahüi Indios vecinos de Quiriego D Fr Javier de Valenzuela B Joaquín Elías González de Zayas

The Spatial and Cultural Creation of *Itom Ania*

Northwestern Mexico constituted a series of overlapping borderlands radiating northward from the tropical environments of the Gran Nayar, Chametla, and Culiacán to the arid lands of the Sonoran Desert (Hers, 2013, pp. 273–312; Sauer, 1935).[2] What the Spaniards first called *petatlán* (land of reed houses) became defined by four major river valleys—the Sinaloa, Zuaque (Fuerte), Mayo, and Yaqui—that flowed through slopes and canyons to water the coastal plains leading from the sierra to the Gulf of California. Petatlán reflected the production of culturally crafted spaces through mixed practices of food procurement and dwelling in the land. Native peoples created these landscapes through their labor, knowledge, and ceremonial cycles across different ecological zones extending eastward from the Gulf of California through coastal beaches and brackish wetlands, networks of streams and floodplains, and the *monte* of thorn forest, cacti, and hardwood trees (Crumrine, 1977: 11–14; Camacho Ibarra, 2017: 18–19; Gentry, 1942: 27–41; Gentry, 1995; Ingold, 2000) (Fig. 1).

Indigenous peoples in this region defended these territories as "our world" (*itom ania*) through their relationship to nature. These mosaics of distinctive geographical features were further inscribed with meaning by the spiritual power that emanates from nature (Camacho Ibarra, 2017: 312; Crumrine, 1977: 98; Shorter, 2009). In addition to the life-giving monte—the source for hunting, gathering medicinal plants, fruits, seeds, and building materials—*itom ania* embraced *kawi* (the sierras) with hills and ranges whose names evoke stands of vegetation, boulders, and oral traditions of *rancherías* (small, seasonal settlements); *wasam*, cultivable farmland in the river valleys; *pueplum,* the villages that mark their domestic spaces, the *ramadas* (arbors) for their rituals and centers of governance; *batwe,* the streams and arroyos that flow into the main river channels; and *bawe*, the maritime estuaries and open waters of the Gulf of California (Lerma Rodríguez, 2011: 53–72).

---

[2] Sauer and Brand (1998) referred to the pre-Hispanic northwestern frontier of Mesoamerica as Aztatlán, extending from Acaponeta to Culiacán.

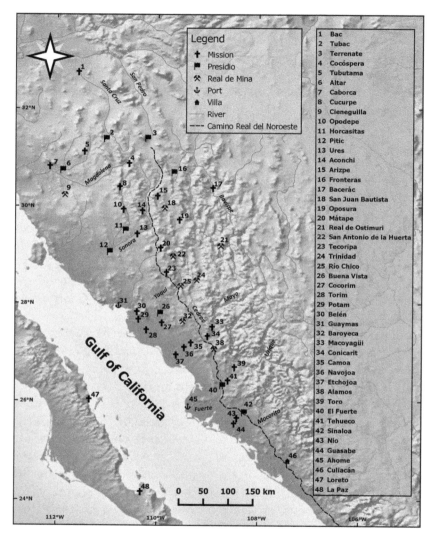

Fig. 5.1 Northwestern New Spain: Provinces of Sinaloa, Ostimuri, and Sonora. Map designed by Javier Etchegaray, University of North Carolina, Chapel Hill. Reprinted with permission from the University of Arizona Press

The peoples of these four river basins adapted Mesoamerican cultigens to these semi-arid but watered valleys, developing local varieties of maize, cucurbits, beans, chilis, amaranthus, cotton, and tobacco as well as root crops like camotes (*Ipomoea batatis*). Beyond the floodplains, along the arroyo beds, and in small clearings throughout the monte, the *Yoremem* and neighboring peoples planted shifting milpas with periods of fallow.[3] Their combined practices of planting, harvesting, and gathering redistributed different species of herbaceous plants, succulents, and cacti; furthermore, all the Indigenous communities of the highlands and river valleys both cultivated and gathered numerous varieties of edible and fibrous plants (Bañuelos, 2013: 403–407; Harriss Clare, 2012: 27, 72–73). The rivers and coastal mangroves and estuaries provided an abundance of fish, clams, freshwater shrimp, mollusks, and amphibians as well as salt. Indigenous peoples lived in these spaces and moved seasonally through different biomes to create a mixed livelihood of farming, foraging, and trade.

Indigenous landscapes were not timeless or unchanging, however, and their practices for planting, hunting, and foraging produced evolving and shifting borderlands. Indigenous skills for survival learned over centuries through experimentation created knowledge that was passed across the generations through quotidian practices, oral traditions, and rituals of dance and song (Shorter, 2009; Yetman & Van Devender, 2002). Recent archaeological findings suggest that the Fuerte river valley was occupied continuously for at least 2000 years BP, and the cultural roots of the historical communities of this region ran deep through changing environments from the Mocorito valley in the south to the Yaqui basin in the north. Analyses of ceramic, lithic, and shell remains for different sites in the Sinaloa, Fuerte, and Mayo drainages support hypotheses of continuous settlement in Petatlán with datable material for over a millennium prior to European contact. Adapting to ecological transformations in these fluvial borderlands, ethnic boundaries shifted within networks of chiefly alliances and rivalries

---

[3] The Mayos, Yaquis, Guarijíos, and Rarámuris constitute living peoples with enduring cultures to the present day. The Tepagüis, Macoyagüis, and Conicaris no longer exist as separate ethnic groups; however, they maintained their languages well into the nineteenth century, and they identify as *Yoreme* or mayo to the present day.

among different bands of rancherías and villages (Carpenter, 2014: 37–52).

The peoples of the floodplains and highland ranges throughout the four major river basins of Petatlán exhibited discernible patterns of livelihood and political organization that differed from the urban settlements of Culiacán and Acaponeta to the south, the terraced villages of the eastern Sonoran highlands, and the hunter-gatherers of the Sonoran Desert. Archaeological surveys and excavations carried out in present-day central and southern Sonora point to shared cultural traditions in the piedmont leading eastward to the Sierra Madre Occidental with cultivation, terracing, and methods for manufacturing ceramic and lithic artifacts even as they suggest regional variations in their time span and settlement sizes (Doolittle, 1988; Pailes, 1972, 1994). A complex of villages known as Huatabampo developed on the coastal marshes of the Mayo and Fuerte river valleys through agriculture and long-distance trade networks from c. 200 BCE to 1450 CE (Álvarez Palma, 1990). On the northern boundary of Petatlán in the mid-Yaqui drainage and its tributaries, archaeological surveys of the Ónavas Valley found evidence of trade with Huatabampo, in the lower Mayo valley, and the Casas Grandes complex centered in Paquimé, in northwestern Chihuahua (Gallaga Murrieta, 2006: 105–252).

Petatlán was connected through these trade routes to Aztatlán and the urban centers on the western and northern borderlands of Mesoamerica. Even after the ceremonial and trading hubs of Huatabampo, Paquimé, Chalchihuites (in Durango) and the Hohokam settlements in the Gila and Santa Cruz river valleys of Arizona began to disperse a century before European contact—due to climate change, internal dissention, and external invasions—clusters of agricultural villages throughout the alluvial valleys and piedmont continued to thrive, producing food surpluses, tools, and prestige insignia like pearls, shells, obsidian, and cloth, that supplied the regional trade networks Spaniards would encounter in the sixteenth century (Berrojalbiz, 2012; Webster et al., 2008). Well-storied sixteenth-century expeditions led by Nuño Beltrán de Guzmán, Álvar Núñez Cabeza de Vaca, Marcos de Niza, Francisco Vázquez de Coronado, and Francisco de Ibarra passed through portions of Sinaloa and Ostimuri, following multi-lingual

Indigenous guides who traveled from one province to another along trails that linked together coastal routes, highland ravines, and mountain passes (Obregón, 1584/1988; Pailes, 1997: 147–157; Sauer, 1932).

Colonial Borderlands in Petatlán

Spanish colonialism came to the *Yoreme* borderlands in the early seventeenth century through Jesuit missions and the mining industry. The Society of Jesus founded a network of mission districts from its base in the Colegio de San Felipe y Santiago in the Villa of Sinaloa, extending through the river valleys of Sinaloa, Ostimuri, and Sonora. In mid-century, the frontiers of mission evangelization advancing northward from the Villa de Sinaloa met the mining frontier proceeding westward from Parral in Nueva Vizcaya. Mining began in Ostimuri with silver strikes in the 1660s deep in the sierra between the Yaqui and Mayo drainages (West, 1993: 44–55). From the late seventeenth to the early eighteenth century, new mines opened in the Reales de los Alamos, Río Chico, Tacupeto, Piedras Verdes, and Baroyeca. Alamos and La Aduana became the principal commercial and administrative centers for Indigenous pueblos and Spanish settlements in the Mayo river basin (Cramaussel, 2012; West, 1993: Fig. 18, "The Alamos mining district, southern Sonora").

The parallel development of missions and *reales de minas* created conflicts involving labor and resources. The mines and surrounding settlements became markets for mission produce and livestock, but at the same time missions became labor repositories for the mines through the supply of rotational workers recruited for specific places and time periods, known as *repartimiento* (BNFR AF Caja 32, 650.1, 1698. Fojas 1–44v.) The Real de los Alamos drew heavily from the Mayo pueblos even as Yaquis, Mayos, and other Indigenous groups from Sonora and Ostimuri crossed the Sierra Madre to work in Parral (Cramaussel, 2014). Parallel to the labor drafts ordered through repartimiento, Indians traveled on their own to the mines, incorporating paid labor into their patterns of seasonal migration. Over time colonists and mission pueblos competed ever more rigorously for Indigenous workers and the productive resources of croplands, pasturage, woodlands, and sources of water in streams and springs.

Colonial boundaries were negotiated over more than three centuries through different practices of defining space and allocating resources in a mixed thorn forest and riparian environment, extending from the basin-and-range topography of the Sierra Madre Occidental to the deltas, wetlands, and estuaries of the Gulf coast. The Jesuit missions were intended to bring scattered Indigenous settlements into permanent villages, but, in practice, they became an interface between the rancherías that supported the Indians' seasonal movements through different biomes and the ceremonial and political centers of the missions. As the Spanish population grew with the expansion of mining and livestock ranching, the institutional status of the missions under colonial law protected Indigenous agricultural lands and portions of the monte, where Native shepherds and cowhands tended mission herds. Yet, mission lands were not necessarily contiguous or clearly defined; rather separate parcels of croplands, dependent on rainfall or the seasonal flow of intermittent streams, and grasslands mixed with thorn forest, were occupied at different times by both Indigenous and Spanish cultivators, pastoralists and foragers. Called *parajes*, which translates simply to "place," these irregularly shaped stretches of monte were named—often with Indigenous toponyms—thus revealing locally recognized histories of occupation and use.

Customary land tenure and usufruct in these northern borderlands of New Spain became more formal under the pressures of population growth and an increase in the size of livestock herds. Over the course of the eighteenth-century colonial authorities in Nueva Vizcaya, Sinaloa and Sonora pressured Spanish *vecinos* (Hispanic resident heads of households) to regularize their occupation of portions of the monte, known as *realengos* (public lands nominally owned by the Crown), through *composición*, a legal process involving the payment of royal fees to carry out the measurement, public auction, and issuance of title that turned open parajes into discreetly bounded pieces of private property

(López Castillo, 2010, 2014; Radding, 1997: 171–206, 2005, 2022: 152–191).[4] Private landholdings evolved from use rights negotiated through verbal agreements to demarcated and titled properties. The legal process for enclosing portions of the monte contrasted with informal arrangements for sharing pasturage and water or transferring use rights through purchase and inheritance.[5] Formal land ownership with measured boundaries and property markers ran counter to Indigenous practices of land use and access based on seasonal occupation of the monte and the resources of water, forests, and wildlife sheltered within it. Table 5.2 lists a selection of archival sources for the legal enclosure of private landholdings in the Provinces of Ostimuri and Sinaloa.

Defending the *Monte*

In the middle portion of the Mayo river drainage, where the floodplains widened to support extensive croplands, the pueblos of Camoa, Tesia, Navojoa, and Cohuirimpo straddled both the northern and southern banks of the river. The Indigenous officers of Navojoa became the principal litigants in a case that represented the entire village in their demand to maintain the riparian wetlands and scrub forest that surrounded their croplands as open monte not to be measured or adjudicated to private owners. Together with the officers of Tesia, they claimed these realengos by right of possession and "because the king, in

---

[4] *Realengos* referred to lands that were nominally under the jurisdiction of the monarch and not alienated to a private owner. See the interesting discussion of *realengos* and *real* in BNFR AF Caja 32, 650.1, 1698, f. 44. Vecinos were Hispanic subjects, identified as residents of a given city or town, with potential landholding rights and the duties to contribute to local defense. The provinces of Sinaloa, Ostimuri, and Sonora were under the administration of the Kingdom of Nueva Vizcaya under 1734, when the governorship of Sonora–Sinaloa was established; with the Intendancy system of 1786, the western provinces from Chiametla to Sonora came under the authority of the Intendancy of Arizpe.

[5] Hundreds of files created through the process of approving land titles through composición, often called *títulos primordiales*, are preserved in the Archivo General del Estado de Sonora, Títulos Primordiales (AGES TP), in the Archivo Histórico General del Estado de Sinaloa, Ramo de Tierras, in the Archivo Histórico de Jalisco, and in the Archivo General de la Nación (AGN), Ramos de Tierras and Indiferente Virreinal.

his mercy, had given them the lands that were left unmeasured"
between the boundaries of neighboring villages and private
landholdings.[6] The contentious arguments recorded in the field and
entered into the court proceedings in the Audiencia of Guadalajara
hinged on the meanings of *possession*; their dissonance arose from
distinct ways of securing a livelihood and dwelling in the land. In the
following paragraphs, the positions upheld by the *Yoreme* village officers
are analyzed in detail, because they illustrate in profound ways both the
material qualities of the land and its resources and the values that
Indigenous peoples ascribed to the landscapes that they and their
ancestors had created.

Don Manuel Ignacio Valenzuela, a vecino of the Villa of Sinaloa,
filed a claim (*denuncia*) to uncultivated lands between his ranch of
Husibampo and the pueblo of Navojoa before the magistrate in charge
of land sales and composiciones in the Audiencia of Guadalajara, Don
Guillermo Martínez de Aguirre y Viana. In June 1787, Judge Martínez
de Aguirre commissioned Juan María de Figueroa, a vecino of Álamos
in the Province of Sinaloa to survey the lands surrounding Husibampo,
ordering Figueroa to measure or estimate (*regular*) the distances
between Navojoa and Husibampo and show their locations in a map;
determine how much land the Indians of Navojoa cultivated and how
much they occupied beyond their croplands; report the number of
Indigenous families living in Navojoa; measure the entire area over
which Valenzuela had filed his claim and determine its "intrinsic
worth." Figueroa fulfilled this demanding commission only partially: he
did not actually measure any of the lands in question, rather he merely
estimated their circumference and extension, nor did he get an accurate
count of the population of Navojoa. If he drew a map, it is not
included in the archive. Despite these omissions, his report revealed key
aspects of the human ecology of the Mayo river basin and surrounding
monte in the late eighteenth century; furthermore, the language he used
to summarize verbal exchanges in the field articulated some of the
conflicting notions of territoriality expressed by Indigenous

---

[6] AGES TP XXVIII, 387, 28 February 1788, f. 70. "… [e]l Gobernador de [Tesia] y sus vocales
… dijeron que respecto a que el Rey les ha hecho limosna del realengo sobrante, no consienten
en la mensura de él".

communities, Hispanic landowners, and officials charged with determining property boundaries, awarding land titles, and collecting the corresponding royal fees.

In late February 1788, Juan María de Figueroa assumed his commission and sent a citation to the claimant Manuel Ignacio Valenzuela, who named a representative to defend his interests for the land survey in Husibampo and the surrounding monte. Likewise, Figueroa summoned the Indigenous officers of Navojoa, naming those who responded to witness the survey and defend their pueblo's lands and boundaries: Governor Nicolás Cubil and council members Ignazio Raqui, José Manuel, Vidal Melchior Usacamea, Juan Ynacio Arze, Ramón Baqui, Fernando Baquivi, José Matías Machiri, Ygnazio Anuamea, José Qroba [Quiroga?], Oní Agustín y Nomamea. Figueroa read to the assembled *Yoreme* officers out loud the order he had received from the Audiencia Judge Martínez y Aguirre, to which they responded that they heard and understood it. Governor Cubil signed this portion of Figueroa's *auto*, as the official record of the proceedings in representation of all the village officers, together with Figueroa and one Hispanic witness, Simón de Santelizer. Nicolás Cubil, literate and bilingual in Spanish and *Yoremnokki*, was named Captain General of the Mayo River in the following decade. He appeared in numerous cases of land measurement and titling as an interpreter and adviser to his own and other *Yoreme* pueblos.[7]

The following day Governor Cubil and his fellow council members walked with Figueroa along the perimeter of all the croplands cultivated by the villagers of Navojoa, accompanied by witnesses Miguel María de Figueroa and Simón Santelizer, the former charged with counting cord lengths and estimating the distances they traversed. In effect, they led him downstream toward the west, following the river bank until they reached the limits that the Mayos of Navojoa recognized between their pueblo and the lands of Cohuirimpo. All along this transect Figueroa observed cultivated land, some of it dependent on seasonal rainfall and other parts watered by the highland streams that replenished the floodplain. Figueroa estimated its length in 125 cords of fifty varas each;

---

[7] AGES TP XXVIII, 387, 26 February 1788, f. 66–67.

this linear distance computed to 6250 varas, roughly equivalent to 5700 meters or 5.7 km. At this point he and the Navojoa officers crossed the river and turned to the north, passing through watered cropland for a distance that Figueroa estimated in fifty cuerdas (2.28 km). Turning eastward on the northern bank of the river, Figueroa, the Mayos, and the two witnesses walked upstream along the river's edge a distance of 250 cuerdas (11.4 km) to reach the limits between the pueblos of Navojoa and Tesia. At this juncture they crossed the river turning south, walking for another fifty cuerdas, then turned westward for an equal transect of 125 cuerdas. In this manner, their survey traced an uneven rectangle along the full width of the floodplain running parallel to the river and surrounding the pueblo. Figueroa estimated the circumference of the entire area in six leagues (approximately 24 square kilometers). He noted, further, all along these cultivated lands short term, seasonal plantings in the washes, where the humidity of the soil would allow crops to germinate and mature.[8] Thus, Nicolás Cubil and the officers of Navojoa guided Judge Surveyor Juan María de Figueroa to confirm their pueblo's possession of the wasam, the lands that were replenished each year by the floodplains in the Mayo river basin and renewed seasonally for planting and harvesting.

The dialogue took a sharper tone when Figueroa turned to the thorn forest that the people of Navojoa enjoyed by right of possession, but without cultivating it as cropland. Figueroa made the striking statement that north of the pueblo, these lands were "limitless," extending as far as thirty leagues to the southern margin of the Yaqui river. Looking south of the pueblo, Nicolás Cubil asked Judge Figueroa to establish the boundaries of Navojoa's lands at the same time that he measured the portion of the realengo that lay adjacent to Valenzuela's holdings in Usibampo. Figueroa verbally assented and issued an auto, stipulating that he would give Governor Cubil a copy of this agreement in writing,

---

[8] AGES TP XXVIII, 387, 26 February 1788, f. 68. The distances were computed into modern measurements on the basis of the conversion of one vara to .912 meters, from the online *Diccionario de la Lengua Española*, https://dle.rae.es/?id=bMH7x5e. The following lines express Figueroa's description of the wasam. "De aquí tomamos el viento poniente por la orilla de tierras de la misma qualidad a cerrar nuestro paseo a este mismo pueblo de donde salimos, regulando por 125 cuerdas: cuya circunferencia es la de 6 leguas: y he visto en las tierras que labraban, sembrados los bajíos que el temporal de verano les permite humedad para ello."

following legal procedure. On February 28, the commission convened for the land survey and set out to walk through the monte leading southward from the pueblo. When they arrived at the boundary marker for the puesto of San Joseph del Retiro, Governor Cubil declared that he did not consent to the measurement of these lands in favor of Valenzuela, because—he declared—they belonged to Navojoa, and the villagers had no other grazing land for their livestock. Pedro Joseph Solano, who represented Valenzuela, insisted that the survey proceed as ordered by Judge Martínez de Aguirre in Guadalajara. Figueroa then "crossed the line" between the lands recognized as belonging to the pueblo of Tesia and those of San Joseph del Retiro. Standing there, he turned to the governor and council members of Tesia, who responded in unison with Governor Cubil that they did not consent to the measurement of these lands, because the king had awarded them to their pueblo. Despite their recorded protest, they deferred to Figueroa's authority on the strength of the order issued by the Audiencia Judge, telling him to act as he thought best.[9]

Overriding the objections of the Indigenous officers of both Tesia and Navojoa, Figueroa proceeded in early March to carry out the survey from the puesto of Husibampo, where he had summoned the officers of the pueblo of Camoa (north of Tesia). When he announced to them the order he had received from the Audiencia, the Camoa officers responded that "they had nothing to say" and acceded to the measurement of the realengo bordering their pueblo. Accompanied by Solano and the witnesses, Figueroa estimated the area that corresponded to Valenzuela's claim as four sitios, but he dispensed with their measurement in order not to increase the costs that would accrue to him for securing title to the land. Two local vecinos, Miguel María de Figueroa and Andrés Armenta, estimated the value of the land at five reales per *caballeria* (43 hectares). The four sitios awarded to Valenzuela corresponded to 162.5 caballerías; when valued at five reales each, the cost of converting them into titled property would have totaled 812.5 reales or 101 pesos and four reales, a sum on which Valenzuela would

---

9 AGES TP XXVIII, 387, 28 February 1788, f. 69–70.

have paid an additional amount in taxes and fees to the royal treasury had he completed the process of securing title to the land.[10]

Immediately following the land survey Governor Nicolás Cubil and the councilmen of Navojoa and Tesia registered their protest over the measurement of the realengo that extended between their pueblos and Husibampo before the Justicia Mayor of the Real de los Álamos, Don Juan Manuel de Zavala, who forwarded their signed statements to the Protector de Indios in the Audiencia of Guadalajara. As they had done in person before Don Juan María de Figueroa, the Indigenous officers defended their right of prior possession according to the primacy awarded to them as Indigenous pueblos (*pueblos de indios*) by the Sovereign, the longevity of their occupation of these lands—as it was commonly known in the region—and their need for these realengos in which to pasture their livestock. Magistrate Zavala halted the process of issuing the land title to Valenzuela, ordering that until the Juez Privativo de Tierras in the Audiencia ruled on the case, the *Yoreme* villagers should continue to exercise their use rights to this portion of the monte according to their custom.[11]

In his role as the commissioned surveyor don Juan María de Figueroa had requested the census corresponding to the Pueblo of Navojoa in order to comply with the orders he had received from Judge Martínez de Aguirre in Guadalajara. Figueroa counted one hundred families in the mission census but doubted its veracity because—he said—he could

---

[10] AGES TP XXVIII, 387, 3–7 March 1788, f. 72–73. Miguel María de Figueroa had served as one of the witnesses throughout the land survey, while Andrés Armenta, who had not appeared previously, was illiterate and could not sign his name. *Caballería*, a common measurement for grasslands or thorn forest, computed as 1104 × 552 varas, or 43 hectares. A *sitio* for bovine cattle, computed at 5000 × 5000 varas, constituted 1747 hectares. Radding, 1997: Table 6.1, p. 177, based on AGN AHH *Temporalidades* legajo 1165.

[11] AGES TP XXVIII, 387, 5 March 1788, f. 75–78. En el Rl de los Álamos a 5.III.1788. Ante mi, D. Juan Manuel de Zavala, Justicia Mayor y Capitán a Guerra de este mencionado real y su jurisdicción, pareció presente el Gobernador del Pueblo de Navojoa D. Nicolás Cubil por sí y a nombre de los demás naturales, haciéndome manifestación ... en oposición de las medidas que ha practicado a pedimento de D Manuel Ygnacio Valenzuela, en lo que dicen ser realengo, y reclaman pertenecer dichas tierras a los dos pueblos de Navojoa y Tesia, de esta comprension ... mediante a la primacía que les concede el Soberano, y a la inmensidad de años que tienen de posesión y reconocimiento por sí y sus antecesores y en el interim se resuelve lo conveniente por el referido S. Juez Privativo, proseguirán los Indios Naturales de los citados pueblos en el proprio método que han gozado, sin innovación alguna.

find no more than ten "poorly built" houses within the village perimeter. This statement does not ring true, however, when compared with the figures in the census carried out four years prior, in 1784, by the first Bishop of Sonora, and because of what we know about Indigenous patterns of settlement and livelihood. The Bishop's census showed 140 families and 350 persons in Navojoa, 95 families with 300 persons in Tesia, and 35 families with 90 persons in Camoa. The families of the fourth pueblo, Cohuirimpo, were dispersed in the monte, while the population of all four villages was mixed among Yoremem, Spaniards, and castas (mestizos). Figueroa opined that the pueblos' scant population undermined their alleged need for the land they so staunchly defended, an argument that Valenzuela's attorney repeated vociferously before the Audiencia. Determined to promote the privatization of the realengos, Figueroa willfully denied the enduring pattern of rancherías in which clusters of families distributed themselves along the length of the floodplain to cultivate their milpas and procure a wide variety of plants and wildlife in the marshes and upland monte. Their jacales of reed mats and puddled adobe provided shelter and served them in their seasonal movements through different ecotones of thornscrub, wetlands, and highland deciduous forest (Dimmitt, 2000: 3–18; López Castillo, 2008: 60).[12]

Maneuvering between deference and defiance the governors and village officers of Tesia and Navojoa expressed the conflicting practices of land use that defined the open quality of the realengos (Deeds, 2003). Leaders like Nicolás Cubil navigated the judicial and administrative institutions of local governance and the Audiencia of Guadalajara to persevere in their assertion of usufruct rights and access to both the cultivated floodplain and the resources of the monte. Their ancient traditions of dwelling between the wasam and the monte clashed, however, with notions of titled ownership of measured pieces of land and its commodification in set fiscal and monetary terms. Indigenous pueblos were caught between the customary rights they defended and the commercialization of the land and its resources, a

---

[12] AGES TP XXVIII, 387, 5 March 1788, f. 72; BNFR AF 34/759 Obispo Antonio de los Reyes, Informe, 1784.

process in which they participated by breeding their own livestock herds and trading their produce and labor in the colonial market. Governor Cubil and his fellow officers forestalled the division and enclosure of the monte surrounding Navojoa and Usibampo with the simple words: "we do not consent." The corpus of documents generated by the process of land titling shows the momentum of issuing land titles through *composición* and the gradual dispossession of Indigenous communal lands in favor of the commercialization of landed properties. To read them only in this way, however, would deny the perseverance of individuals and communities among the Yoremem and neighboring peoples who intervened in the contentious histories of enclosing the open territory of the monte.

North of Camoa in the mountainous territory of the headwaters of the Mayo and Cedros rivers, the Pueblo of Tepagüi defended its longstanding possession of Bacusa, a stretch of land with sufficient water for agriculture some two leagues north of their village.[13] Faced with the threat of its measurement and conversion into private property in favor of Don Gabriel Felix, a vecino from the mining real of Baroyeca, the governor and village council of Tepagüi mobilized to demonstrate to local authorities their effective use of Bacusa. In the spring of 1791, Governor Pedro Misquy and his fellow officers walked the land with the Subdelegado of the Province of Ostimuri, Don Cristoval Gimenes Moyano, taking care to point out the milpas they had cleared for planting and the irrigation canals—*acequis*—they had dug to water their crops. Located northwest of the main channel of the Cedros river, the Tepahuis harnessed the downward flow of arroyos that fed into both the Cedros and Mayo rivers to irrigate small fields immediately adjacent to their pueblo and in separate puestos like Bacusa. Furthermore, the Tepahuis maintained reciprocal relations with several vecinos from the highland ranching settlement of Quiriego, upstream on the Cedros river, who used portions of Bacusa but acknowledged that it belonged to Tepagüi. It was not clear whether the

---

[13] AGES TP Tomo VI, 69, 1791–1820, ff. 431–469. Don Gabriel Felix y el Pueblo de Tepagüi por el sitio de Bacusa.

vecinos paid rent or offered a different kind of material Exchange (AGES TP Tomo VI, 69, 1791–1820, f. 435–436).

When Subdelegado Gimenes Moyano forwarded the file to Intendant-Governor Enrique de Grimarest, he dismissed the Tepahuis' strenuous arguments, referring disparagingly to the acequia they had dug as nothing more than a shallow ditch. In August 1791, he commissioned Gimenes Moyano to survey Bacusa in favor of Felix's bid to gain title to the puesto. Having summoned the neighboring landholders Don Francisco Xavier de Valenzuela, the priest Don Joaquín Elías González de Zayas—both members of prominent families in the mining center of Álamos—and the Pueblo of Tepagüi, Gimenes Moyano began the measurements in late October. Using a twisted ixtle rope of fifty varas' length, Gimenes Moyano established the center of the property in a corral built on the top of a low mesa. The work began by counting 48 cordeles to the north, ending in the boundary marker for Quiriego, where the subdelegado placed another marker for this northern limit of Bacusa at the foot of a *teso* tree (*Acacia occidentalis*; Martin et al. eds., 1998: 110). Returning to the center, the survey team walked in the other three cardinal directions to establish the boundaries of the puesto. Gimenes Moyano extended the perimeters of the surveyed land toward the east and south, by directing an additional set of measurements that reached the northern limits of Tepagüi, where Governor Misquy and his fellow officers placed the boundary marker that separated their village lands from this portion of Bacusa that—until that momento—had been part of their communal patrimony. Gimenes Moyano's elongated survey rendered this measured portion of Bacusa in 2.5 sitios "and three cordeles more," surrounded on its southern and eastern limits by the village lands of Tepagüi and to the north and west by the properties of González de Zayas and Valenzuela. The subdelegado valued Bacusa at 69 pesos, considering that it was forested and intersected by arroyos that provided alluvial lands for planting maize and natural water holes for livestock (*abrevaderos*). Don Gabriel Felix paid the fees to obtain title to Bacusa two decades later, in 1820, on the eve of Independence.

The testimonies recorded in the survey of Bacusa in support of the Tepahuis as well as their detractors offer pointed observations regarding different systems of cultivation and water management in the foothills surrounding the Mayo river. Witnesses called to defend or refute the Tepahuis' claim to Bacusa described canals that conducted water to cleared fields, and separate plantings *de humedad* and *de verano*—that is, crops dependent on groundwater at the margins of arroyos or on summer rains without irrigation. Even those witnesses who argued against the Tepahuis' effective use of Bacusa, emphasizing the small scale of their visible cropland, named a series of puestos—Choymoco, Techueca, Supabampo, Saposero, Los Chinos, and Suybampo—across the stream beds of tropical deciduous forest to which the Tepahuis had access for hunting, gathering, and cultivos de humedad. Don Juan María Figueroa and Don Victorino Gil, vecinos who purported to have long-term knowledge of the región, calculated the number of *fanegas* of wheat or corn that could be planted in Tepagüi village lands communally or by individual families. Despite their intention to undermine Indigenous claims to this puesto and promote its enclosure as a titled property, Figueroa and Gil revealed the cyclical or seasonal nature of land use in the monte with different kinds of vegetation, soils, and sources of water.

Language and Histories of Cultural Landscapes

For the Indigenous peoples of today, land titles and the documents they generate provide a wealth of information about the natural environment and the landscapes their ancestors created, adding to their own stores of knowledge and opening pathways to future possibilities for restoring their ecological patrimony. The histories of land enclosures through the process of composición to grant land titles upon payment of a fee do more than merely document a bureaucratic legal procedure under colonial rule; rather they illustrate the persistence of Indigenous communities in the face of contested power relations and the disputed control over vital resources. The testimonies included in these files provide the names of Indigenous officers, who represented the institutional presence of community governance in the northwestern borderlands of New Spain. The principles they defended in specific instances of disputed claims to portions of the monte were centered in the material resources of vegetation, water, and cultivable land. Moreover, they expressed the very identity of community linked to these commonly held spaces in the wider cultural meaning of territory. Their words are echoed by the *Yoreme* citizens of Cohuirimpo gathered in Los Nachuquis, who articulated the significance of *itom ania* in their affirmation: *we claim our natural right, because we are part of the territory* (Radding, "Jornada de trabajo en torno al Pueblo de Cohuirimpo," field notes, 23 October 2021).

Orality and the written word are integral to the narratives that are woven into the colonial land titles analyzed for this chapter. Verbal confrontations and negotiations among the historical actors—the vecinos and *Yoreme* peoples who appear in the recorded proceedings—can be inferred from the written documents, even in the formal language of the scribes' annotations. The Indigenous officers who were called to witness land measurements and defend the boundaries of cropland and monte surrounding their villages deliberated among themselves in yoremnokki and then directed their words to the colonial authorities either directly or through interpreters. Regionally based Spaniards and Indians shared a common lexicon for geographical markers relating to place names and species of plants and animals, even when they disagreed on the meanings of possession, usufruct, and ownership of the vital resources of forested land and water. Governor Nicolás Cubil and the council officers of Navojoa, Tesia, and Camoa contested the measurement of Husibampo in the middle Mayo river drainage during the land survey in the field and before the magistrates of the Province of Sinaloa and the Audiencia of Guadalajara. With the words "we do not consent," they alleged their legal right to the territory as pueblos de indios and subjects of the king, asserting their need for this portion of the monte as grazing land. In a similar confrontation, Governor Misquy and the council officers of Tepagüi defended their corporate possession and usufruct of Bacusa against the encroachments of private landowners in the forested monte and watered cropland that surrounded their pueblo. They called witnesses to attest to the ways that Tepahuis had turned Bacusa into a cultivated landscape through their labor to dig acequias, gather the forest, and clear brush for planting milpas.

The full meaning of the conflicts narrated in this chapter emerges from a careful reading of the internal evidence generated by the títulos primordiales, interpreted in the context of the ecological settings for these territories and the ethnohistorical patterns of both cultural continuity and historical change. In this way, histories of cultural landscapes can be woven from the verbal descriptions and numerical measurements of portions of the monte. The language that gives them meaning is deeply rooted in the spatial environments of Petatlán, connecting the colonial past with the present in the living histories of *Yoreme* communities.

References

Archives

AGES Archivo General del Estado de Sonora.
TP Títulos Primordiales.
AGN Archivo General de la Nación, Mexico.
AHH Archivo Histórico de Hacienda.
BNFR Biblioteca Nacional, Mexico, Fondo Reservado.
AF Archivo Franciscano.

Bibliography

Álvarez Palma, A. M. (1990). Huatabampo: Consideraciones sobre una comunidad agrícola prehispánica en el sur de Sonora. *Noroeste de México, 9*, 9–93.

Bañuelos, N. (2013). Etnobotánica, una ventana hacia la concepción de los mundos mayo y guarijío. In J. Luis, M. Zamarrón, & A. A. Zeleny (Coords.), *Los pueblos indígenas del Noroeste. Atlas Etnográfico* (pp. 403–407). Instituto Nacional de Lenguas Indígenas, Instituto Nacional de Antropología e Historia, Instituto Sonorense de Cultura.

Berrojalbiz, F. (2012). *Paisajes y fronteras del Durango prehispánico*. UNAM Instituto de Investigaciones Antropológicas, Instituto de Investigaciones Estéticas.

Camacho Ibarra, F. (2017). *El sol y la serpiente: el pajko y el complejo ritual comunal de los mayos de Sonora* (M.A. Thesis). Universidad Nacional Autónoma de México, Programa de Maestría y Doctorado en Estudios Mesoamericanos.

Carpenter, J. (2014). The pre-hispanic occupation of the Río Fuerte Valley, Sinaloa. In E. Villalpando & R. H. McGuire (Eds.), *Building transnational archaeologies/Construyendo arqueologías transnacionales* (pp. 37–52). University of Arizona Press.

Cramaussel, C. (2012). Poblar en tierras de muchos indios: La región de los Álamos en los siglos XVII y XVIII. *Región y Sociedad XXIV, 53*, 11–53.

Cramaussel, C. (2014). The forced transfer of Indians in Nueva Vizcaya and Sinaloa: A hispanic method of colonization. In J. Barr & E. Countryman (Eds.), *Contested spaces of early America* (pp. 184–207). University of Pennsylvania Press.

Crumrine, N. R. (1977). *The Mayo Indians of Sonora: A people who refuse to die*. University of Arizona Press.

Deeds, S. M. (2003). *Defiance and deference in Mexico's colonial North: Indians under Spanish Rule in Nueva Vizcaya*. University of Texas Press.

Deeds, S. M. (2000). Legacies of resistance, adaptation, and tenacity: History of the native peoples of Northwest Mexico. In R. E. W. Adams & M. J. MacLeod (Eds.), *The Cambridge history of the native peoples of the Americas*, Vol. 2, Mesoamerica, Part 2 (pp. 44–88). Cambridge University Press.

Dimmitt, M. A. (2000). Biomes and communities of the Sonoran Desert Region. In S. J. Philips & P. W. Comus (Eds.), *A natural history of the Sonoran desert* (pp. 3–18). Arizona Sonora Desert Museum.

Doolittle, W. E. (1988). *Pre-hispanic occupance in the valley of Sonora, Mexico: Archaeological confirmation of early Spanish reports, anthropological papers University of Arizona, No. 48*. University of Arizona Press.

Gallaga Murrieta, E. (2006). *An archaeological survey of the Onavas Valley, Sonora, Mexico. A landscape of interactions during the late prehispanic period* (Ph.D dissertation). University of Arizona.

Gentry, H. S. (1942). *Río Mayo plants: A study of the flora and vegetation of the valley of the Rio Mayo, Sonora*. Carnegie Institution.

Gentry, H. S. (1995). "Caminos of San Bernardo." Special Issue: Explorations on the Río Mayo. *Journal of the Southwest, 37*(2), 135–141.

Harriss Clare, C. J. (2012). *Wa?ási—kehkí buu naaósa-buga* = *"Hasta aquí son todas las palabras"*: *la ideología lingüística en la construcción de la identidad entre los guarijío del alto mayo*. Instituto Chihuahuense de Cultura, PIALLI, Consejo Nacional para La Cultura y las Artes.

Hers, M.-A. (2013). Aztatlán y los lazos con el centro de México. In M.-A. Hers (Coords.), *Miradas renovadas al Occidente indígena de México* (pp. 273–312). Universidad Nacional Autónoma de México, Instituto de Investigaciones Estéticas; Instituto Nacional de Antropología e Historia, Centro de Estudios Mexicanos y Centroamericanos.

Ingold, T. (2000). *The perception of the environment: Essays in livelihood, dwelling and skill*. Routledge.

Lerma Rodríguez, E. (2011). *El nido heredado*: *Estudio sobre cosmovisión, espacio y ciclo ritual de la tribu yaqui* (Ph.D. Dissertation). Universidad Nacional Autónoma de México, Posgrado en Antropología, Facultad de Filosofía y Letras, Instituto de Investigaciones Antropológicas.

López Castillo, G. (2008). El territorio cahita, de frontera chichimeca a frontera misional. In A. Fábregas Puig, M. A. Nájera Espinoza & J. F. Román Gutiérrez (Coords.), *Regiones y esencias: Estudios sobre la gran Chichimeca* (pp. 57–68). Seminario de Estudios de la Gran Chichimeca, Universidad de Guadalajara, Universidad Autónoma de Zacatecas, Universidad Autónoma de Aguascalientes, Universidad Intercultural de Chiapas, El Colegio de San Luis, El Colegio de Michoacán, El Colegio de Jalisco, Universidad Autónoma de Coahuila.

López Castillo, G. (2010). *El poblamiento en tierra de indios cahitas*. Siglo XXI, El Colegio de Sinaloa.

López Castillo, G. (2014). *Composición de tierras y tendencias de poblamiento hispano en la franja costera: Culiacán y Chiametla, siglos XVII y XVIII*. Instituto Nacional de Antropología e Historia-Centro INAH Sinaloa, H. Ayuntamiento de Culiacán-Instituto Municipal de Cultura.

Martin, P. S., Yetman, D., Fishbein, M., Jenkins, P., Van Devender, T. R., & Wilson, R. K. (Eds.). (1998). *Gentry's Río Mayo plants: The tropical deciduous forest and environs of Northwest Mexico*. University of Arizona Press.

Moctezuma Zamarrón, J. L., & Aguilar Zeleny, A., (Eds.) (2013). *Los pueblos indígenas del noroeste: atlas etnográfico*. Instituto Nacional de Lenguas Indígenas, Instituto Nacional de Antropología e Historia, Instituto Sonorense de Cultura.

Obregón, B. (1584/1988). *Historia de los descubrimientos antiguos y modernos de la Nueva España escrita por el conquistador en el año de 1584* (M. Cuevas, Ed.). Editorial Porrúa, S.A.

Pailes, R. A. (1972). An archaeological reconnaissance of Southern Sonora and reconsideration of the Río Sonora Culture (Ph.D. Dissertation). University of Southern Illinois, Carbondale.

Pailes, R. A. (1994). Relaciones Culturales Prehistóricas en el Noroeste de Sonora. In B. Braniff & R. S. Felger (Eds.), *Sonora: antropología del desierto. Noroeste de México* (pp. 117–122). Centro INAH Sonora.

Pailes, R. A. (1997). An archeological perspective on the Sonora Entrada. In R. Flint & S. C. Flint (Eds.), *The Coronado expedition to Tierra Nueva. The 1540–1542 route across the Southwest* (pp. 147–157). The University Press of Colorado.

Radding, C. (1997). *Wandering peoples. Colonialism, ethnic spaces, and ecological frontiers in Northwestern Mexico, 1700–1850.* Duke University Press.

Radding, C. (2005). *Landscapes of power and identity. Comparative histories in the Sonoran Desert and the forests of Amazonia from colony to republic.* Duke University Press.

Radding, C. (2022). *Bountiful deserts. Sustaining indigenous worlds in Northern New Spain.* University of Arizona Press.

Sauer, C. O. (1932). The road to Cíbola. In *Ibero-Americana.* University of California Press.

Sauer, C. O. (1935). Aboriginal population of Northwestern Mexico. *Ibero-Americana, 10,* 1–33.

Sauer, C. O., & Brand, D. D. (1998). Aztatlán, Prehistoric Mexican Frontier on the Pacific Coast, 1–66. I*bero-Americana 1.* University of California Press.

Shorter, D. D. (2009). *We will dance our truth: Yaqui history in Yoeme performances.* University of Nebraska Press.

Webster, L. D., McBrinn, M. E., & Carrera, E. G. (Eds.). (2008). *Archaeology without borders: Contact, commerce, and change in the U.S. Southwest and Northwestern Mexico.* University Press of Colorado, CONACULTA, INAH.

West, R. C. (1993). *Sonora: Its geographical personality.* University of Texas Press.

Yetman, D., & Van Devender, T. R. (2002). *Mayo ethnobotany: Land, history, and traditional knowledge in Northwest Mexico.* University of California Press.

6

Gender Disparities in Guaraní Knowledge, Literacy, and Fashion in the Ecological Borderlands of Colonial and Early Nineteenth-Century Paraguay

Barbara A. Ganson

During my memorable interview and visit in 1991 to the village of Acaray-mí in a remote area of Alto Paraná on the Paraguayan-Brazilian border, an Avá-Chiripá shaman and cacique emphasized to me that "he has entire books in his head." My former Paraguayan professor of philosophy of history at the Universidad Católica Nuestra Señora de la Asunción and Director of the Centro de Estudios Antropológicos, Dr. Adriano Irala Burgos, had arranged for me to interview two shamans and village elders with the assistance of a graduate student who had resided two years among them and who spoke their same language variant, which is distinctive from the Guaraní (known as *yopará*) spoken by most Paraguayans today. While on other occasions, I had traveled by automobile, bus, or on horseback to visit twenty-two of the former thirty Jesuit missions in the borderland region of Paraguay, northern Argentina, and southern Brazil, on this trip, I walked over two hours through the forests and across shallow muddy creeks and fields in

B. A. Ganson (✉)
Florida Atlantic University, Department of History, Boca Raton, Florida, USA
e-mail: bganson@fau.edu

© The Author(s) 2024
J. Olko and C. Radding (eds.), *Living with Nature, Cherishing Language*,
https://doi.org/10.1007/978-3-031-38739-5_6

heavy rain and thunderstorms first to set up an appointment for the interviews, which took place under the same climatic conditions the following day. From this experience, I realized the importance of Native religion, artisanry, and mythology in the everyday lives of the Guaraní, rooted in their natural surroundings, from the lighting of their campfires in their modest wooden homes, carving of sacred animals, to the location of their villages constructed near pindó palm trees. The Avá-Chiripá belong to the linguistic family of the Tupí-Guaraní.

Anthropologists recently uncovered evidence that the Tupí-Guaraní among other Indigenous peoples of the Amazonian region and tropical lowland areas of South America had vast knowledge of plant and landscape domestication, including village gardens, orchards, and forests well before the arrival of Europeans in South America in the early sixteenth century (Clement & Denevan, 2015; Moore, 2014). While the Tupí-Guaraní societies were sophisticated, they did not develop a writing system independently of Europeans. Instead, strong oral traditions were an integral part of their culture. Traditional knowledge has been passed down from one generation to another through the telling of stories and education, both in a Native school where an Indigenous teacher instructs in accordance with the Paraguayan Ministry of Education, and by parents, village elders, traditional healers, caciques, and other family members.

This essay analyzes the gender disparities in Guaraní education and literacy in the province of Paraguay based on original Guaraní texts and other archival sources, including school censuses and Guaraní schoolwork, during the colonial era through the mid-nineteenth century (see Fig. 6.1). Native sources are vital for helping us understand which social values, technologies, and material goods the Indigenous peoples accepted or rejected from the Europeans. For the nineteenth century, in addition to these sources, traveler accounts provide impressions of Paraguayan women's education. Fashion is an element in this analysis as well because the types of clothing worn reflected the Guaraníes' changing sense of identity. Decisions over what the Guaraní wore became a critical part of the strategies employed by missionaries and Spanish officials to make the Indigenous people conform to their notion of what was proper behavior and dress. By learning Spanish in the schools rather than their Native language and adopting European-style clothing, the Guaraní became negotiators of change in their identities and gender norms. Their adoption of the technology of writing was part of a complex process of transculturation, meaning that the Guaraní invented new traditions from the materials or elements introduced to them by a more dominant European culture, and selected those which served to be beneficial (Ganson, 2003; Wilde, 2019). Their use of the Native language, whether in written form or spoken, however, proved to be a contentious issue in Paraguay, reflecting the cultural resiliency of the Guaraní, the forces of cultural domination, and the relationship between education, literacy, and gender.

Early Franciscan and Jesuit missionaries first introduced reading and writing using a Roman or Latin alphabet to the Guaraní (with emphasis on the sons of chiefs and members of the nobility) in the early seventeenth century. However, only Native boys attended mission schools for instruction (see Fig. 6.2). Scribbling of the alphabet on wet clay tiles that were made in workshops for church roofs and floors, and the inscription of the names, ages, and dates of the deceased on gravestones in Guaraní and Spanish may indicate that missionaries educated male children in general, although the missionaries concentrated on the education of the sons of elites (Ganson Fieldwork Notes, 1990–1991).

Patriarchal beliefs that men were superior lay at the core of Catholic teachings in these missionary enterprises, which served as primary institutions of European colonization. Missionaries concentrated on the education of male children who would then teach their parents and play influential roles in their communities. Boys learned how to read and write and do mathematics, including multiplication and the adding of fractions. Peruvian Jesuit Antonio Ruiz de Montoya (1585–1652) observed how the Guaraní were highly capable in mechanical matters, and served as excellent carpenters, blacksmiths, tailors, weavers, and shoemakers. He also found them "remarkably attached to the music in which the fathers instructed the chiefs' sons, along with reading and writing" (Ruiz de Montoya, 2017, p. 229). Every Sunday, the Guaraní in the missions celebrated the Mass with choral music and fine musical instruments. The Guaraní, Ruiz de Montoya noted, carried harps and

Fig. 6.1 Map of colonial Paraguay, 1733, *Source*: Library of Congress, Geography and Map Division. Ovalle, A. D., Techo, N. D. & Homann Erben. (1733) Typus geographicus, Chili, Paraguay Freti Magellanici &c. [Norib. i.e. Nuremberg: Editoribus Homannianis Heredibus]. Map retrieved from the Library of Congress, https://www.loc.gov/item/2004629179/

other instruments with which they celebrated the spiritual deity, personified as the Christian God, on their festive days. The Guaraní played a range of instruments, including violins, clarinets, bassoons, cornets, and oboes, which helped to attract new converts and make them desire to have the Jesuits enter their lands so that they could teach their children. Early Jesuits themselves learned vital survival skills and knowledge about the environment, such as herbal medicine and cures

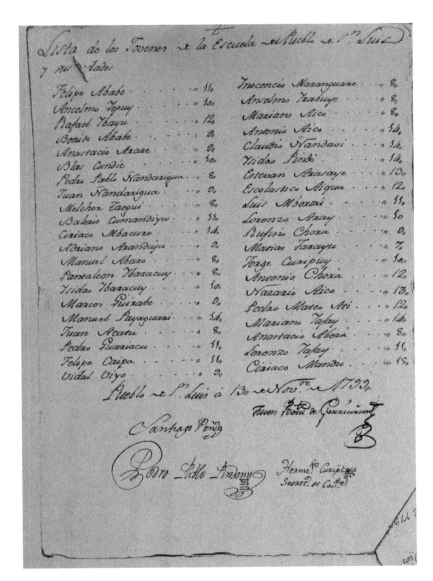

Fig. 6.2 List of pupils in the school in the town of San Luis and their ages, November 13, 1799. Source: "Lista de los jovenes de la Escuela del Pueblo de San Luis y sus edades, 13 de noviembre de 1799, AGN Buenos Aires, IX 18-2-4)

for snakebites, which the Native peoples had acquired through trial and error and passed down through oral traditions over the course of several centuries in the tropical lowlands of the greater Paraguayan basin. Missionaries trained Guaraní men to serve as militia officers and soldiers, river boat sailors, printers, sculptors, scribes, blacksmiths, pottery-makers using a potter's wheel, as well as ranchers in the handling of livestock, among many other skills. The division of labor by gender required that men often work further afield, such as in the gathering of yerba maté before its domestication in the mid-seventeenth century and on the ranches, in workshops, and in other tasks that required their periodic absence from the missions.

Jesuit and Franciscan missionaries limited the educational experiences of Guaraní women and girls, apart from religious instruction and the making of thread, clothing, and in some villages, Paraguayan lace (*ñandutí* meaning "spiderweb" lace) using European-style looms. While Guaraní women planted and harvested a wide variety of crops, formal female education and labor were largely relegated to performing domestic chores, taking care of children, along with transporting water and wood, and the spinning of cotton and wool thread for the making of clothing for their families, and pottery-making using the coil method but without use of a pottery wheel at a workshop. The term *hilandera* (a female weaver) appears in the dictionaries and vocabularies, which suggests that the proper work for women was weaving of cloth and making of clothing in the missions. Guaraní women learned to spin thread on spinning wheels, winding off bobbins of yarn from the spool into a ball (Ganson, 2003). Spanish women who resided in Asunción, "were known to be experts with the needle" by an early chronicler and noted how Guaraní and mestiza women learned how to make lace (*ñandutí or spiderweb lace*) using fine threads and needles brought from Europe. They designed lace patterns fashioned after Native floral and animal designs, such as coconut flower, fish bone, guava flower, and corn flower (Sixteenth-century chronicler Ruy Díaz de Guzmán quoted in Félix de Azara, and Plá, 1969). Women's roles became evident in the early phase of Spanish settlement in Paraguay, providing a context for examining female literacy among both Hispanic and Guaraní populations. In 1556 in Asunción, Doña Isabel de Guevara who was literate and one of the most influential women in the viceroyalty of Peru petitioned the crown for her *encomienda* (grant of Indigenous laborers) to remain in perpetuity. She described how Spanish women had served as sentinels, cooked, and fed the troops what little food that was available, washed clothes, collected firewood, nursed the sick, and buried the dead when the men became so weak. After one thousand of the original 1500 Spaniards on Pedro de Mendoza's expedition had perished, mainly due to starvation, the remaining Spaniards traveled in two brigantines to Asunción in 1537, where initially Spanish women cleared fields, and planted and harvested the crops, according to her account (Lockhart, 1994; Lockhart & Otte, 1976). More than a hundred Spanish women joined the early conquerors by leaving Seville in 1550 on a Spanish expedition headed by Doña Mencía de Calderon and her eighteen-year-old son, Diego. Their caravel ran aground off the Brazilian island of Santa Catalina. Shipwrecked, the Spaniards constructed a new vessel and proceeded north to the port of São Francisco do Sul where after being held in prison, a Portuguese governor released them and provided them with food, cattle, clothing, guns, and other supplies, which enabled them to arrive at their destination in Asunción in 1556 (Cruz, 1960). Spanish women dedicated a good part of their day to spinning cotton, weaving cloth, and making lace. In addition to weaving the rough fibers of a variety of wild pineapple (*caraguatá*), cocoanut trees, and other Native fibers (*yviraí*), as Indigenous women had practiced, Spanish women instructed the Guaraní and mestiza women how to make a lace using fine threads.

There is some controversy in the historical literature over the extent of education and literacy of Guaraní women and girls. Paraguayan historian Olinda Massare de Kostianovsky in *La instrucción publica en la época colonial* (1975) contends that boys learned to read and write and do mathematics, while girls were educated in cooking and weaving and other subjects apart from the boys and at different times. While it may be clear that the missionaries taught catechism, there appears to be no evidence of literate Guaraní women and girls who were instructed to read and write and do mathematics. On the other hand, French scholar Capucine Boidin and Argentine authors Leonardo Cerno and Fabián R.

Vega, in an engaging new article "This Book is Your Book: Jesuit Editorial Policy and Individual Indigenous Reading in Eighteenth-Century Paraguay" (2020), suggest women and girls may have learned to read devotional books in their Native language based on an introduction in a text written in Guaraní by a Jesuit that was to be made available in print for the Indigenous people. This evidence appears thin, but these authors did very carefully qualify their statement. This, of course, does not preclude Guaraní women and girls from learning to read on their own by taking an interest in the schoolwork of their respective sons, brothers, or spouses or receiving instruction from their male children in the evenings, but there is no hard or direct evidence of Guaraní female literacy.

It is important to examine the disparity in Guaraní women's and girls' education within the broader context of female education in the province of Paraguay. Indeed, the absence of formal education of Guaraní women and girls was not all that distinctive from that of mestiza, black, and mulatta women and girls in the province. In 1598 Dominican abbess Francisca Jesusa Pérez de Bocanegra founded an orphanage and a women's shelter (a type of boarding school or housing for women that was distinctive from a nunnery in that women did not take strict religious vows) in Asunción called the Casa de Huérfanas y Recogidas. The residence was a workshop rather than any type of school. The residence was constructed in Asunción on land belonging to the Governor Hernando Arias de Saavedra. Between sixty and one hundred orphans and young women resided there. Many had been the "daughters of honorable Spaniards who lost their lives in the conquest of the province," but mestiza and mulatta women and girls lived there, as well. Residents spun thread and made cloth to earn a living and could increase their dowry, which may have improved their chances for marriage. When the abbess died in 1617, the institution was closed because no one replaced her. In 1650 there was an attempt to reestablish the orphanage or women's shelter, but it failed due to an epidemic (type not indicated) in Asunción. Some Spanish women nonetheless learned to read and write on their own in their homes (Olinda Massare de Kostianovsky, 1975).

Lacking a formal education, Indigenous, mulatta, and mestiza women were not without influence of their own, however, and their many voices appear in judicial records or through those who wrote letters on their behalf or confirmed their oral testimonies. In an extensive judicial record referring to a love triangle and cases of adultery in a former Jesuit town of Jesús, Margarita Arandí, a married Guaraní woman whose husband had abandoned her and fled from the missions, had Gregorio Pahaya confirm the accuracy of her testimony by signing his name on her behalf. Guaraní women clearly understood the power of the written word but had to use literate Guaraní men to express their points of view or testimonies (Ganson, 2020).

A request by a Guaraní woman to have a letter written in the Native language in defense of her late son further illustrates the literacy gap between men and women. On September 20, 1803, María Rosa Aripuy, along with her daughter Rosa Guarapey, on behalf of her late son Juliano Tariqui from Mission Corpus Christi (now Misiones, Argentina) explained to the members of the *cabildo* (town council which was composed of Guaraní), including the *corregidor* (head official), that she expressed great concern and regret over what happened to him. Juliano Tariqui had insulted a Spaniard, Don Teodoro Flores, while rounding up cattle, having been replaced by the new overseer. She also expressed concern about what may happen to Teodoro given the grief or nightmare he now caused his parents. However, she did not hold him entirely responsible for the death of her son, perhaps having knowledge of her son's strong will. María Rosa Aripuy explained she did not know how to sign her name and asked the secretary of the town council to sign in her place (Carta al Corregidor y Cabildo de Corpus, 1803, María Rosa Aripuy, in Guaraní and transliteration in Lenguas generales de América del Sur, https://www.lan gas.cnrs.fr/#/consulter_corpus/liste/1).

#1
Candelaria 19 de abril de 1806.
Mission Candelaria, April 19, 1806.

2

Hae che María Rosa Arĭpĭi co taba Corpus rayĭ hae Libre cherecobae;
Ha'e che María Rosa Arypýi ko Táva Corpus rajy ha'e libre che reko va'e;
Yo María Rosa Arypýi, natural de este pueblo de Corpus, libre
I, María Rosa Arypýi, of the town of Corpus, a free person

3
Juliano Tariquĭ cĭ, amĭrĭ
Juliano Tariquy sy, amyrĭ
y Madre del finado Julián Ty[by] riqui
and mother of the late Julián Ty[by] riqui

4
ambohupiguaha co quatiapĭpe che ñeē, cherubicha Corregid.or hae Cavildo robaque
ambohupiguaha ko kuatia pype che ñe'ē che ruvicha Corregidor ha'e Cabildo rovake
hago presente en este Papel mi verdad ante mis superiores el Correg.or y Cabildo
present this truthful letter before my superiors, the Corregidor and the members of the town council.

5
ndarecoihaba mbae amo ayerure baerāma Caray Teodoro Flores amboajebaecue rehe
ndarekói háva mba'e amo ajerure va'erāma karai Theodoro Flores amboaje va'ekue rehe,
diciendo qe no tengo que pedir contra D. Teodoro Flores,
saying that I ask nothing against Don Teodoro Flores

6
hey haguerami chebe chemembĭ amĭrĭ ymanoeӯmobe
he'i hague rami chéve che memby amyrĭ imano e'ӯmove
por lo que este Cometio como melo dijo mi difunto hijo antes de morir diciéndome que
because of the crime he had committed against my deceased son who before he died said

7
aniche hagua maramobe che mandua hagua co teco rehe
aníche haguā maramove che mandu'a haguā ko teko rehe;

no me acordase de lo que pasa respecto a que
he had no recollection with regards to what had happened

8
Caray raỹ raco nomboayei che reche hemimbota rupi, co teco oyehuha
che rehe;
karai ra'y rako nomboajéi che rehe hemimbota rupi, ko teko ojehuha che
rehe;
el moso referido no hizo lo que ha hecho conmigo por su propio motivo,
the young man in question did not do what he did for his own reasons.

9
che raco aheca baecue tenondebe,
che rako aheka va'ekue tenondeve,
pues yo fui quien lo busque,
well, I was the one who was went looking (for trouble?).

10
hae rãmo tupã ñandeyara ynĭro hagua chebe, che angaipapague rehe,
ha'e ramo Tupã Ñandejára iñyrõ haguã chéve, che angaipa pague rehe
y para que Dios me perdone todos mis pecados,
and may God forgive me for all my sins.

11
hae Caray raỹ Teodoro upe animo Justicia óàhoce co teco rehe
ha'e karai ra'y Theodoro upe anímo justicia oahose ko teko rehe
y para que las Justicias no caigan sobre el moso Dn Teodoro
and so that the authorities do not fall hard on the young man Don
Teodoro.

12
hae tinĭrõte chebe ayehu hagua tupã robaque amboyequaa ndebe
ha'e tiñyrõte chéve ajehu haguã Tupã rovake ambojekuaa ndéve
le pido perdon paraq.e asi pueda aparecer ante Dios, lo que asi le hago
presente,
I ask pardon so that when God appears, this will occur.

13

ndoyehui ramo yepe che rembireco co ára che robaque
ndojehúi ramo jepe che rembireko ko ára che rovake
sin embargo de no hallarse hoy presente mi Muger
Nonetheless, not finding my woman is present.

14

ahenduca yoapĭ hagua ychupe hae ramo omboaye hagua
ahenduka joapy haguã ichupe ha'e ramo omboaje haguã,
para que asi se lo haga presente paraq.e ambas hagan presente esto mismo
so that in this manner in that both being present this same thing occurs.

15

ñanderubichaupe pemee hagua co quatia Caray Theodoro ruba upe yñĭrō
hagua chebe, hae rāmi; cherubicha reta orōmboayebo chemembĭ ñēengue
oromeebo co quatia penderobaque tupā ñandeyara tiñĭrote chemembi upe
hae ñandebe abe
ñande ruvicha upe peme'ē haguã ko kuatia karai Theodoro rúva upe,
iñyrō haguã chéve, ha'e rami; che ruvicha reta oromboajévo che memby
ñe'ēngue orome'ēvo ko kuatia pende rovake Tupā Ñandejára tinyrōte che
memby upe ha'e ñandéve ave
al padre de Teodoro para que esté igualm.te me perdone y para cumplir
con lo que mi hijo nos dejo dicho damos ante Vuestra Merced este Papel
to the father of Teodoro so that equally forgive me and so fulfill what my
son had said in presence of your grace, this letter.

16

hae rami oroyerure ñande Rey recobia upe omoiha corupi ñandebe
yporoquaita marāngatu omboayehagua
ha'e rami orojerure ñande Rey rekovia upe omoĭha kórupi ñandéve
iporokuaita marāngatu omboaje haguã;
y pedimos al que representa al Rey en estos destinos
and we ask that of who represents the king for this purpose.

17

aniche hagua oahoce hagua coteco rehe Caray Theodoro Flores upe,
aniche haguã oahose haguã ko teko rehe Karai Theodoro Flores upe,
para que por esta Causa no sea perseguido el Español Teodoro Flores,

so that this does not lead to the Spaniard Teodoro Flores being pursued.

18
orohecha yepe Che rubicha orōmboacĭ guazuha coteco oyehuhague chemembĭrehe,
orohecha jepe che ruvicha oromboasy guasuha ko teko ojehu hague che memby rehe,
sin embargo, de que hemos sentido mucho este suceso que le ha pasado à mi hijo,
Nonetheless, we have greatly felt awful about what had happened to my son.

19
haete hae rami abe oromboacĭ guazuete Caray Theodoro rehe
ha'ete ha'e rami ave oromboasy guasuete karai Theodoro rehe
pero mucho más sentimos el que se bea así el Español Teodoro
but even more so, we feel the same way as to what happened to the Spaniard Teodoro.

20
oyehu hague coteco hece ombopĭa poriahubo ocĭ hae ytuba upe
ojehu hague ko teko hese ombopy'a poriahúvo osy ha'e itúva upe
por la pesadumbre que ha dado a su Padre y a su Madre
for the nightmare he has caused his father and mother

21
co iporerequaetei bae ñanderehe hae ñanderubarāmo yarecobae
ko iporerequa eteí va'e ñande rehe ha'e ñande rúva ramo jareko va'e,
que tanto nos cuida
and those who look after us.

22
hae cobae mbohobaibo che rubicha oromee co quatia ymopĭata hagua ore ñee ndoroiquai ramo yepe Orerera oromboguapĭhagua tomboaye Orerehe Secretario de Cavildo co ara 20 de sept.e 1803.
ha'e ko va'e mbohováivo che ruvicha orome'ē ko kuatia imopyatā haguā ore ñe'ē ndoroikuaái ramo jepe ore réra oromboguapy haguā tomboaje ore rehe Secretario de Cabildo ko ára 20 de septiembre de 1803.

y para que tenga más valor y fuerza lo q
ue decimos dando este papel y porque no sabemos firmar pedimos lo
haga el secretario a nro nombre, hoy 20 de septiembre de 1803.
So that with courage and strength, we state this, presenting this letter and
because we do not know how to sign, we ask the secretary to do so in
our name, today 20 September 1803.

23
Oromboyequaa Ore Correg.or hae Cavildo
Oromboajekuaa ore Corregidor ha'e Cabildo
Y nos el Corregidor y Cavildo Certificamos
And the Corregidor and cabildo certify.

24
hupiguarete heco haba opacó quatia pĭpe
hupigua rete heko háva opa ko kuatia pype
ser cierto quanto expone en este papel
be certain with regards to what is exposed or pointed out in this letter.

25
hey orerayĭ
he'i va'e ore rajy
nra hija
our daughter

26
yyayebaecue Caray Theodoro Flores rehe
ijaje va'ekue karai Theodoro Flores rehe
en lo sucedido con Teodoro Flores
with regards to what happened with Teodoro Flores.

27
oromondo ramo ore Estanciape Baca pĭcĭbo hae Capatas pĭahu
mboguapĭbo
oromondo ramo ore estánciape vaka pysývo ha'e capataz pyahu
mboguapývo
quando lo mandamos à nuestra estancia à coger ganado y a poner un
capataz nuebo,

when we sent him to the ranch to round up cattle and install a new foreman.

28

hae uperãmo yyaye co teco oñemotĩe'ỹbo ore raỹ Juliano Caray Theodoro upe hae cobae orombohupiguarãmo

ha'e upe ramo ijaje ko teko oñemotie'ỹvo ore ra'y Juliano karai Theodoro upe ha'e ko va'e orombohupigua ramo,

en cuyo tiempo fue quando nro hijo Julián le insultó a dho Teodoro, y sea verdad todo lo referido damos la presente q.t

at which time our son Julián insulted Teodoro, and this be the truth to what we present.

29

orome̅E co quatia hae omboguapĩ orerera heche ore Secretario de Cavildo ndoroiquai ramo oroyapo hagua.

orome'ē ko kuatia ha'e omboguapy ore réra hese ore Secretario de Cabildo ndoroikuaá ramo orojapo haguã.

firmamos ante Nuestro Secretario de Cavildo los que sabemos fha ut supra.

We sign before the secretary of the town council, those who know how to.

30

M.a Rosa Arĭpĭi ymembĭ Rosa Guarape

Maria Rosa Arypýi, imemby Rosa Guarape

A ruego de María Rosa Arypýi y de su hija Rosa Guarapey

At the request of María Rosa Arypýi and her daughter Rosa Guarapey

31

Norberto Añengara [Al]cale de 2º voto

Norberto Añengara [Al]calde 2º voto

Norberto Añengara, Alcalde de Segundo voto

Norberto Añengara, mayor of second vote.

32

Fermín Aguaiy, secreo de Cavildo

Fermín Aguaiy Secretario de Cabildo.

Fermín Aguaiy Secretario de Cabildo
Fermín Aguaiy Secretary of the town council.

33
Antonio Morales
Translation from Guarani and transliteration by Bartomeu Melià, S.J.,
Capucine Boidin and Angelica Otazu Melgarejo, (April 2013); https://
www.langas.cnrs.fr/#/consulter_corpus/liste/1
Guaraní letter presented by Barbara Ganson and English trans. (2012).
Manuscrito de la traducción al castellano reproducido en Archivo
Nacional de Asunción, Documentos en guaraní, 2006. Source: SCyJ.
Vol. 1388. Fol. 63–64.

Colonial Legacies for Literacy in the Early Republic of Paraguay

A cursory reading of women's wills from the Archivo Nacional de Asunción indicates that Paraguayan women often did not know how to sign their names. The 1838 will of Manuela Ayala, a childless widow who may have been a mestizo from the village of Luque near Asunción indicated that her late husband had left some land to orphan nieces that the couple had raised and educated since the age of eight. However, she had no money but left them some additional land and each a pair of oxen. Since she did not know how to sign her name on her will, three witnesses signed on her behalf. Another female head of household, on the other hand, María Dorotea Aguero, a widow of a rancher in Aparipí and mother of four children, described her property, including more than 800 head of cattle, a female African slave, furniture, linen, dishes, and clothing in her will that she signed (ANA, vol. 501, no. 6, 1840). This pattern of women's illiteracy in Paraguay is not unlike other parts of Spanish America, such as in Yucatán, Mexico, where an extensive collection of women's writings in Spanish, including petitions, letters, and notarial documents from Mexico's Archivo General de la Nación has been recently uncovered. More letters came from the upper echelons of society, but some letters were written on behalf of women of modest means and backgrounds (Lentz & Goode, 2022).

It is still unknown how many or what percentage of the Guaraní population became literate in the Guaraní language or Spanish. The evidence suggests that literacy rates were extremely low. Caciques and members of Indigenous town councils often indicated that they did not know how to sign their own names; they signed documents with an X. Nonetheless, among Guaraní migrants who lived in Buenos Aires and Santa Fé in the 1750s Jesuits noted how the Guaraní kept in touch with events in the mission towns "through the letters they received" from other Guaraní in the missions (Nusdorffer, 1753).

Changes in the language policy of the Spanish crown undoubtedly had a negative impact on the Guaraní learning how to read and write in their Native language. In 1743 the Spanish crown ordered Spanish to be taught, encouraging the eradication of Native languages spoken in the Spanish American Empire. The king of Spain specifically ordered the provincial of the Society of Jesus in Paraguay to teach Spanish to Indian children that same year. Nevertheless, the Jesuits found it nearly impossible to put this decree into effect because the Tupí-Guaraní language was so widely spoken, not only in the missions but in the entire province of Paraguay, including Spanish towns, and served as a *lingua franca* in Brazil. Following a tour of the province in 1744, the Bishop of Paraguay Jose Cayetano Parvicino, observed: "The customs and language of the Spaniards born here as well as that of Negros and mulattoes of which there are many, is that of the Indians, with few differences" (Cayetano Paravacino, 1744). A Jesuit observed: "In Paraguay, many have forgotten the Spanish language and adopted that of the Indians, which they use in their homes in the towns and in the rural areas where many live, as in the towns, and where no one knows another language other than that of the Indians" (Cardiel, 1994). Boys, he noted, studied Spanish in the schools, but did not know it well and were punished if they spoke their Native Guaraní language. As soon as school was out, the boys would again speak Guaraní. Few females, he observed, knew how to speak Spanish because girls never studied the language in the schools.

The crown circulated a similar decree in 1760, encouraging the eradication of the diverse languages spoken in the Spanish Empire and requiring that only Spanish be spoken. Charles III ordered that the "Indians be taught the dogmas of our religion in Spanish and ... to read and write in this language only...in order to improve administration and the spiritual well-being of the natural ones (Indians) and so that they can understand their superiors, love the conquering nation, rid themselves of idolatry, and become civilized" (Cédula Real, 1760). The Guaraní resisted this change in language policy. The missionaries explained that the Guaraní "expressed a certain love toward what was their own," their Native language. The Guaraní in Franciscan missions in the central region of the province also were reluctant to learn Spanish. In 1744, according to José Cayetano, the bishop of Paraguay, following a visit to the missions, the Guaraní "prefer to be punished, rather than learn the rational language (Spanish)" (Visita de Obispo José Cayetano, 1744).

Following the Jesuit expulsion from Spanish America in 1767, when they were administering 88,864 souls in thirty missions among the Guaraní in the province of Paraguay, the administration of the missions no longer remained entirely in the hands of Catholic missionaries. Franciscans, Mercedarians, and Dominicans were assigned to look after the spiritual welfare of the Guaraní. Beginning in the early 1780s, creole administrators hired Spanish teachers to instruct male children in the Christine doctrine, how to count, and how to read and write in the Spanish language in the mission schools. The creole administrator, the Guaraní *corregidor,* and members of the Indigenous town council were to provide a large house for the schoolteachers, along with the same provisions that administrators received, aside from their salaries. They were supposed to be given a list of the boys in the town over the ages of four or five, and their type of occupation they may have been employed, such as the weaving of textiles or whether they were altar boys or the priests' helpers or assistants. The male children of caciques were to receive instruction in reading and writing and learning how to count with preference and with greater urgency because they would be expected to assume positions within their Indigenous communities. There were Guaraní teachers of primary education as well, who were responsible for the teaching of music and dance under the guidance or supervision of the Spanish teacher and the Catholic priest.

In 1783, the viceroy in Buenos Aires determined that instruction was to be given in Spanish but with the use of Guaraní under the notion that Guaraní instruction was necessary to understand Spanish and with the intention that the male children would speak Spanish with more frequency in the schools. The viceroy also insisted that prizes be handed out every six months to those pupils who speak Spanish. These prizes consisted of various yards of cloth that could be made into clothing, most likely by their mothers. One of the duties of the schoolteacher was to ensure that male children wore clothing and not run around naked. Schoolteachers were to ensure girls who came to recite prayers and attend Mass, especially on religious holidays, were to be dressed decently (with "honestidad") be washed or clean. For this reason, it was important for the schoolteacher to maintain a list of all the girls and the boys as provided by the *corregidor* and the secretary of the town council. The Spaniards placed emphasis on cleanliness of the Guaraní, including looking after their appearance and being well groomed (Pueblo de Concepción, 1786). The use of the Guaraní language to facilitate the comprehension of Spanish is notable; there was no mention of children being punished for speaking their Native language. Beginning in the late eighteenth century, however, Spanish became the language of the schools.

Teachers kept track of the ability of their male Guaraní pupils who knew how to read and write in long hand, mastered the art of printing, or were still learning and practicing the alphabet (see Fig. 6.3). As far as mathematics was concerned, teachers taught the Guaraní boys the basics of addition, subtraction, and multiplication. Finally, the Guaraní studied Christian doctrine, and recited their prayers in two languages, Spanish and Guaraní (Bernardo Gonzalo, Pueblo Real de Nuestra Señora de la Concepción, November 12, 1799, AGN, Buenos Aires, IX 18-2-4). In part, when it came to religious doctrine, therefore, these were bilingual schools, although there was preference for the use of Spanish. Interestingly, Spanish creole and members of Indigenous town councils learned that it was not beneficial to establish or open schools in certain missions in times of epidemics, which seemed to appear in clusters of different towns.

In 1786, we learn from Gonzalo Doblas, the lieutenant governor of the Department of Concepción, that at the mission of San Carlos, there were 25 boys between the ages 6 and 13 who attended school. Each pupil received a notebook (*cartilla*) and the mission had 84 notebooks in their warehouse. No teacher had been appointed to teach at San José (Josef); no school had been established there, even though the mission had an equal number of boys as San Carlos; nor were there any notebooks available in their warehouse. An epidemic had broken out in Apóstles, which prevented the establishment of a school until the virus subsided; there were no notebooks in their warehouse. Mission Concepción had 50 or 60 boys but no Spanish schoolteacher. An Indian teacher was present; their warehouse had 168 notebooks. Mission Santos Mártires had 31 boys between the ages of 6 and 13 who were provided with notebooks; 84 notebooks remained in their warehouse.

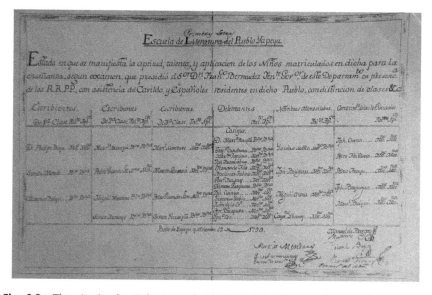

Fig. 6.3 The Aptitude, Talent, and Ability of Pupils Enrolled in the Primary School in the town of Yapeyú, November 12, 1799, "Escuela de Primera Letras del Pueblo Yapeyú," AGN Buenos Aires IX 18-2-4

At Santa María la Mayor, the epidemic took the lives of many of the schoolboys; there were 49 boys between ages 5 and 11, but their warehouses lacked notebooks. In the town of San Francisco Xavier, there was a school with 50 pupils between ages 5 and 12; there were 23 notebooks (Bernardo Gonzalo, Pueblo Real de Nuestra Señora de la Concepción, November 12, 1799, AGN, Buenos Aires, IX 18-2-4).

Schoolwork from the 1790s provides further evidence that only Guaraní boys were learning to read and write in Spanish. Female children did not author or sign any of the schoolwork from Santa María la Mayor, Santa Rosa, Santo Angelo, San Luis, San Cosme, San Borja, San Francisco Xavier, Nuestra Señora de la Concepción, Santos Mártires del Japón, San Carlos, San José, Santo Tomé, San Borja, and San Luis submitted to the governor in Buenos Aires in 1795 for first, second, and third prize, usually consisting of samples of fine penmanship and occasionally mathematical problems. Lists of Guaraní pupils from 1799 and of those who could print or write in long hand indicate that no female children attended mission schools. All the names on the lists of pupils in mission schools in Apóstoles, San Nicolás, Santo Angelo, San Luis, and San Juan Bautista in 1799 were those of male Guaraní children.

For those Guaraní pupils who competed for prizes demonstrated fine penmanship, along with some knowledge of Christine doctrine, proper etiquette, and the willingness to praise King Charles IV and the royal family. Miguel Yaribu, first prize winner from mission San Carlos, expressed good wishes to King Charles IV (1748–1819) for his birthday and hoped the king lived a long and happy life, along with Doñā Luisa de Borbón (1780–1846) and the Prince of Asturias Don Fernando (1784–1833) (Miguel Yaribu al Rey, November 11, 1799, AGN, Buenos Aires, IX 18-2-1.). His schoolwork underscores the process of transculturation, that is, accepting selected elements of European culture that proved beneficial. First, second, and third prize winners received yards of cloth or actual items of clothing, such as wool ponchos and a straw hat, for their submissions. Others who demonstrated knowledge of the Spanish language but did not receive a prize, still received items of clothing or rosaries, as did the Native teacher's aides (*ayudantes de la escuela*).

All the mission towns had ample numbers of books from the libraries that had belonged to the Catholic missionaries for advanced pupils who knew how to read. There were pupils, Gonzalo Doblas insisted, who could read well and demonstrate their ability to write through exercising good penmanship. Guaraní boys initially practiced learning how to print on small tablets made of wood. Later they wrote on leather hides and then on paper, if they wrote in long hand or using script; the supply of paper was low in the missions (Provincia de Buenos Aires, 1786).

By the end of the eighteenth century, several Guaraní had become competent writers in Spanish. One became a Catholic priest, Francisco Javier Tubichpotá, the son of a cacique from the Mercedarian mission of Santiago. At age thirteen, he entered the Real Colegio Seminario de San Carlos in Asunción, as the legitimate son of lieutenant corregidor Estanislao Tubichapotá and María Solome Araquí. His father brought him to study there and offered to pay for his own room and board, using funds from the mission. The bishop of Paraguay decreed that he should be admitted to the *colegio* (seminary) "to serve as an example." Tubichapotá studied Latin, philosophy, and theology from 1789 to 1800. He was probably better educated than most of the Spaniards and mestizos in Paraguay. To attend the Real Colegio Seminario, one had to demonstrate one's *limpieza de sangre* (purity of blood), meaning that one had to be of the Indigenous nobility, baptized, confirmed, a legitimate son, and not mixed with a "*baja esfera*" (lower social group or section of society) (ANA, NE 844, Nicolás Obispo de Paraguay, 1806). In 1803, Tubichpotá was ordained a priest, having taken an examination to be fully ordained. He was assigned to work in a Native parish to work among neophytes. In 1800, Andrés Araro became a certified teacher of primary letters in his town of Apóstoles, having demonstrated his ability to read, write, and count, and knowledge of the Christian faith and the Spanish language. He earned a monthly salary of six pesos (Letter of Feliciano Cortez to the Ten. Govr.de Concepcion, April 22, 1800, AGN, Buenos Aires, IX 18-2-3.). Andrés Araro was one of two pupils from Apóstoles (the other was Juan Antonio Paraye) who had demonstrated and provided samples of their mathematical ability, including fractions, on November 10, 1795, for the governor of Buenos Aires (see Fig. 6.4).

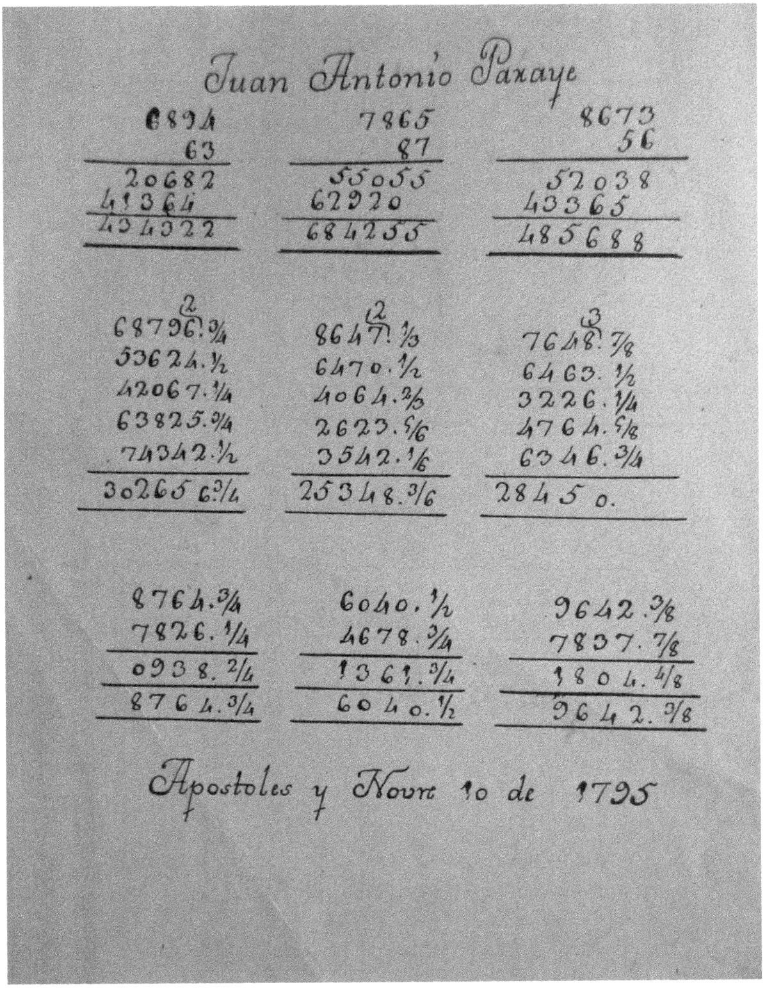

Fig. 6.4 Mathematical Problems of Juan Antonio Paraye, Mission Apóstoles, "Trabajos de matemática," November 10, 1795, AGN Buenos Aires, IX 18-2-1

Fashion and Social Standing Among the Guaraní

Fashion further underscores social differentiation and the patriarchic nature of colonial society. Lists of items of clothing from the Guaraní missions indicate that Guaraní men wore trousers, a loose shirt, a wool poncho in the winter, and maybe a straw hat. From an eighteenth-century Jesuit illustration of a mission town, including a drawing of Guaraní women in the *coty guazú* (women's shelter or asylum), we know Guaraní women wore long white dresses or a chemise (typoí) or loose-fitting smock with a sash. A 1781 text in Guaraní, with a Spanish translation, provides further details about types of clothing by certain members of the Guaraní who resided in the former Jesuit mission of La Cruz (now in province of Corrientes, Argentina). The document indicates how social differentiation for males was made visible, based on their standing within their communities.

> Documento de data de 40 pesos razón de las prendas de vestuarios subm. A los Comisarios y Peiones en el citado Pueblo de La Cruz, January 12, 1781. AGN IX 12-1-4 (Buenos Aires) A Guaraní text regarding clothing of the Comisarios and peones of the town of La Cruz, 1781, Provincia General de los Pueblos de Guaranís.

> Exmo. Sr. Virrey:

> Ore curuzu tava rehegua, orogue rubae cue hazienda oretavahegui oro moi mami, Administr.or General pope hae Oroicotebe anga Señor o neñemde-hagua mini rehe, tupa hae Rey marangatu rerapipe, o no yerunendebe, enemeeucahagua orebe. Oro ha ano nedepohegui co tupa mbae tupa nandeyara tandeo basa aranabo. Alcalde de primer voto y comionado, Miguel Chauy. Secretario Manuel Ñema.

> Señor. El Administr.or General de los Pueblos en las misiones del Uruguay y Paraná en cumpliento del Ley decreto de V.E dize que el ....ha remite el Corregidor y Cavildo del Pueblo la Cruz se dirije a suplicar a se Digna mandar seles..prendas de vestuario....

January 11, 1781. The general administrator of the missions purchased the clothing needed for the town of La Cruz.

Five ponchos cordobeses
Five yards of blue cloth
Five dozen buttons
Five caps or hats made of plain cloth
Two caps or hats of Braga (?) superior.
Five knives
Six yards of cloth (pañete de Zapallanga?)
Nine yards of baieta de la tierra (local cloth?)

Those individuals who received more elaborate clothing or items, such as blue cloth, wool ponchos, tin buttons, and caps were the Alcalde de Primer Voto, Miguel Chau, and El Cacique Don Manuel Ñena. Others who received less elaborate clothing, cloth, and a poncho but no tin buttons were Norberto Chauy, Pedro Guachuca, and Miguel Guayaré. As could be expected, the more elaborate clothing was a sign of their position in society. We do know that from the Jesuits period, especially from the illustrations by Jesuit missionary Florian Baucke (or Paucke), those Guaraní who were militia soldiers and officers wore elaborate uniforms during military exercises in the main plaza of the missions. The list of items of clothing for these individuals reflect the basic hierarchal and patriarchal nature of mission society. There is no mention of elaborate clothing needed for the wives of the caciques or women in general, who wore long white cotton dresses from neck to foot. Mid-eighteenth-century drawings by Jesuit Florian Baucke, in contrast, show examples of elaborate headdresses, feathered caps, worn mostly like by the Indigenous nobility of the Rio de la Plata, and Indigenous commoners who wore no clothing while gathering wild honey from trees. Either Indigenous women had a tradition of wearing skirts or the Catholic missionary depicted them as wearing skirts and not much else, as they made hand-made pottery or used looms for the making of cloth (Paucke, 1973, 1999).

Fig. 6.5 Map of South America, 1822. This map was delineated and engraved by mapmaker Alexander McPherson (1796-1824). It was published February 1, 1822, by Sherwood Neely and Jones Paternester Row. Author's Collection

Public Education in the Republic of Paraguay

Following the independence of Paraguay from the domination of Buenos Aires and Spanish royal authority in 1811, the first public schools were established during the reign of dictator Dr. José Gaspar Rodríguez de Francia (1814–1840) (see Fig. 6.5). Nonetheless, only male children received free elementary education. Boys studied arithmetic, grammar, and spelling in the Spanish language. In the *pueblos de indios* (Native towns) administrators paid teachers' salaries using the funds belonging to the community. Various religious institutions provided a higher quality education for the sons of elites in Asunción, the male children of landowners, public officials, and merchants. The Real Colegio Seminario de San Carlos that trained boys to enter the priesthood offered three years of philosophy, four years of theology, and Latin. Religious education however suffered under Dr. Francia's administration. In 1823, Dr. Francia, who scorned the religious orders and perceived the clergy as a threat to his power, closed the Seminario de San Carlos. Its teachers were fired, exiled, or imprisoned. On the other hand, in 1828, Dr. Francia required mandatory education for all young males, including free blacks and mulattoes. Thus, gender was more of a limiting factor or obstacle in securing a free public education than race, and ethnicity in Paraguay. The National Treasury assumed most of the responsibility for paying of teachers' salaries and upkeep of the primary schools throughout the country. Dr. Francia himself sent supplies of paper and primers to the elementary schools, and even clothing for needy male pupils (Cooney, 1983). While Spanish was the language of instruction, its spread was reduced to the benefit of the Guaraní language by the isolationist policies of Dr. Francia, who disallowed foreign immigration and discouraged Spanish-speaking merchants from entering the country (Nickson, 2009).

Paraguay's second prominent dictator Carlos Antonio López (1840–1862) expanded the number of primary schools to 408 and 16,755 pupils by 1857. By 1862, there were 435 schools with 24,524 pupils. Boys studied philosophy, civics, mathematics, history, geography, Spanish grammar, Latin, religion, and biology. Carlos

Antonio López also sent several talented pupils for advanced study in engineering, education, law, medicine, and military science to Europe. One recalled having received four or five lashes for having spoken Guaraní at school (Centurión, 1948 quoted by A. Nickson). Female education remained largely nonexistent in Paraguay during this period. The writings of nineteenth-century travelers allow us to discern something of the specificities of gender difference, intertwined with social class in Paraguay. Foreign travelers noted that Paraguayan women were "utterly devoid of education, beyond reading and writing" (Mansfield, 1971). George F. Masterman in *Seven Eventful Years in Paraguay: A Narrative of Personal Experience Amongst the Paraguayans* (1870) observed that "women, except a few of the higher classes, are quite uneducated, so much so that it is rare to find one who can read and write. The men, however, nearly all do so" (Masterman, 1870). Masterman also described how every morning some four-to-five hundred women were engaged as vendors in the marketplace of Asunción. They road donkeys or drove carts loaded with corn, manioc, oranges, melons, wood, molasses, eggs, chickens, and other produce that might find a sale in the capital. The market women were said to have a smart and clean appearance, having exchanged their snowy white dresses (*typóis*) for ones that were white and clean, having walked some twenty miles the night before. Marguerite Dickins, the wife of a sea captain who visited cities along the coast of Brazil, the interior of Argentina, and Asunción described Native Paraguayan women who, along with their children, sold gourds or feather dusters, dressed in white cotton skirts and manta from head to foot (Dickins, 1893). Clearly, Paraguayan market women and street vendors played essential roles in the nineteenth-century economy (Graubart, 2007; Mangan, 2005). Given the widespread presence of mestizas and Indigenous women in the market towns of Paraguay, the basic teaching of mathematics that would have been of great value to the community. Basic accounting would have been a tool that the mestiza and Indigenous market women and street vendors would have had learned on their own through experience to carry out their transactions.

Educational reforms to benefit women and girls took place only during the second half of the nineteenth century. An 1856 census of the pupil population indicates that all the pupils who attended the Academia Literaria de la República del Paraguay (which later became the Colegio Nacional) were boys. All teachers and principals were men (UNESCO Microfilms of the Archivo Nacional de Asunción, vol. 226, 1844, December 11, 1843, documents 60–68). In 1856 Carmen Encina petitioned to open a public school for girls in the village of Caacupé (UNESCO Microfilms of the Archivo Nacional de Asunción, Vol. 320, documents 124-48). Of a list of more than 150 schoolteachers in 1863, one can recognize the names of three women (UNESCO microfilms of the Archivo Nacional de Asunción, vol. 336, 1863, document 440-07).

Conclusions: Knowledge, Education, and Gender

In conclusion, despite the dependence of Spanish conquerors upon the Guaraní women's agricultural labor and ecological knowledge for their survival during the conquest and early colonization of Paraguay, Guaraní women held subordinate status during the long period of colonial rule; they were often traded for metal tools and objects, enslaved, and even served as currency (Austin, 2020; Owensby, 2022). European colonization increased the significance of male dominance, as reflected in the division of labor and educational system in the missions. Religious instruction in colonial times took place through religious sermons, music, and the liturgy at church services, and catechism was taught to both boys and girls, but at separate times. Yet there appears to be no direct evidence of Catholic missionaries teaching Guaraní women and girls how to read and write and do mathematics at mission schools. Only one document written in the Guaraní language on behalf of a Native woman and daughter, but not signed by them, appears among the abundance of Guaraní letters uncovered in archives from Argentina, Paraguay, Brazil, Spain, the United States, England, and France, which date from the mid-eighteenth through the early nineteenth century.

Sermons and other religious texts published in the Guaraní language, such as the *Explicación de el Catecismo en lengua guaraní* (1724) by cacique and musician of mission Santa María, Nicolás Yapuguay, nonetheless were intended for the conversion of entire communities (Furlong, 1947). Religious teachings may have been printed not only to be read by those Guaraní who were literate but could be read out loud to members of religious congregations, including its female members. Clerics apparently did not see the benefit of literacy for women and girls, although the missionaries recognized the significant role of Guaraní women in religious congregations (see Ruiz de Montoya, *The Spiritual Conquest*).

By examining a selection of Guarani sources and other texts, we have learned how Spain attempted to assimilate the Native population by imposing the use of the Spanish language in the schools beginning in late eighteenth Paraguay. We can also discern the transcultural process by which certain Guaraní came to read and write in the Spanish language. We still lack details about the extent and manner in which the Guaraní school children were punished by authorities for speaking their Native language. Parental influence over their children's education is unclear but it undoubtedly declined with the instruction provided by Catholic missionaries and Spanish teachers. It is notable, however, that several Guaraní became teachers and one an ordained priest, and that the Native language endured despite the assimilation attempts. Finally, the impact of Spanish colonialism on Native culture becomes evident in that the Guaraní came to adopt European-style clothing that was acceptable by the standards of Catholic missionaries, although they still retained and cherished their Guaraní language. The cotton *typói,* the white chemises worn by Indigenous and mestiza women and girls, which were flounced with lace at the neck and arm, were not all that distinct from the long cotton dresses with some embroidery worn by women and girls in the Chiquitos and Moxos missions of present-day Bolivia and in other parts of the Americas, such as the *huipiles* of Mexico. Notwithstanding the undeniable demographic and cultural presence of Indigenous women, the disparities of women and girls' education and literacy, regardless of race, ethnicity, and social class, continued well into the nineteenth century, because women lacked full citizenship following Paraguayan independence.

Bibliography

Archivo Nacional de Asunción (ANA). (1806, December 5). NE 844, Nicolás Obispo de Paraguay.

Austin, S. M. (2020). *Colonial Kinship: Guaraní, Spaniards, and Africans in Paraguay*. University of New Mexico Press.

Azara, F. d. (1969). Geografía y física y esférica de la provincial del Paraguay y misiones guaranies. In J. Plá (Ed.), *Las artesanías en el Paraguay*. Ediciones Comunes.

Boidin, C., Cerno, L., & Vega, F. R. (2020). This book is your book: Jesuit editorial policy and individual indigenous reading in eighteenth-century Paraguay. *Ethnohistory, 67*(2), 247–267.

Cardiel, S. J. (1994). *Compendio de la historia del Paraguay* (1780). Secretaría de Cultura de la Nación.

Carta al Corregidor y Cabildo de Corpus. (1803). María Rosa Aripuy, in Guaraní and transliteration in Lenguas generales de América del Sur, provided by Barbara A. Ganson. https://www.langas.cnrs.fr/#/consulter_cor pus/liste/1

Cayetano Paravacino, J. Carta a Su Majestad dando informe de la visita pastoral hecha a todos los pueblos de la provincial, Asunción, November 14, 1744, MG 1035a, Gondra Collection, Benson Latin American Collection, University of Texas at Austin.

Cédula Real disponiendo se ponga en práctica lo propuesto por el Arzobispo de México, a fin de conseguir que se destierren los diferentes ídiomas que se usan en los dominios solo se habla el castellano, Yo el Rey, Aranjuez, May 10, 1760, ANA, Nueva Encuadernación, Vol. 62, folios 99–104.

Centurión, J. (1948). *Memorias o reminiscencias históricas sobre la Guerra del Paraguay*. Editorial Guarania.

Cooney, J. W. (1983). Repression to reform: Education in the Republic of Paraguay, 1811–1850. *History of Education Quarterly, 23*, 413–428.

Clement, C., & Deneven, W. (2015). The domestication of Amazonia before the European Conquest. *Proceedings of the Royal Society*.

Cruz, J. (1960). *Doña Mencía: La Adelantada*. Editorial Le Reja.

Dickins, M. (1893). *Along shore with a man-of-war*. Arena Publishing Company.

Documento de data de 40 pesos razón de las prendas de vestuarios subm. A los Comisarios ... Pueblo de La Cruz, January 12, 1781. Archivo General de la Nación, (AGN), Buenos Aires, IX 12-1-4.

Furlong, S. J., (1947). *Origenes del Arte Tipografico en America, especialmente en la República Argentina* (pp. 144–145). Editorial Huarpes, S.A.

Ganson, B. A. (2003). *The Guaraní under Spanish Rule in the Rio de la Plata*. Stanford University Press.

Ganson, B. A. (2020, April). A Patriarchal Society in the Río de la Plata: Adultery and the double standard at mission Jesús de Tavarangue, 1782, In L. Newson (Ed.), *Cultural worlds of the Jesuits*. Institute of Latin American Studies, School of Advanced Study, University of London. https://ilas.sas.ac.uk/publications/cultural-worlds-jesuits-colonial-latin-america

Gonzalo, B. (1799, November 12). Pueblo Real de Nuestra Señora de la Concepción, AGN, Buenos Aires, IX 18-2-4.

Graubart, K. B. (2007). *With our labor and sweat: Indigenous women and the formation of colonial society in Peru, 1550–1700*. Stanford University Press.

Lentz, M., & Goode, C. T. (2022, February 23). The personal and the political: Women's letters in Late Colonial Yucatan. Paper presented online, AHA.

Letter of Feliciano Cortez to the Ten. Govr.de Concepción, April 22, 1800. AGN, Buenos Aires, IX 18-2-3.

Letter of Isabel de Guevara, July 2, 1556, in James Lockhart and Enrique Otte, eds. and trans. *Letters and Peoples of the Spanish Indies: Sixteenth Century* (1976). Cambridge University Press, 14–17.

Lockhart, J. (1968/1994). *Spanish Peru, 1556–1560* (2nd ed.). University of Wisconsin Press.

Mangan, J. E. (2005). *Trading roles: Gender, ethnicity, and the urban economy in colonial Potosí*. Duke University Press.

Mansfield, C. B. (1971). *Paraguay, Brazil, and the Plata: Letters written in 1852–1853*. Macmillan 1856, reprint ed., AMS Press.

Masterman, G. F. (1870). *Seven eventful years in Paraguay: A narrative of personal experience amongst the Paraguayans*. Sampson, Low, Son and Marston.

Moore, J. D. (2014). *A prehistory of South America: Ancient cultural diversity on the last known continent*. University of Colorado at Boulder.

Nickson, R. A. (2009). Governance and the revitalization of the Guaraní language in Paraguay. *Latin American Research Review, 44*(3), 5.

Nusdorffer, B. (1753). Segunda parte de lo sucedido en las Doctrinas después que salió de ellas el Pe. Luis de Altamirano para Buenos Ayres. Biblioteca Nacional de Rio de Janeiro, Coleção de Angelis, Seção de Manuscritos, Roll 31846, Defensa de los Jesuítas, 1.2.34, Guerra de los Guaraníes, 8.2.25, folio 17.

Olinda Massare de Kostianovsky. (1975). *La educación pública en la época colonial*.

Owensby, B. P. (2022). *New world of gain: Europeans, Guaraní, and the global origins of modern economy*. Stanford University Press.

Paucke, F. (1999). *Hacia allá y para acá: una estada entre los indos Mocobíes, 1749–1767*. 4 vols. Editorial Nuevo Siglo.

Paucke, F. (1973). *Iconografía colonial rioplatense, 1749–1767: costumbres y trajes españoles, criollos e indios*. Elche.

Plá, J. (1969). *Las artesanías en el Paraguay*. Ediciones Comunes.

Provincia de Buenos Aires, Departamento de Concepción sobre que el Teniente governador de dicho departamento a razón a los libros y demás utensilios que necesitan y deven tener los Maestros de Escuela, 1786, AGN, Buenos Aires, IX 22-8-2, Legajo 3, Expediente 27.

Pueblo de Concepción, Departamento de Concepción, Sobre que el Teniente Governor a la Departamento de razon a los libros y demás utensilos que necesitan y deben tener los maestros de Escuela, 1786 IX 22-8-2 AGN, Buenos Aires, Signed Gonzalo de Doblas, Candelaria, September 19, 1783 and copied on November 15, 1786.

Ruiz de Montoya, A. (2017). *The Spiritual Conquest: Early Years of the Jesuits Missions in Paraguay* (1639) trans. Barbara A. Ganson and Clinia M. Saffi. Institute of Jesuit Sources, Boston College.

"Trabajos de matemática," November 10, 1795. AGN, Buenos Aires, IX 18-2-1.

UNESCO Microfilms of the Archivo Nacional de Asunción, vols. 226, 320–336.

Visita de José Cayetano, Obispo del Paraguay, November 21, 1744, Archivo General de la Indias, Audiencia Charcas 76, 4, 49; unpublished notes of Carlos Leonhardt, S.J., Archivo Provincial de Buenos Aires de la Compañía de Jesús.

Wilde, G. (2019). Frontier missions in South America: Impositions, adaptations, and appropirations. In D. A Levin Rojo & C. Radding (Eds.), *The Oxford handbook of the borderlands of the Iberian world* (pp. 545–570). Oxford University Press.

Yaribu, M. R. (1799, November 11). AGN, Buenos Aires, IX 18-2-1.

Part II

Language, Environment, and Well-Being: Contemporary Challenges

7

The Generational Gap and the Road to Well-Being: Language and Natural Environment in the Discourses of Indigenous Old and Young in Tlaxcala

Gregory Haimovich

Introduction

At the scale of major societies, or nations, the problem of intergenerational gap is often considered so trivial that misunderstandings between the young and old generations are mostly perceived as a natural order of things. Margaret Mead, for example, in her book about generation gap (apparently the first notable study of this subject in anthropology), argues that only relatively isolated, "primitive" societies exhibit "postfigurativeness"—a model, in which the main source of learning for the young generation is their oldest relatives and compatriots, while in the industrialized societies members of the same generation are much more prone to learn from their contemporaries and the adults also learn from their children (Mead, 1970, pp. 1, 26,

---

In the memory of Candelario Bautista Galicia (1944–2022), farmer, musician and storyteller.

G. Haimovich (✉)
University of Warsaw, Warsaw, Poland
e-mail: grhaimovich@al.uw.edu.pl

© The Author(s) 2024
J. Olko and C. Radding (eds.), *Living with Nature, Cherishing Language*,
https://doi.org/10.1007/978-3-031-38739-5_7

189

51–52). The acculturation of migrant and colonized ethnic groups into the mainstream societies is thus seen as a progressive turn in the universal "cultural evolution" (Mead, 1970, p. 51). Certainly, this paradigm is characterized by a considerable degree of eurocentrism and overgeneralization (see e.g. Lohmann, 2004, on the latter tendency in Mead's research). But with the rise of the value of multiculturalism and dismissal of the evolutionary discourse in anthropology, the very notion of postfigurativeness has acquired a new significance.

The study of ethnic minorities and the process of their assimilation brought forward such issues as intergenerational conflicts and psychological trauma, present not only among those members of an ethnic group who were directly colonized or themselves migrated to a new country but also in the subsequent generations (Feir, 2016; Krieg, 2009; Lee, 2004; Marino, 2020; O'Neill et al., 2018). Although the focus of assimilation studies has long been on migrant groups (e.g. Alba, 2005; Alba & Nee, 1997; Gans, 1992), it is obvious that minoritized Indigenous groups globally experience the same ruptures in the transmission of language and cultural traditions (religion, clothing, cuisine) from one generation to another. The ecological deterioration of natural surroundings is also interconnected with the loss of traditional ecological knowledge in Indigenous societies, as this type of knowledge becomes less relevant for the youth who strive to improve their socioeconomic status in the wider society (Cristancho & Vining, 2009; Russell & Tchamou, 2001).

The recent global revalorization of Indigenous identities, however, has also led to situations, in which representatives of younger and more acculturated generations of Indigenous peoples themselves strive to reconnect to older ethnic group members in order to regain their ties with the ancestral cultures, languages, and ways of knowing (Iseke & Moore, 2011; Nishita & Katsutani, 2022; Pace & Gabel, 2018; Saba, 2017). Although this drive embraces only a part of Indigenous youth, and unlikely changes their lifestyles sharply, it is emblematic in the sense that it underscores the value of traditional Indigenous knowledge in the modern industrialized world. Furthermore, it can boost the self-esteem of the aged populations who still preserve this knowledge.

In this respect, Indigenous "communities in transition", i.e. those experiencing a shift towards industrialization and assimilation during the last hundred years or less, deserve particular attention. One side of the change that takes place in such communities is visible to the eye: change in the landscape, the ways people dress, and the language(s) spoken in the community on a daily basis. The other side is an inner side of the change, which is manifested in the attitudes and ideologies of community members. If views on the practices that distinguish the character of the community—such as language, customs, and natural environment—vary significantly between the young people and the elderly, it indicates that the transition has been virtually completed.

Since September 2017, when I started conducting fieldwork in Mexico, I was involved in research on the sociolinguistic situation and subjective well-being in historically Nahuatl-speaking communities in Tlaxcala, primarily in the municipality of Contla de Juan Cuamatzi. In addition to a sociolinguistic survey in the beginning of the research, I conducted a number of semi-structured interviews, from half-an-hour to one-hour length, with local residents who were eager to share their own perspectives on well-being and the future of Nahuatl. One part of the interviewees, among whom women and men were both present, belonged to the age group of younger than 35, while another part were people older than 55 years old. In this manner, the study aimed to find out to what extent the views of the older generation, most of whom were Nahuatl speakers, differed from those of the young people, none of whom learned Nahuatl systemically. At first, the primary focus of the interviews was Nahuatl in the lives of interlocutors, but as our conversations were developing, this topic evolved into discussions about the past and the present, and also on well-being—not only in the individual but also in the broader, social dimension of this concept. Most of the interviews were conducted in Spanish and only one was conducted in the local variety of Nahuatl, since in the beginning of the study I did not yet obtain the necessary level of proficiency in spoken Nahuatl.

This work unites the material collected in these interviews and examines how the topics of language and natural environment, the two aspects that have experienced a profound change in Contla during the last half of a century, are viewed by younger and older members of the community. It also refers to how these views are connected, directly and covertly, to their perception of personal and communal well-being. In this manner, the chapter helps to examine the true depth of the intergenerational gap that developed in Contla communities due to their gradual assimilation during the past century.

San Bernardino Contla Then and Now: Overviewing the Environmental and Sociolinguistic Change

Sixty years have passed since Hugo G. Nutini, who dedicated most of his life career to the research of Nahua communities in Mexico, undertook a study of the social organization of the municipality of Contla, the results of which were later published in his monograph *San Bernardino Contla: Marriage and Family Structure in a Tlaxcalan Municipio* (1968).[1] Nowadays, the book, apart from being an example of a classical anthropological study embedded in the structural functionalist tradition, also represents an important historical document, providing the reader with a description of Contla communities as they appeared in the mid-twentieth century—the look that no longer exists. Other researchers of Nahuatl-speaking communities in Tlaxcala, such as Hill and Hill (1986), Robichaux (2005), and Messing and Rockwell (2006), attest the fundamental changes that took place in the socioeconomic and sociolinguistic profile of southeastern Tlaxcala communities, including Contla de Juan Cuamatzi, during the last quarter of the past century.

Nutini writes that except for the *cabecera*, the municipal center, located in the lower part of the municipality, Contla presents a "semi-dispersed type" of settlement, "in which the houses are located in the middle of cultivated plots of land" (1968, p. 28). Most of the private houses, according to the author, "are one-story adobe buildings

---

[1] Although a productive ethnographer of rural central Mexico, Nutini also has a controversial reputation due to his active involvement in the ill-famed Camelot project (Solovey, 2001).

on foundation of stone and mortar" (1968, p. 36). However, at present, if we visit such sections of the municipality as San Miguel Xaltipan[2] and San José Aztatla, we find the majority of houses built along the streets and not in the middle of *milpas* (corn fields). Many of these houses are two-storied and even three-storied buildings, depending on the landowner's wealth, and are predominantly made of brick and concrete, with flat rather than slanted roofs. There are a very few houses made of adobe that can still be found in Contla; some of them are preserved as family landmarks and some of them at some point became a part of larger buildings, where the newer adjustments were again built of brick and concrete. Most of the households are now enclosed behind concrete walls. The roads, even in the remote sections of the municipality, became asphalted and have also expanded significantly (Fig. 7.1).

These changes also reflected the substantial growth of the population of Contla during the last sixty years: if at the time of Nutini's research

Fig. 7.1 A view on the Malintzin volcano from San Miguel Xaltipan, Contla. November 2018. Copyright: Author

---

[2] Nutini (1968) calls it San Miguel Xaltipac in his book, and it is difficult to determine when and how this toponym changed its ending.

the number of residents was 10,699 (Nutini, 1968, p. 24), according to the national census of 2020 it reached 38,579 people.[3] During two decades, 1970s and 1980s, according to INEGI data, the number of residents of the municipality almost doubled (González Jácome, 2009). Obviously, this growth required an expansion of living space and infrastructure, including roads, and the latter meant the easier delivery of cheaper building materials, such as concrete and bricks, from the centers of their production. Thus, not only the improvement of infrastructure led to the increase in labor opportunities and greater influx of material goods due to the faster connections to urban Spanish-speaking centers; it also caused local communities to gradually obtain an appearance of urban zones (Hill & Hill, 1986, p. 11; Robichaux, 2005, p. 65). Foreign investments in the municipal economy (primarily in textile industry), unheard of in the mid-twentieth century, comprised about 92 million USD in the first half of 2021.[4] The wells, mentioned by Nutini (1968, pp. 35–36), have been largely abandoned in favor of tap water in 1970s (see Hill & Hill, 1986, p. 13). Currently, private households in Contla usually have water reservoirs located on the roof, but their filling now relies more on centralized supplies by the regional water company than on rain water.

Nutini himself noted certain environmental and cultural changes that took place in Contla in the past century. First, it was the draining of *barrancas*, or ravines that historically divided the municipality of Contla into sections. Whereas contemporary elderly residents affirm that in the times of their parents and grandparents there was always water in the *barrancas* (see below), Nutini (1968, p. 25) wrote that they become filled only in the periods of torrential rains. Two decades later, Hill and Hill (1986, p. 9) mentioned that people still "occasionally drowned" in the streams coming from the slopes of Malintzi (hispanicized Malinche) volcano during rainy seasons, which is unthinkable today. Both Nutini (1968) and Hill and Hill (1986) attest the decline of traditional Indigenous clothing practices in Contla and other communities of southeastern Tlaxcala. Nowadays it has become

---

[3] https://datamexico.org/es/profile/geo/contla-de-juan-cuamatzi#population-and-housing.

[4] https://datamexico.org/es/profile/geo/contla-de-juan-cuamatzi#economia-inversion-extranjera.

even more evident, as even the elderly women do not wear traditional blouses and skirts in Contla (see Hill & Hill, 1986, p. 14) and only the oldest male residents wear *huaraches*, which were common at the time of Nutini's (1968, p. 42) research.

The main aspect of Indigenous culture that has experienced a dramatic shift during the last sixty years is the language. Nutini (1968, p. 24) gives the figure of 8688 speakers of Nahuatl, or Mexicano, in Contla, which at that time constituted 81.2% of the whole population; the vast majority of monolingual Spanish speakers lived in the *cabecera*. Furthermore, 419 of the Nahuatl speakers of Contla (about 4% of the population) were monolingual (Nutini, 1968, p. 24). In comparison, the results of the 2020 census showed that there were only 4373 Nahuatl speakers (including persons with a minimum of three years of age) living in the municipality, whereas the overall population of the municipality has increased nearly fourfold during the same period.[5] There are hardly any Nahuatl-only speakers left. In other words, in the beginning of 1960s the community of Contla was largely bilingual, whereas now Nahuatl became extensively substituted by Spanish.

This decline conforms to the evidence presented in Hill and Hill (1986), Robichaux (2005), and Messing and Rockwell (2006), as it shows that the major language shift began in 1960s–1970s, when the first generation of Nahuatl speakers who had gone through the public school system grew up and, upon becoming parents, did not teach the language to their children. One of the main reasons for this interruption of language transmission was most likely the trauma experienced by that generation as a result of the forceful assimilation practiced at schools, where any sign of the Indigenous culture, and particularly the language, was mocked and persecuted by the teachers, up to corporal punishments (Robichaux, 2005, pp. 76–78). Many members of that generation justified their decision not to speak Nahuatl to their children by pragmatic considerations, seemingly accepting or succumbing to the dominant assimilation discourse, which associated Indigenousness with backwardness and socioeconomic misfortune (Messing, 2007). In recent years, however, when the ideas of

---

[5] https://datamexico.org/es/profile/geo/contla-de-juan-cuamatzi#population.

revalorization and revitalization of Nahuatl as well as other Indigenous languages have become actively promoted (in both the mainstream public discourse in Mexico and Indigenous communities where a language shift has been ongoing for decades), many elderly Nahuatl speakers, who did not transmit Nahuatl to their children decades ago, started reassessing their life experience and advocating the maintenance of their language in the community. The conversations I had with Contla residents confirmed the existence of this tangible change in local attitudes towards language transmission.

Attitudes of the Elderly and Young People Towards Language and Mutual Communication

When I started conducting my research in Tlaxcala, the first generation of local residents who were not taught Nahuatl in their family settings were already adults in their fifties and sixties. I collected multiple testimonies from this generation, in which they recalled that their parents avoided speaking Nahuatl to them despite using it between themselves, or even directly prohibited their children to speak the language at home. These persons, who still had certain proficiency in Nahuatl, argued that they themselves did not face much discrimination at school, but their parents, who were now in their eighties or late seventies, were scolded and even beaten at school, when they studied in 1940s–1950s. I also met speakers of above seventy years' old who described that experience. One of them recounted that while he studied the three grades of the primary school in Contla, he freely communicated in Nahuatl with his peers, and although the teachers did not like it, the children still continued speaking Nahuatl among themselves. But once he moved to a school in San Pablo Apetatitlán (see Messing & Rockwell, 2006, p. 258), he and his *paisanos* met significant pressure from the predominantly Spanish-speaking environment:

CB (m, 73): So I, there in San Pablo, since here I had already finished the third year, I already could read. And I already could talk to those who didn't speak Nahuatl there, I didn't speak Nahuatl there, [I spoke] Spanish. I spoke Nahuatl with my fellow countrymen who went there, there were eight of us from here, eight kids who went to San Pablo, and wherever we met there, we spoke Nahuatl [to each other]. And they, the people there, like kids and teachers [told us]: "Hey, you, what are you saying? What do you say about us? Are you lying to us or what?" [And one of us] said: "No, we are just talking to each other, me and my fellow..."—"But what do you say?" So we explained to them, and they said: "Aaa, so no-no-no-no, we don't like the way you speak, no-no-no-no-no... Here you speak Spanish, we don't want you to speak that *huanhuanguinguinhuanguentzin*, no." They shunned us. So we had to speak Spanish between us, who were all friends from here, too. We couldn't speak Nahuatl.[6] (04.09.2017)

In September 2017, during the above-mentioned introductory survey, a number of people in their seventies maintained that they now tried to pass some of their knowledge of Nahuatl to their grandchildren of school and pre-school age. One man of 61 years old admitted that although his parents spoke to him solely in Nahuatl, he spoke only Spanish to his children when they grew up, but at the same time he was determined to speak the language to his grandchildren when he would have them Another man, of fifty years old, lamented the loss of Nahuatl in his community and the lack of desire of his own adult children to learn it, whereas later he responded that he had not spoken Nahuatl to his children at all when they were young, even though it was his main language of communication with his peers and older people, including his parents. Rather than the lack of understanding of the cause-and-effect relations in this case, his bitterness was probably caused by the inability to overturn the course of time and raise his children in a different manner.

There were other people, for example, a few women in their fifties, who accepted the break in intergenerational transmission as inevitable; one of them said that she had tried to speak Nahuatl to her sons when

---

[6] All excerpts from the interviews cited in this chapter are translated to English by the author.

they were children, but they had not liked it, and now, when they were adults, they asked her to teach them the language, but now she was reluctant to do so. Many respondents, mostly above fifty years old, stated that they would like to see more Nahuatl taught in schools but they doubted that the children themselves would like it. It should be noted that such conciliatory views came from the people who clearly expressed their support for the revitalization of Nahuatl in general, when they were asked whether they wanted Nahuatl to be used more widely in various public domains, such as education, mass media, and health care.

In my broader interviews with elderly persons, I came upon similar expressions of sadness regarding the contemporary youth and children who neither wanted to learn Nahuatl nor showed respect to the elderly when they spoke it in public. As one of the interviewees said, "kids make fun of us on their road to school", when they saw him and other old people talking Nahuatl between themselves. Similar experience was recalled by another woman in her seventies, who expressed regret about the fact that contemporary adolescents could not even recognize Nahuatl when they heard old people talking it:

NF (f, 75): Yes, in the past [all the people] knew Mexicano, but now the girls already… You address them, they don't understand it. Sometimes I go somewhere with her (a grandchild—GH) and we talk Mexicano, [me and other people] whom I know. Yes, all Mexicano. And she says, she's looking at us from aside and says: "What are you talking? English?" I'm telling her: "And this is because you don't understand, you expect English?" English… she just laughs, that's what it is. (03.08.2021)

Nevertheless, it appears that the elderly feel more comfortable to teach some Nahuatl to their grandchildren, if the latter are still in pre-school age and thus are more attached to them and less affected by the larger society, than to share their linguistic knowledge with their adult children and other younger adults, as the elderly do not hold any expectations about their motivation to learn. For example, I met two male Nahuatl speakers in their seventies, each of whom had a notebook where for many years they had been writing down Nahuatl words and phrases that they remembered and wanted to preserve. Both of them did it independently of any language revitalization activities taking place in their community—it was their personal endeavor that they cherished but were hesitant about making it public. They were more keen to share their work with me, an external researcher with an interest in Nahuatl, than with their children, who probably were even unaware of the existence of these notebooks. As one of these elderly speakers told me, he wanted to leave his dictionary to his grandchildren, not mentioning his children, who were already in their thirties and forties.

For the elderly, their knowledge of the language is not only about the language itself. When we discussed with them the role of Nahuatl in their life, both past and present, they often emphasized the practices and customs that had their proper discourse and terminology in Nahuatl Among those practices, they mentioned scraping magueys (agave plants) for an alcoholic beverage called *pulque* and sponsoring a religious fiesta as a part of *mayordomía*—a system of the veneration of holy images, part of a larger social-religious system of *cargos*, which has played an essential structural role in the life of Indigenous communities in Mexico since the early colonial period (Robichaux, 2005).[7] The ceremonial practices related to *mayordomía* have not only been preserved across Tlaxcala but, due to the population growth, they have become richer and more complex (Robichaux, 2005, pp. 66–67); however, in the eyes of the older generation of Contla, the decline of the language causes them to lose a crucial element, something that constituted their uniqueness and intimacy.[8] This could even be related to the very names of celebrations, as Day of the Dead, which in the recent period has been influenced by North American Halloween tradition. One of the interviewees explained it in the following way:

> BM (m, 57): For us *Miccailhuitl*, or All Saints' Day, or Day of the Dead, is very different from what is called Halloween in other places, yes. For us *Miccailhuitl* is a holiday of respect to our dead ancestors, a holiday

---

[7] For more information about *mayordomía* and its place in the system of *cargos* in Tlaxcala and Mexico, see Romero Melgarejo (2009) and González de la Fuente (2011).

[8] These testimonies question Robichaux's approach, according to which the system of cargos itself is the main indicator of Indigenous identity in Mexico (Robichaux, 2005, p. 69).

of… of love, of care for our deceased relatives. We leave them bread, fruit, *molli*, everything we can, we offer them the best we can, although we know that they won't come, we still offer it to them as a way to remember them. […] And what is going on with what they brought to us, with Halloween? A holiday of witches, holiday of vampires, ghosts and so on. Demons and all things like that. So, it isn't our culture, our culture is respect, it is care, yes, to all the things. And this is also achieved on the basis of the Nahuatl language. As long as we try to save Nahuatl and are aware of these values, I think the things could be better. (08.09.2017)

Unlike the elderly interviewees, who were either Native speakers of Nahuatl or at least spoke it freely since childhood, among the younger people whom I interviewed only one woman of 22 years old, spoke Nahuatl as her mother tongue. She was not Native to Contla but to San Francisco Ixquihuacan, a village in Sierra Norte de Puebla, from where a number of residents migrated to San Miguel Xaltipan in late 1990s in search for work opportunities. Although the variety of Nahuatl spoken in Ixquihuacan differs little from the variety spoken by elderly people in Contla, she attested that she had never tried to communicate with any Nahuatl speaker in Xaltipan in Nahuatl. Moreover, an older co-worker of the woman, who proceeded from the same village as she did, affirmed that she never talked to him in Nahuatl as well, saving this language only for communication with her younger sister and, presumably, her parents at home. Asked in the interview whether she wanted to teach her own children Nahuatl, when she might have a family in the future, she contemplated a bit and then replied: "Well yes, it is indeed important… Well, I don't know, but since it's being lost, and there are few people who speak it, and so the more people will speak it, the better…".

Other young interviewees, who were born in Contla, as a rule, could be characterized as language "rememberers" (see the classification of endangered language speakers in Grinevald & Bert, 2011), i.e. who knew only a few words or phrases in Nahuatl. According to them, their parents or grandparents occasionally taught them a word or two, or, upon hearing Nahuatl words in conversations of their older relatives, the young themselves asked to explain what this or that word meant, and the elderly explained it to them. The common declarative attitude among the young people whom I met was that, while they did not at all avoid learning Nahuatl, they were not presented with such an opportunity on the part of speakers from their families or were too shy to show initiative and ask their parents or grandparents to teach them. One of them did say that if she had asked her father to teach her, he would have most likely agreed. It is, of course, impossible to verify how conversations would develop in a real situation if an elderly member of the family were to offer a younger relative to teach her/him Nahuatl and/or a younger relative would ask an elderly one to teach her/him the language on a regular basis. According to the young people, an important factor here is the level of intimacy (or, as another young interviewee characterized it, *límite de la confianza*, 'confidence limit') with elderly persons who speak Nahuatl:

> YA (f, 34): There are indeed moments, in which we say "What are they talking about?", we get this curiosity… to know, "What is that they're saying?" And well, maybe we don't dare [to ask] at that moment—if I know the lady, I may ask her, but if I don't know her, then like no, better not to ask her, she may be annoyed because… so I just keep listening to their conversation, although maybe I don't understand what they are talking about at that moment, but… it would still be important. (28.11.2019)

Afterwards, my interlocutor said that in order to learn Nahuatl from the elderly, the youth should project an interest not only in the language itself but in the persons who speak it. But this is again where the generational gap, according to the testimonies of other young people, created a considerable obstacle—even if a young person respects the elderly (and not only members of their own families), there are differences in both the manner of communication and the worldview that can discourage them from approaching an older person for advice or even a mundane conversation. Therefore, the youth refrain from making the first step in communication.

A notable testimony came from a teenager, who, discussing the traditions of his community, emphasized the role of the language component in them:

EA (m, 17): There are quite a few things that the old people discussed among them [in the past], and most of them they didn't discuss in Spanish. Because then, with my grandparents, they spoke together in that manner, and I was there, but I couldn't understand them, it was necessary to ask them what they were saying or so. [...] Without the language, we won't do the things as they did, or how can I say, we can't do [them] as we won't know the things of the past. (05.10.2018)

Surprisingly, this corresponded to the deliberations of elderly residents, whom I interviewed earlier, who put a clear emphasis on the linguistic aspect of tradition.

As can be seen above, the young interviewees in general expressed concern about the fate of Nahuatl, but I also encountered an opinion that the language would persist no matter what. The fact that the opinion's holder did not speak Nahuatl himself did not worry him, as he still believed that he would learn it in the future, so that not only he but his future children would also learn to speak the language. It appeared that for the majority of young interviewees Nahuatl was a matter of interest but not a crucial element of their identity. However, I also met a young man who, despite being a language "rememberer" himself, argued that knowing and pronouncing words in Nahuatl, learned from his parents and grandparents, made him feel special against the mainstream society, which has not been sympathetic to Indigenous languages and their speakers:

EB (m, 28): So, when you speak Nahuatl, you feel superior in something else. When you speak a native-native language, something that is yours... I, for example, as I'm telling you, when I come and speak it there, at SAMS or WalMart (refers to the stores in the city of Tlaxcala—GH), like that, I feel superior in something. This is my ego. And something that I feel very well. It's my pleasure, I like it very much to speak like that. And the people didn't look at me well when I spoke like that. But it doesn't matter to me. I am doing this because I like breaking rules. Because I like it, because I don't want to lose it. (17.09.2018)

He was also proud to pass the little knowledge that he had to his daughter, although he was divorced with her mother, who did not approve any use of Nahuatl. This remarkable perspective, although uncommon, demonstrates that attitudes to Nahuatl among the young adult generation in Contla can vary and can surprise the elderly speakers who are skeptical of the youth's desire to learn and maintain the language.

Attitudes of the Elderly and Young People Towards Natural Environment and Its Change

In the beginning of my research in Tlaxcala, the ecological transformation of the community did not especially interest me, as I wanted to focus primarily on the sociolinguistic transformation. But in our discussions about past and present with elderly members of Contla communities other than the municipal center, where the environmental change was more recent and experienced by still living residents, it became clear that this subject evoked equally strong emotions and reflection as did the language. As in the case of language, the prevalent feeling among elderly residents was that something inherently valuable had disappeared from their lives with the change of the landscape, an aspect that the younger generation could not even properly value as they were born too late to witness it.

The first concern focused on the industrialization of the water supply. Although elderly residents acknowledged that the installation of water pipes prevented their community from suffering from water shortages, and that it was necessary in light of the population growth, in their eyes it was accompanied by the abandonment of a part of their heritage, towards which they harbored nostalgic feelings:

NF (f, 75): There are forty years since we have tap water... Earlier, we had water from the well, we had a number of wells, very deep ones, of seventy-five meters, eighty meters. We drew water with a rope, people drew water with a bucket, like that, in primitive way. [...] Earlier, we weren't so many people as today, before we grew up, our grandparents told that in their time they went to take water from the streams, even some non-Indigenous persons[9] did so [...] and the water collected [from there] was clean-clean-clean... (03.08.2021)

The wells, of which nowadays only the upper part is preserved at best, in this discourse were more than just a water source—they were a monument of human, communal effort that required a long period of time to reach the alleged depth by digging, and likely a popular spot for social interactions. When industrialization came to San Miguel Xaltipan and the roads in the community, according to the words of another interviewee, were widened, state authorities also sent a team of engineers with modern equipment, community members carried pipes, and the wells were instantly dug up, with pipes installed, and then the holes were covered up with earth. At that time, none of the residents opposed this change—it was perceived as the progress through technology—but in effect, it also marked the moment when water became the property of a water supply company and not of the community.[10]

It is notable that the memory of crystal clean water in the *barrancas*, which already was the thing of the past at the time of Nutini's research, has been preserved through generations, and even more notable that this memory is contrasted with the contemporary experience of tap water. The allegedly pure fresh water that was available to everybody, from the perspective of the elderly, symbolizes the pre-industrial epoch, while industrialization and contamination transformed the once clean streams into waste and mud containers, in which the appearance of

---

[9] In this case (the interview was conducted in Nahuatl), the interviewee used the word *coyohtzitzin*, a plural form of *coyotl*, "coyote", the historical denomination of a Spanish-speaking person in Nahuatl, but using the reverential form, which should emphasize the high(er) social status of the people referred.

[10] As regards the changes of ownership of water resources in Mexico during the past century from local (community) to centralized (federal or regional) management and to private companies, as well as inequalities and conflicts associated with each stage, see e.g. Wilder & Romero Lankao (2006) and Mendoza García (2013). In this context, the case of Contla de Juan Cuamatzi also deserves a more substantial study on the subject.

water during rainy seasons can only cause mud slides and soil erosion. The contamination, gradually accumulated throughout the years, was noticed by elderly residents not only in the water but also in the air and in the sky itself:

CB (m, 73): Now, the manner of our life, in comparison to what I [remember], it's very different. Why? Because when there was, let's say, sixty or sixty-five years ago, we saw the blue, clean sky. And now we don't see it as blue as it was, we see it grey, it has changed its color. And I recall, in that epoch, when I was a kid sixty-five years ago, there were *mesurcos*, or boundaries. We call it *mesurcos* in Nahuatl,[11] and a boundary, that corresponded to a boundary [between land properties], was a row, and these rows were all of magueys. For example, every two meters there was a maguey… And now there aren't any. (04.09.2017)

The shrinkage of magueys, or agave plants, in Contla thus did not only mean the decline of traditional practices for extraction of agave juice, but also the loss of green hedge method for marking boundaries between lands. This method needed a replacement, which h was provided then by concrete and cement. Apart from magueys, the diminishment of vegetation in the local communities included also fruit trees, such as apple, pear, peach, and plum. In some households, they constituted an additional source of income, but it also turned out that for Contla residents, the value of fruit and fruit trees was not merely economic. When asked whether her community looked more attractive today or when she was young, one of the elderly interlocutors answered:

NF (f, 75): Earlier I think it was more beautiful. Because there were many trees. We had *mesurcos*, magueys. […] And after one maguey was used up, people planted another one, so that the *mesurcos* remained intact. Earlier it was different. There was a hell of a lot of fruit. So much that my father sold it, he carried it in baskets, carried it to Santa Ana, who knows where dad sold it. At that time there were so many various fruits, and I planted them at my property, so besides *mesurcos*,

---

[11] In fact, the word *mesurco* is a loan blend, consisting of the Nahuatl word *me-tl*, "maguey", and Spanish *surco*, "furrow".

there were many fruits, pears, apples, all cultivated at home. Now you saw, there's nothing. Everything has been used up, now all have become people of jobs, everybody lives from a job. Earlier was one thing and today is another thing. (03.08.2021)

In general, elderly residents do not deny the positive developments—they acknowledge that was harder in the past than it is today. In past times, everybody had to work in the field and only a few could make a living from working in the textile industry (which started to develop in Contla only in the mid-twentieth century); now the choice of occupations is much broader and profits are higher. But the sentiments about greener and cleaner community life remain strong, and the prevalent view among the older generation is that these qualities can never return.

The younger residents of Contla were already born and bred in the new, densely populated and industrialized environment, but that does not mean that they are not able to value things, both material and spiritual, associated with the pre-industrial epoch. Those of them who used to hear stories about the past from their parents and grandparents or those who have visited and stayed in other, less urbanized Indigenous communities, can express a feeling of deprivation that unites them with the elderly. This feeling, for example, can be provoked by a piece of tradition that they discover in other places and cannot find in their own community of origin. For example, one of the interviewees spent a significant amount of time working in San Isidro Buensuceso, a community on the southwestern slope of Malintzi, where the majority of population still speak Nahuatl (see Nava Nava, 2016), and recalled the local practices of steam bathing, or *temazcal*:

EB (m, 28): I asked them: "And why do you lay out *ocoxal* (dried, sand-like resin of *ocote* pines—GH)? Because in my community I bathe in a *temazcal* and there we don't lay out anything, we only put some sacks, cartons, so that you don't get burnt…" They tell me: "No, actually *ocoxal* gives you a rich aroma." So when I got inside… indeed, it emitted a rich aroma. Very rich. And in the case of mothers, when they are pregnant or have recently given birth, they bathe them seven times in a *temazcal*, with *ocoxalito*… It's like a ritual for the Lord, and [the head of the family] goes to collect *ocoxal* to the mountain, and then arranges a bath for his daughter or daughter-in-law, so that the baby would be healthy. So I learned it there, and during vacations, when I came back home [...], I said to my dad and mom: "Listen, mom, and why don't we put *ocoxal* here?" She says: "Well, actually, [this custom], it's already lost." It was completely lost. (17.09.2017)

One aspect of this story is a regret for the lost tradition, but under a closer look, it becomes clear that this loss occurred due to an environmental change as well. San Isidro Buensuceso is not only a more conservative community in terms of the retention of the Indigenous language and culture in comparison to Xaltipan, the interviewee's home village. It also borders the conifer forest on the upper slopes of Malintzi volcano, where *ocote*, or Mexican pine, is the prevalent species. This forest is currently threatened by illegal logging (Jiménez Laguna, 2021). In pre-industrial times, the same forest also reached the upper part of the Contla municipality, but after the forest boundary moved upwards and a whole day's journey was required to collect *ocoxal*, this practice, as it seems logical to assume, died out. For an elderly community member, who can still recall how the wood was used as a construction material (Nutini, 1968, p. 36), the connection would be clear, but not so for a younger one, who has seen much less interaction with the forest.

As far as I can judge from my conversation with young people in Contla, a significant part of them are not inclined to reflect on what has been lost as a result of the municipality's industrialization. In interviews, when we started talking about the environment, they usually expressed concerns over the roads that had not been paved and their maintenance. The positive aspects of the industrialization overweighed the negative ones for many. One of the interviewees, when I asked her to compare the older times and modern times with regard to the communal well-being, answered:

YA (f, 34): We can say that now we have better services. As I told you, before there were surcos, o mesurcos, or lanes, and now we have streets, so... so there is a progress from [what was] fifty years ago comparing to now, yes, because there is already a system of roads, there is centralized tap water, because earlier, I still remember it as a kid, the water was very scarce... [...] [There were] persons who had wells, and we went to buy water from them, so the water was very scarce... [...] The persons who had wells, they were like upper class, because they had water, right? and in that case the people who didn't have it, had to go to bring [water] for cooking, for drinking, in order to wash clothes they also had to go to another place... [...] So as others tell me and as I myself see it, it has been a progress, maybe not at a large scale, but it has been a progress, and we have benefited from it, because now we have roads, means of communication... [...] There are still some roads in the upper part [of the municipality] that are made of dirt, but I actually feel there has been certain progress, because the principal things, like roads, water, electricity, now we have it all. (28.11.2019)

Such an obviously more practical view of industrialization, in comparison to those disclosed above, also presents a critical perspective on the traditional organization of water resources, mentioning the issue skirted by the elderly interlocutors: the inequality that existed between those who had a well (or had resources to dig it) and the rest of the population. Hence, it is also evident that sentiments about the old times can eclipse certain social issues that underline the need for change. But at the same time, the push for progress does not prevent all the youth from noticing the pitfalls of industrialization as well as its benefits (Fig. 7.2).

Thus, another young interviewee expressed an opposite view of the environmental change that has affected Contla, focusing on the damage done to the ecology and economical sustainability of the local communities. His deliberations on this subject came forward right after I asked him how he thought the community's well-being could be improved. In his view, the current landscape presented the evidence why people should be educated to understand that well-being is generated not only by consuming resources, but also by creating them.

RS (m, 29): Before, I remember that I was [told], that this part from here and up to there was still sown, with what was corn, certain vegetables, fruits, and... and we lived on this, actually. And now we don't. No, because this part became like they built it up, they made... so I think that, talking about education, [it is important to tell people] that

Fig. 7.2 A road in San José Aztatla, Contla. September 2017. Copyright: Author

if I build, if I do something, I also have to think how to create, how to produce my food as well. And without relying on the industry, without relying on… for example, here, instead of growing your own tomatoes, you go and buy them, or corn. Earlier you had this here, but now you go and buy canned corn. […] My mom told me that here, where the chapel stands now, there was flowing water. And there were many trees. And what I got to know is that the landlord of that place decided to cut down the trees, and then the water stopped flowing. So I believe that the lack of water is caused by the lack of trees." (02.08.2021)

For the interviewee, these issues not only indicated that the community needed to shift towards the greener and more sustainable environment, but, moreover, he had a will to contribute to that shift himself:

RS: I also talked to my brother and said… I explained to him the same thing, that the lack of water is created by the lack of trees and… many things, but trees are my priority. And he said "Ok", he said to me: "So I don't know, brother, how do you want to do things?", he tells me, [and I answer] "Well, we have to aim for a group, we form a group, and we start with cleaning the road, and then after we look what can be done, we plant trees and... do what was done before and save all this, right?" So that we can live as before in terms of resources, but with the abilities that technology gives us and with the ability to purchase stuff. [...] It is necessary to look for persons of our generation, more or less of our age. (02.08.2021)

The seeming idealization of the state of environment in the past is secondary here—the more important is the desire to aspire to that ideal and try to regain the ecological benefits of the pre-industrial environment. Such a proactive approach among the local youth is probably unique and it is uncertain if it indeed results in regular and organized planting of trees, as my interlocutor envisioned. However, what is also remarkable here, is the understanding that the work of that sort should be done by involving a number of people rather than individually. It is also possible that the motive for organizing a group of activists lies not only in a desire to pursue more significant practical outcome, but also in the ability to stimulate a social change together with an ecological one, by disseminating the ideas of sustainability around your peers and persuading them to take part in a constructive work for the common good.

The accent on the young, made by the interviewee, is also notable: it may reflect plans to seek participation among the more active part of the community, but it may also indicate the aforementioned difficulty to cooperate with the older generation, as well as the expectation that they would not accept young people's ideas due to the worldview differences. It did not come up in our conversation how the older generation might contribute to such a project, but since it was their memories that gave incentive to the idea of re-greening the community, and in general they had more experience in horticulture than Contla's youth, it would be appropriate for me to ask then if the elderly people could also play an active role in that project.

Discussion: Combining the Opposites

The qualitative study that lies in the basis of this work is not without certain limitations. First, the number of interviewees recorded hardly constitutes a representative picture of how the local society generally views the issues of language, environment, and well-being. Another limitation comes from the fact that I interacted with the persons who were willing to share their experience and thoughts with me and trusted me enough to give me their consent for recording our conversations. I have been perfectly aware that there are plenty of people in Contla who would not regard me, a white foreign researcher, as trustworthy, or for whom the issues that interested me are of no real concern. In addition, there is a possible limitation caused by what Bernard (2006: 241) calls "the deference factor"—when the interviewee tells the interviewer what the latter expects to hear, or what the former *thinks* that the latter expects to hear.

Here it should be noted that I found the elderly residents in general more open to converse with me and inclined to share their stories. My younger interlocutors could be divided to those in their late twenties—middle thirties and those in late teens or early twenties. While the former manifested more socially and ecologically oriented thought about well-being, the latter showed reluctance to talk about complex social topics. In addition to shyness and possible lack of trust towards me, their aspirations were more orientated towards outside the community. They put more emphasis on individual well-being, which in their eyes was associated with opportunities to study and live an independent life, with traveling to new places and making acquaintances with people from those places. Proceeding from the socially disadvantaged groups, they could perceive an incentive to focus on issues immanent to their community as an attempt to keep them within the circle from which they mostly wanted to leave. In this situation, the elderly, feeling that the youth seek to move away from them and neglect their authority,[12] may be more eager to share their

---

[12] See e.g. Merali (2004) on similar intergenerational gaps in migrant and refugee contexts.

knowledge with an outsider if the latter shows an interest in them and their experience.

However, in the interviews my younger interlocutors also wanted to show that the lack of connection with the elderly was a two-sided problem. According to their words, their older relatives also tended to look at them critically and were somewhat reluctant to share their knowledge with the youth. One of the interviewees (EB, m, 28) even said that the elderly "do not want to share [traditional knowledge] because they feel inhibited... because they feel that we are going to take something from them". But sometimes, as can be seen from the example of YA (f, 34), young people can look positively at this issue, admitting that it is also their responsibility to show their respect and curiosity towards the elderly and their knowledge (here, notably, it was also the oldest of the younger interviewees who expressed that view).

In essence, we have two virtually opposing age groups who are apt to reproach each other for mutual indifference. This situation acquires the form of a vicious circle, in which the elderly do not place much trust in the young, since the latter, in the former's opinion, hold very different values from them. Indeed, the contemporary young residents of Contla were socialized in a different language, different environment, and therefore in a different culture, but it is also important to take into account that the older generation themselves contributed to the change that created this rupture. The young people, feeling the distrust on the part of elderly, are also wary to approach them for any kind of communication, thus persuading the elderly in the young's lack of respect. Hence, it is not fruitful to evaluate the fairness of each group's allegations to each other. Rather, it is useful to look at possible points of convergence in the discourse of the young and the elderly, which would show that the generational gap, despite its obvious existence, is not as solid as it seems to many representatives of the opposing generations. Of course, the opinions demonstrated in the interviews are quite diverse with contrasting points of view within each generational group, but this is precisely what shows that one generation's views of another are too generalized.

Referring to the attitudes towards the heritage language, we can see that although the vast majority of young residents do not speak Nahuatl (and even those few who are proficient in Nahuatl do not speak it outside the family), the interviewees from the respective age group maintain that they would like to learn it if they were given the opportunity. Even if the "deference factor" played a role here, such ideas as the reasoning that Nahuatl should be learned in order better to understand the tradition or the use of linguistic knowledge as a means to "break the rules" could hardly be provoked by any of my questions or my own views known to the interviewees. This shows that regarding the local youth as entirely disinterested in the linguistic heritage can block the opportunity of reconciliation between the distinct generations for the purpose of language transmission. The young people, in their turn, do not take into consideration the experience of forced assimilation at schools decades ago that undoubtedly contributed to the elderly's unwillingness to transmit the language—a traumatic experience of which the contemporary youth have been spared. None of the interlocutors who spoke about the reluctance of their elderly relatives to share linguistic or other traditional knowledge with them, mentioned the discrimination that the past generations had suffered.

As regards the attitudes towards the natural environment, the misunderstanding between the generations in this case is not so evident, as the elderly people mainly contemplate this issue in romantic terms, considering the change of the landscape as irreversible, and the young people rarely reflect on the environmental loss that occurred before they were born. However, in those rare cases, when the feeling of loss is shared by the young people, and more remarkably, when this feeling prompts them to take action, it indicates that for such young persons the greener environment constitutes no less a value than for the elderly, who remember it from their life time. Therefore, it is also a possible point for collaboration between community members of both generations, if the aim is to preserve and revitalize what is valuable to them all.

The question is how such collaboration could happen in practice. A suitable model could be drawn from those community-based participatory projects, in which Indigenous elders are reunited with those representatives of Indigenous youth who are interested in learning from them and work together for the maintenance of traditional cultural practices and sustainable future of their communities (Kahn et al., 2016; Pace & Gabel, 2018; Saba, 2017; Wexler, 2011). In Contla there are probably no "Indigenous elders" in the ordinary sense of this word, but there are many knowledge holders of Indigenous origin and Nahuatl speakers who would like to pass their knowledge to the next generations. Thus, a space may be established where these knowledge holders could communicate with curious and sympathetic young people, who would not necessarily be relatives of the elderly participants, on a regular basis. It could function as a sort of workshop, or a meeting space, where any young person who would like to learn Nahuatl, could receive lessons from a Native speaker. As for the environmental knowledge, it could represent a platform, where the elderly could share the knowledge that has fallen out of use, which could help younger activists to plan and conduct ecological projects as the one described in the end of the previous section. But of course, such a platform can be created only if the initiative comes from community members themselves, and if there are enough representatives of both generations who would like to participate and support it.

The interviews recorded in Contla presented me with a multifaceted discourse about well-being to show different ways in which the residents of different age and gender viewed the better life for themselves and their communities. Two principles arose in the discourse of elderly and young persons alike, when our conversations focused on the well-being of the local society and what was needed to achieve it—unity and education. They expressed a common discontent about the adherence of community members to their personal material interests, which prevented them from joining efforts for the common good, even in such minor cases as the maintenance of a piece of road. Education in this context was understood not as the instruction of school subjects, but the communication of values—moral and cultural, and not exclusively to children but also to adults.

In this context, both old and young residents sometimes look towards a more traditional way of life, where the society is more knitted together, Indigenous language is still spoken and natural environment is greener and more diverse. While admitting the benefits for material well-being brought by industrialization, the majority of interlocutors share a feeling that some of the essential components of well-being have been lost or are hard to achieve. A model that can show that things could be different is not necessarily sought in the past, as an inspiration can be found in a contemporary and not so distant community, which has been less acculturated than Contla communities, even if the economic sustainability there remains at a lower level. Such was the case of the young interlocutor who had worked in San Isidro Buensuceso and was helped there by people who were poorer than he, which he later revealed during our interview:

> EB (m, 28): But the teaching of that place gave me a lot to understand life. That is, you come and do something not in order to earn money, not because you want something from the people, no. The teaching that it gave me is that all of us are humble and all of us are on the same road, which is well-being. And if we don't understand it well, we may fall, we may panhandle, we can ask someone for food, and we'll be given it, there will be always people who will give us and those who won't give you. And with them [residents of San Isidro Buensuceso—GH] I learned it all. (17.09.2017)

Thus, the language, ecology, and traditional values become associated with well-being when they are associated with live people around you— people who are inclined to share that knowledge and give you a helping hand. Although not explicitly expressed by other interlocutors, this view, which places knowledge-sharing and mutual help in the center of well-being, it can characterize their distinct narratives, of the old and young alike.

Conclusions

The considerations of Contla residents about language and ecology, past and present, should be viewed in light of profound social and economic changes that took place in Mexico since the mid-twentieth century. The narratives cited in this chapter show the individual dimension of these changes, including the reflections of both the generation that witnessed them directly and the generation that grew up with their results. While most of research on the process of urbanization and industrialization in Tlaxcala and Mexico focuses on broad social, economic, and environmental implications, these individual narratives reveal an additional, esthetic aspect of the change—nostalgia for the lost beauty and colorfulness of the landscape that featured Contla communities half a century ago. This nostalgia can be shared not only by the elderly residents but also by the young ones, who took to heart stories about the past.

This study aims to demonstrate that an intergenerational gap, which becomes especially significant in ethnic minority societies that experience intense assimilation, should not be considered merely a given and unchangeable fact. An inquiry into the human experience and discourse that form but at the same time bridge that gap is essential. The example of Contla shows that the need in what Mead called "postfigurativeness" can be found even among disadvantaged and acculturated Indigenous youth, who need to concentrate their efforts on finding better economic opportunities than their community can offer. We can see that while the dividing factors of the generational gap in the historically Indigenous community are clear and familiar to both sides of the gap, the factors that still unite or are able to unite the old and young generations remain covert but present nonetheless. In order to reveal them, it is necessary to analyze comparatively the discourses of representatives from both generations.

Thus, in studying the views of community members on subjects related to local Indigenous identity it is essential to include the perspectives of different age groups. The exclusive reliance on the perspective of the elders, or knowledge holders, however productive it could be in terms of ethnographic data, does not take into consideration the future directions of community's development, which will take place after the current generation of knowledge holders passes away. Suggesting the bridging elements of different community discourses, as is made here for the language and the natural environment, is one step. Another one is to make these elements known to community members themselves, so that they could reflect on the findings and use them for their own benefit, such as enhancing the dialog and collaboration between the two allegedly opposite age groups. From the ethnographic perspective, such a dialog can test the statements expressed by community members in individual interviews and the assumptions that can be made on their basis. In the social dimension, it can help the elderly generation to deal with their sense of loss and can open a new opportunity for the younger generation to acquire knowledge has been difficult to learn in other settings.

List of Interviews

AV, 21. 04.10.2018. San Miguel Xaltipan, Contla de Juan Cuamatzi.
BM, 58. 08.09.2017. San Miguel Xaltipan, Contla de Juan Cuamatzi.
CB, 73. 04.09.2017. San Miguel Xaltipan, Contla de Juan Cuamatzi.
07.09.2017. San Miguel Xaltipan, Contla de Juan Cuamatzi.
EB, 28. 17.09.2017. San Miguel Xaltipan, Contla de Juan Cuamatzi.
EA, 17, 05.10.2018. San Miguel Xaltipan, Contla de Juan Cuamatzi.
NF, 75. 03.08.2021. San Miguel Xaltipan, Contla de Juan Cuamatzi.
RS, 29. 02.08.2021. San Miguel Xaltipan, Contla de Juan Cuamatzi.
YA, 34. 28.11.2019. San Miguel Xaltipan, Contla de Juan Cuamatzi.

References

Alba, R. (2005). Bright vs. blurred boundaries: Second-generation assimilation and exclusion in France, Germany, and the United States. *Ethnic and Racial Studies, 28*(1), 20–49. https://doi.org/10.1080/0141987042000280003.

Alba, R., & Nee, V. (1997). Rethinking assimilation theory for a new era of immigration. *International Migration Review, 31*(4), 826–874. https://doi.org/10.2307/2547416

Bernard, H. R. (2006). *Research methods in anthropology: Qualitative and quantitative approaches* (4th ed.). AltaMira Press.

Cristancho, S., & Vining, J. (2009). Perceived intergenerational differences in the transmission of traditional ecological knowledge (TEK) in two indigenous groups from Colombia and Guatemala. *Culture & Psychology, 15*(2), 229–254. https://doi.org/10.1177/1354067X09102892

Feir, D. (2016). The intergenerational effect of forcible assimilation policy on school performance. *International Indigenous Policy Journal, 7*(3), 1–44.

Gans, H. J. (1992). Second-generation decline: Scenarios for the economic and ethnic futures of the post-1965 American immigrants. *Ethnic and Racial Studies, 15*(2), 173–192. https://doi.org/10.1080/01419870.1992.9993740

González de la Fuente, I. (2011). Comunidad, sistema de cargos y proyecto social. Una propuesta analítica de sociedades locales en México. *AIBR. Revista de Antropología Iberoamericana, 6*(1), 81–107.

González Jácome, A. (2009). Las faldas de la Malinche: el paisaje de tierras temblado-frías y sus pueblos. In F. Castro Pérez & T. M. Tucker (Eds.), *Matlalcuéyetl: visiones plurales sobre cultura, ambiente y desarrollo* (vol. 1, pp. 257–282). El Colegio de Tlaxcala, A. C.

Grinevald, C., & Bert, M. (2011). Speakers and communities. In P. Austin & J. Sallabank (Eds.), *The Cambridge handbook of endangered languages* (pp. 45–65). Cambridge University Press. https://doi.org/10.1017/CBO9780511975981.003

Hill, J. H., & Hill, K. C. (1986). *Speaking Mexicano: Dynamics of syncretic language in Central Mexico.* University of Arizona Press.

Iseke, J., & Moore, S. (2011). Community-based indigenous digital storytelling with elders and youth. *American Indian Culture and Research Journal, 35*(4), 19–38. https://doi.org/10.17953/aicr.35.4.4588445552858866.

Jiménez Laguna, D. K. (2021, March 12). Tala de árboles en la Malinche, preocupación de unos pocos. *Escenario Tlaxcala.* Retrieved August 6, 2022, from https://escenariotlx.com/tala-de-arboles-en-la-malinche-preocupacion-de-unos-pocos/.

Kahn, C. B., Reinschmidt, K., Teufel-Shone, N. I., Oré, C. E., Henson, M., & Attakai, A. (2016). American Indian elders' resilience: Sources of strength for building a healthy future for youth. *American Indian and Alaska Native Mental Health Research (online), 23*(3), 117–133. https://doi.org/10.5820/aian.2303.2016.117

Krieg, A. (2009). The experience of collective trauma in Australian indigenous communities. *Australasian Psychiatry, 17*(1_suppl), S28–32. https://doi.org/10.1080/10398560902948621.

Lee, J.-S. (2004). *Intergenerational conflict, ethnic identity, and their influences on problem behaviors among Korean American adolescents* (Ph.D. Dissertation). University of Pittsburgh, School of Social Work.

Lohmann, R. I. (2004). Sex and sensibility: Margaret Mead's descriptive and rhetorical ethnography. *Reviews in Anthropology, 33*(2), 111–130. https://doi.org/10.1080/00938150490447439

Marino, S. (2020). *Intergenerational ethnic identity construction and transmission among Italian-Australians: absence, ambivalence and revival.* Palgrave Macmillan; Springer Nature.

Mead, M. (1970). *Culture and commitment: A study of the generation gap.* The American Museum of Natural History.

Mendoza García, J. É. (2013). El manantial La Taza de San Gabriel Chilac (Puebla) y los manantiales de Teotihuacan (Estado de México) ante la federalización: un análisis comparativo entre 1917 y 1960. In A. Escobar Ohmstede & M. Buttler (Eds.), *Mexico in transition: New perspectives on Mexican Agrarian history, nineteenth and twentieth centuries* (pp. 225–259). Centro de Investigaciones y Estudios Superiores en Antropología Social (CIESAS).

Merali, N. (2004). Individual assimilation status and intergenerational gaps in Hispanic refugee families. *International Journal for the Advancement of Counselling, 26*(1), 21–32. https://doi.org/10.1023/B:ADCO.0000021547.83609.9d

Messing, J. (2007). Multiple ideologies and competing discourses: Language shift in Tlaxcala, Mexico. *Language in Society, 36*(4), 555–577. https://doi.org/10.1017/S0047404507070443

Messing, J., & Rockwell, E. (2006). Local language promoters and new discursive spaces: Mexicano in and out of schools in Tlaxcala. In M. Hidalgo (Ed.), *Mexican indigenous languages at the dawn of the twenty-first century* (pp. 249–280). Mouton de Gruyter.

Nava Nava, R. (2016). La socialización infantil bilingüe en San Isidro Buensuceso, Tlaxcala, México. *Revista Española De Antropología Americana, 46*, 29–47. https://doi.org/10.5209/REAA.58286

Nishita, C., & Katsutani, W. (2022, February 1). Maintaining cultural identity and community: The need for intergenerational programming in native Hawaiian and Pacific Islander populations. *Journal of Intergenerational Relationships.* https://doi.org/10.1080/15350770.2022.2033146.

Nutini, H. G. (1968). *San Bernardino Contla: Marriage and family structure in a Tlaxcalan Municipio.* University of Pittsburg Press.

O'Neill, L., Fraser, T., Kitchenham, A., & McDonald, V. (2018). Hidden burdens: A review of intergenerational, historical and complex Trauma, implications for indigenous families. *Journal of Child & Adolescent Trauma, 11*(2), 173–186. https://doi.org/10.1007/s40653-016-0117-9

Pace, J., & Gabel, C. (2018). Using photovoice to understand barriers and enablers to southern Labrador Inuit intergenerational interaction: Research.

Journal of Intergenerational Relationships, 16(4), 351–373. https://doi.org/10.1080/15350770.2018.1500506

Robichaux, D. (2005). Identidades cambiantes: "indios" y "mestizos" en el suroeste de Tlaxcala. *Relaciones. Estudios De Historia y Sociedad, 26*(104), 58–104.

Romero Melgarejo, O. A. (2009). Poder y cultura: el sistema de cargos, su vigencia en las comunidades nahuas tlaxcaltecas. In F. Castro Pérez & T. M. Tucker (Eds.), *Matlalcuéyetl: visiones plurales sobre cultura, ambiente y desarrollo* (vol. 2, pp. 97–130). El Colegio de Tlaxcala, A. C.

Russell, D., & Tchamou, N. (2001). Soil fertility and the generation gap: The Bënë of Southern Cameroon. In C. J. Pierce-Colfer & Y. Byron (Eds.), *People managing forests: The links between human well-being and sustainability* (pp. 229–249). Resources for the Future; Center for International Forestry Research.

Saba, R. (2017, August 23). Indigenous youth bridging generation gap with elders to help save cultures. *The Globe and Mail*. Retrieved August 6, 2022, from https://www.theglobeandmail.com/life/relationships/indigenous-youth-bridging-generation-gap-with-elders-to-help-savecultures/article36068068/.

Solovey, M. (2001). Project Camelot and the 1960s epistemological revolution: Rethinking the politics-patronage-social science nexus. *Social Studies of Science, 31*(2), 171–206. https://doi.org/10.1177/0306312701031002003

Wexler, L. (2011). Intergenerational dialogue exchange and action: Introducing a community-based participatory approach to connect youth, adults and elders in an Alaskan native community. *International Journal of Qualitative Methods, 10*(3), 248–264. https://doi.org/10.1177/160940691101000305

Wilder, M., & Romero Lankao, P. (2006). Paradoxes of decentralization: Water reform and social implications in Mexico. *World Development, 34*(11), 1977–1995. https://doi.org/10.1016/j.worlddev.2005.11.026

8

"¡Amo kitlapanas tetl!": Heritage Language and the Defense Against Fracking in the Huasteca Potosina, Mexico

Elwira Dexter-Sobkowiak

Abbreviations

COCIHP	Coordinadora de Organizaciones Campesinas e Indígenas de la Huasteca Potosina (Coordinating Committee of the Rural and Indigenous Organizations of the Huasteca Potosina)
EIA	The U.S. Energy Information Administration
INEGI	Instituto Nacional de Estadística y Geografía (The National Institute of Statistics and Geography)
PEMEX	Petróleos Mexicanos (Mexican Petroleum)
PROCEDE	Programa de Certificación de Derechos Ejidales y Titulación de Solares (The Program for Certification of Ejido Rights and Titles of Lots)

E. Dexter-Sobkowiak (✉)
University of Warsaw, Warsaw, Poland
e-mail: elwira.sobkowiak@al.uw.edu.pl

When the last tree is cut
and the last fish killed,
the last river poisoned,
then you will see that you can't eat money.
Native American saying (Speake, 2009, p. 177)

Introduction

The aim of this chapter is to explore the relationship between language and environment loss and the use of heritage language in resistance related to the defense of natural resources. This link is shown by analyzing the case of the Nahua and Ténec communities in the Huasteca Potosina in northeastern Mexico, which between 2013 and 2018 were involved in a range of collective actions against the threat of gas and oil extraction in the area using hydraulic fracturing (fracking). One particularly interesting aspect of these activities was the spontaneous use of the Indigenous languages of the region—Nahuatl and Ténec—in spoken discourse as well as on banners, T-shirts, and in artwork expressing the opposition of the local population to this damaging technique of fossil fuel extraction. Against the backdrop of several centuries of gradual erosion of Indigenous language use, the new extension of use of Indigenous languages in this new domain is a surprising but very welcome development for language revitalization.

In this chapter, I explore the reasons behind the use of Nahuatl and Ténec in the anti-fracking movement, using insights from participant observation and from interviews with activists and members of local Indigenous communities. I demonstrate that the motivation to use Indigenous languages was not only to facilitate communication, but also to strengthen the Indigenous regional identity as a tool for solidarity building and Indigenous resistance against damaging development projects. I argue that the use of the two languages in—precisely—the opposition to fracking (and water pollution) is extremely significant, because water and other natural resources that would be destroyed as a result of fossil fuel exploitation, apart from being essential to survival, also form an integral part of the local

Indigenous worldview. Therefore, the danger of fracking is perceived as a threat not only to the environment but also to the Native cultures and languages. I postulate that this relation could be utilized in designing future sustainable development projects to support the integrity of Indigenous communities that would include the protection of natural resources, local cultural heritage, and also language revitalization initiatives.

The structure of this chapter is the following. First, I provide a brief description of the Huasteca, including its history, biodiversity, and linguistic situation. Then, I proceed to explain what fracking is and what the energy sector reform of 2014 in Mexico involved. I then move on to characterize the anti-fracking efforts that emerged in the Huasteca and I list activities in which Indigenous languages were used. What follows is a discussion in which I interpret the use of Nahuatl and Tének in the context of anti-fracking resistance, leading to my concluding remarks.

Huasteca Potosina—Geography, History, Demographic, and Linguistic Situation

The Huasteca is a geographical and cultural region in the northeastern part of Mexico. It extends from the Sierra Madre Oriental mountain range eastward to the Gulf of Mexico. Its northern border is the Sierra de Tamaulipas and its southern border is demarcated by the Cazones River. The Huasteca is a predominantly rural region and it includes the northeastern part of the state of Hidalgo, southeastern San Luis Potosí, southern Tamaulipas, northern Veracruz, northern Puebla, northeastern Querétaro and a very small portion of the state of Guanajuato. The Huasteca Potosina, which is the focus of this study, refers to the part of the Huasteca region located in the state of San Luis Potosí.

The Huasteca has a humid/subhumid climate, a varied landscape, and considerable biodiversity. Whereas the northern and eastern parts of the region are quite flat, the rest of it is mountainous. One of the characteristics of the Huasteca is the presence of numerous caves and *sótanos* (underground openings up to several hundred meters deep) located on the hillsides. The region is also abundant in water springs and rivers, which are often located in deep canyons. A common sight in the Huasteca is the waterfalls, the biggest one being Tamul. Most of the rivers empty into the Pánuco or the Cazones rivers. These, in turn, empty into the Gulf of Mexico. The Huasteca is covered mostly with tropical cloud forest with an abundance of wild animals and plants. The region is rich in wildlife including birds (e.g. macaws, owls, eagles), mammals (e.g. deer, jaguars, armadillos), and many species of reptiles, amphibians, and insects. The region is also abundant in agricultural crops such as maize, beans, squash, and various chili peppers. Apart from the traditional Mesoamerican crops, other crops introduced after the Spanish conquest are also harvested, including coffee, sugarcane, bananas, and citrus fruits.

The Huasteca Potosina region is also characterized by its substantial ethnolinguistic diversity. The two most widely spoken Indigenous languages today are the Western Huasteca variety of Nahuatl (Uto-Aztecan, ISO 639–3: nhw) and a Mayan language called Téneek (Teenek, Huastec, Wastek; Mayan, ISO 639–3: hus). A small part of the Huasteca located where the state of San Luis Potosí borders Querétaro is also home to the third ethnolinguistic group, namely the Pame people, known in their own language as the Xi'iuy (Otomanguean, ISO 639–3: pbs, pmq). The Spanish conquest of Mexico at the beginning of the sixteenth century introduced into the picture another language—Spanish—which gradually became the dominant language of Mexico and the Huasteca.

The European exploration of the Huasteca beginning in 1523 (Pérez Zevallos, 2010, p. 47) constitutes a turning point in the history of the region. The colonizers quickly recognized a potential economic opportunity in the rich natural resources and agricultural lands of the area, its large population, its established trade routes between the south and the north, and its proximity to the Gulf of Mexico (Palka, 2015). One of the initial consequences of the Spanish colonization was demographic change: a significant decrease in the Indigenous population as a result of deadly diseases brought by the Europeans to which they had no immunity. Depopulation also occurred as a consequence of slave trafficking. In 1525, as many as 10,000

Indigenous people of the Huasteca were sold and sent to the Antilles as slaves in exchange for cattle (Escobar Ohmstede, 1997, p. 34; Márquez, 1986, p. 203). The Spanish colonizers also imposed social and political changes, including new ways of using the land (Pérez Zevallos, 1983, p. 134), that included organizing the local population into *encomiendas*, in which Indigenous people were considered vassals of the Crown and were forced to pay tributes with agricultural produce or labor shifts. Due to evangelization, which was mostly performed in the Huasteca by the Augustinians, many Indigenous people had to abandon their original settlements and were forced into congregations (Pérez Zevallos, 1983, p. 99ff.). Another result of evangelization was the introduction of Christian elements into the traditional Indigenous cosmovision.

Although struggles regarding the division of land and natural resources between different *cabeceras* in the Huasteca had their origin in pre-Hispanic times (Pérez Zevallos, 1983, p. 93), they worsened after the arrival of the Spanish. The conflicts became especially intense in the eighteenth century due to political changes and new taxes introduced in New Spain by the Bourbons (Escobar Ohmstede, 1998, p. 55). Moreover, during the eighteenth century, territorial conflicts were provoked by the expansion of private properties required for increased agricultural production of such crops as sugarcane, maize, and animal farming (Escobar Ohmstede, 1997, p. 40ff.). It was during those times that the most fertile flatlands were often taken over by the owners of large *haciendas* and the Indigenous populations were forced to move to mountainous terrain (Escobar Ohmstede, 1998, p. 60ff; Pérez Zevallos, 1983, p. 61). Among several forms of Indigenous resistance to the exploitation of resources, violent protests were often regarded as the only way to recuperate lands that had been lost to non-Indigenous owners (Escobar Ohmstede, 1997, p. 65).[1]

Significant legal changes regarding land ownership occurred in Mexico after the Mexican Revolution (1910–1917), which redistributed the land from many large *haciendas* back into the possession of local

---

[1] An example of such conflict includes the event at the Vaquerías hacienda near Huayacocotla in Huasteca Veracruzana in 1784 when the local Otomí people rose against the decision of Real Audiencia to end their right to use land for wood collection (Escobar Ohmstede, 1997, p. 63).

Indigenous populations. As a result of these reforms, Indigenous land tenure arrangements are now based on membership in local communities in the form of *comunidades agrarias* or *ejidos*. Indigenous peoples have collective and communal ownership of the lands on which they reside, and both *ejidatarios* and *comuneros* make important decisions regarding land use in official community assemblies. The official titles to the land are held by the community and usually cannot be sold to outsiders, although, as a result of the PROCEDE land certification program in the 1990s, some of the indigenous lands were privatized and divided into individual parcels (see Smith et al. [2009] for more details).

Among many other consequences of the Spanish conquest of Mexico was the decline of the Indigenous languages. The replacement of Nahuatl with Spanish as a new lingua franca of Mexico was, however, rather gradual. In the early part of the colonial period Spanish was only spoken in the Huasteca by an extremely limited number of people (Escobar Ohmstede, 1998, p. 45), and Nahuatl and Tének were used in several official contexts. In particular, Nahuatl and Tének were used for evangelization, as confirmed by the existence of Christian doctrines and catechisms translated to both of these Indigenous languages (de la Cruz, 1571; de Quirós, 2013 [1711]; de Tapia Zenteno, 1753, 1767). This situation officially changed at the beginning of the seventeenth century when Philip IV imposed the policy of Castilianization, which lasted until the end of the colonial period (Fountain, 2015; Heath, 1972). Mexican independence in 1821 brought freedom from the Crown, but it actually resulted in the reinforcement of the use of Spanish in Mexico. Ironically, the colonial language was considered one of the unifying factors in building a new nation. The dominant position of Spanish was further consolidated in the second part of the twentieth century when it became the language of obligatory schooling. The project of alphabetization was carried out throughout the country, including remote Indigenous communities which prior to that time had little exposure to Spanish.

Language Domains in the Huasteca Potosina

Today, the two main ethnolinguistic groups in the Huasteca Potosina are the Nahuas and the Tének people. The former speak a local variety of Nahuatl, which, together with other closely related varieties of Nahuatl spoken in the Huasteca Veracruzana, Hidalguense, and Poblana, as well as in other states of Mexico, is the most widely spoken Indigenous language in the country. According to data provided by Instituto Nacional de Estadística y Geografía (INEGI, the National Institute of Statistics and Geography), the total number of speakers of Nahuatl in Mexico over three years old is 1,651,958 (INEGI, 2020). The principal municipalities of the state of San Luis Potosí where Nahuatl can be heard are Tamazunchale, Matlapa, Xilitla, Huehuetlán, Tancanhuitz, Tampacán, Coxcatlán, Axtla de Terrazas, San Martín Chalchicuautla, and Tampamolón Corona. The total number of Tének speakers over three years old is estimated to be 168,729 (INEGI, 2020). Almost 60% of all Tének speakers live in the state of San Luis Potosí in the following municipalities: Aquismón, Tancanhuitz, Tanlajás, Xilitla, San Antonio, San Vicente Tancuayalab, Ciudad Valles and Tampamolón Corona. The remaining 40% of Tének speakers live in the state of Veracruz. As a result of ongoing migration, there are also speakers of Nahuatl and Tének in such urban centers as Monterrey, Mexico City, and Guadalajara in Mexico, as well as in various parts of the United States. Almost all speakers of Indigenous languages of the Huasteca are bilingual with Spanish.

There are several forms of official recognition of Indigenous languages in Mexico, but their value is mainly symbolic. One of these is bilingual education offered in some Indigenous villages where preschool and primary school children receive reading and writing classes in their respective heritage language. This model does not educate through the medium of the Indigenous language, however, and its de facto aim is to gradually introduce Spanish, which becomes the sole language of education starting from secondary school. Nahuatl and Tének, as well as other Indigenous languages of Mexico, are recognized as national languages by the *Ley General de Derechos Lingüísticos de los Pueblos Indígenas* (General Law of Linguistic Rights of Indigenous Peoples,

2003). This law states that Mexico is a multilingual country in which the Native languages should be protected and promoted in official contexts and no person should be discriminated against for speaking their heritage language. However, the linguistic rights of the Indigenous peoples of the Huasteca (and other regions) are mostly not respected. As an example, during my stays in Xilitla between 2014 and 2020 it was almost impossible to communicate with local government officials, apart from those from the Department of Indigenous Affairs, in Nahuatl or Ténck. Similarly, medical centers, hospitals, and courts usually lack staff who would be able to attend Indigenous people in their Native languages.

The visibility of Nahuatl and Ténck in the Huasteca is also limited. In Xilitla, for instance, there are no official signs in Indigenous languages in the local government buildings or in the streets. However, a number of the local businesses, including several shops and hotels, have adopted names in Nahuatl, most likely for commercial reasons. Certain tourist attractions located near Indigenous communities sometimes have multilingual information panels in Spanish, English, and a respective Indigenous language. Although there are no newspapers published in Indigenous languages in the Huasteca, nor a TV station, there is a local radio station XEANT—Voz de la Huasteca (Voice of the Huasteca) which broadcasts portions of its programs in Nahuatl, Ténck, and Pame.

Spanish is taking over more and more domains, nevertheless, Indigenous languages are used in families—especially with older members—and in the local community life. In certain villages Nahuatl and Ténck are still employed in communication among neighbors and in community gatherings. In some households heritage languages are used for everyday communication, but there is a growing number of homes where Nahuatl or Ténck are solely used when speaking with the elderly. Many parents, despite being able to speak Nahuatl or Ténck themselves, decide not to transmit their heritage languages to their offspring and opt for Spanish instead. As reported by several Nahuatl and Ténck language activists I spoke to during my fieldwork in the Huasteca, the official language is associated with prosperity and a desired lifestyle, and many people consider the heritage languages backward and associate them with poverty and discrimination (see also Grin, 2007; Harbert, 2011; Ladefoged, 1992; Sallabank, 2013).

Indigenous Economies and Resources

The everyday life of the Nahua and the Ténék peoples revolves around the place where they live and rely on the resources that can be found in their community and nearby. Many Indigenous peoples have collective and communal ownership of lands on which they reside. Apart from small-scale farming they also practice foraging and they collect wild fruit, edible wild plants and mushrooms, and engage in animal hunting. Many families also keep domesticated animals including chickens, turkeys, ducks, pigs, rabbits, as well as dogs and cats. Occasionally they sell their crops, such as coffee or squash, the surplus of eggs, homemade local foods, or fruit gathered in the wild to buyers at the weekly market in the *altepetl*, the municipal capital. The region has practically no industry and unemployment is high. Lack of jobs has forced many Indigenous peoples to leave and find jobs in urban centers, particularly in Monterrey, Nuevo León. Those members of the family that stay in the Huasteca often rely on financial help from the relatives who work in cities or abroad. Many Indigenous peoples also receive modest, but regular social help from the government, such as the Social Inclusion Program PROSPERA.

The Nahuatl and Ténék speakers mostly reside in remote communities and their access to utilities is limited. Although most Indigenous communities in the Huasteca are connected to the electricity grid, few residents have landline phones or connectivity with a mobile phone network. It is unusual for an Indigenous community to be linked to a municipal water supply. In many cases the only sources of water are springs located outside the villages from which the residents fetch water to use for drinking, cooking, and other needs. Some communities have a local system of pipes that transport water from a nearby spring to the households. Many people also gather rainwater in plastic or concrete cisterns and use it for daily needs as well.

Natural resources, and water in particular, play a crucial role not only in daily survival but also in the spiritual and cultural life of Indigenous communities. Caves, mountains, and water sources are sacred places for both the Nahua and the Ténck people and the local landmarks are present in their oral tradition (see e.g. van't Hooft & Cerda Zepeda, 2003; Sobkowiak, 2016; Trejo Arenas, 2015). Several centuries of co-existence among the Ténck and the Nahua people in the Huasteca resulted in the emergence of similar themes in their oral traditions (including the origin of fire, the origin of corn, or the arrival of floods). These tales have been passed on from generation to generation in an intimate space of the family in order to secure the integral development of the speakers, which comprises "ecological integrity, life quality, and a management of resources" (van't Hooft, 2009, p. 67). The significance of water for the local people of the Huasteca is summarized by van't Hooft:

> Water paved the way for the present world to come into existence; time started in a new pace after the retrieval of the universal flood waters that wiped out the former world. Today, the cosmos, as created after the flood, is divided into different realms, one of which represents the water world called Apan, which runs across communities in the form of lakes, rivers and wells. [...] Water is also a core element in the foundation of a village, providing it with a right to exist. Huastecan Nahua communities are built around the sacred hill that harbours water as one of its main components. (van't Hooft, 2007, p. 253)

The rich natural resources of the Huasteca, and water especially, have attracted attention from external national and international companies that seek to use them in a number of projects. A current project involves, for example, a thermoelectric power plant near Tamazunchale, and potential projects include, among many, the construction of an aqueduct that would transport water from the Pánuco river to the arid area of Monterrey (the so-called "Monterrey VI" project), damming of the Coy river or development of the Ciudad Valles-Tamazunchale highway. Despite the fact that these initiatives directly affect the Indigenous communities and are located in proximity to Indigenous land and water sources, very rarely are the Indigenous people consulted in the authorization of the projects. What is more, many believe that the infrastructure that has been developed in the Huasteca so far is, in fact, related to plans to begin large-scale extraction of oil and gas in the future. This prospect generates fear among the local population as it would cause further contamination of air, water, and soil. These speculations became more justified in 2013 when the Mexican government initiated a reform of the energy industry that was aimed at opening Mexico's fossil fuel resources for exploration by foreign companies, including in the Huasteca. One of the methods of exploitation that was contemplated involved fracking.

Fracking in Mexico

Hydraulic fracturing, also known as fracking, is a technique used to recover gas trapped in shale, a type of sedimentary rock located deep below the surface of the earth. It involves horizontal drilling and pumping of large quantities of water mixed with a number of chemicals and sand at very high pressure into a well in order to cause fracturing of the shale rock and a release of natural gas.[2] Fracking can be a highly effective method of exploration and has even changed the United States from a gas importing to a gas exporting country (Bertram, 2019). The success of fracking in the United States has led to the promotion of this method of gas and oil exploitation across the world.

Mexico is one of the countries that are richest in shale gas. The U.S. Energy Information Administration (EIA) estimated in 2013 that Mexico's technically recoverable shale gas was in sixth place in the world and that its technically recoverable shale oil reserves were in eighth place in the world (EIA, 2013). These reserves are accumulated in marine-deposited, source-rock shales located along the onshore Gulf of Mexico region located in five geologic provinces that span over eleven Mexican states: Coahuila, Nuevo León, Tamaulipas, San Luis Potosí,

---

[2] While the term 'fracking' is typically associated with the use of the practice in order to recover shale deposits, in Mexico this term is often used in a broader meaning. It may refer to any drilling of wells so that conventional deposits of oil and gas, not only those located in shale rock, can be accessed (Pskowski, 2020).

Querétaro, Veracruz, Hidalgo, Puebla, Oaxaca, Tabasco, and Chiapas (see map in: EIA, 2013, Chapter 2, Part II, p. 1). One of the regions rich in shale resources is the Tampico Basin which extends partly into the Huasteca.

Mexico's shale resources have been explored since 2010 but it was not until 2013 that the Mexican national oil company, Pétroleos Mexicanos S.A. de C.V. (PEMEX), reported its first successful shale gas test well in the Burgos Basin (Alire García, 2013), where there are already approximately 3,500 active conventional gas wells (Pskowski, 2020). Subsequently, several other test wells were drilled by PEMEX (Burnett, 2015), but the Mexican shale reserves remain unexploited despite a very optimistic prognosis.

Developing of fracking in Mexico has, in fact, proven rather slow and problematic. One early obstacle involved a shortage of water, for example in the Burgos Basin located in arid northeastern Mexico (Godoy, 2013). In addition, many areas rich in shale gas are controlled by drug cartels which have demanded that foreign companies performing preliminary drilling pay for access to prospective well sites. Moreover, the history of theft of oil from pipelines has cost PEMEX huge losses (González, 2015) and is also highly discouraging for prospective foreign investors (Meyers, 2015). A more complicated problem involves the high costs of fracking and a lack of infrastructure that impedes PEMEX from performing the exploration on its own without the cooperation of foreign investors. Under the Mining Law of Mexico, the only company allowed to exploit and develop gas and oil resources was PEMEX. Although PEMEX was allowed to sign contracts with private companies, the existing regulations in Mexico did not guarantee oil companies to receive incentive payments based on a percent of production (PEMEX, 2012). Because of this, there was little incentive to develop fracking in Mexico under the existing legislation (Miroff & Brooth, 2013).

This problem could not be solved without a reform which would allow private and foreign investors to get more involved in Mexico's fossil fuel sector. Thus, the energy reform started in December 2013 with the Mexican Congress voting to amend the Mexican Constitution. In August 2014 the amendments were signed by President Enrique

Peña Nieto, and this step opened Mexico's fuel industry to private and foreign companies (Villers Negroponte, 2014). The next step occurred when the Comisión Nacional de Hidrocarburos (National Hydrocarbons Commission) announced in December 2014 the guidelines for bidding by private companies. The first stage, or the so-called *Ronda cero* (Round zero) involved PEMEX identifying the projects it was interested in pursuing. Subsequently, the auction of development blocks was announced, which comprised the so-called *Ronda uno* (Round one). Despite significant initial interest, only nine oil and gas auctions were conducted in Mexico and approximately 90 contracts for onshore and offshore work were awarded until July 2018 (Barrera, 2018). The election of Andrés Manuel López Obrador as the new president of Mexico in 2018 brought a temporary ban on fracking for three years (Bertram, 2019). Investors interested in Mexican oil and gas reserves are waiting for further administrative reforms, as well as more favorable economic conditions that would guarantee the security of their investment.

Controversies Regarding Fracking

The plans to develop fracking in Mexico are controversial for several reasons. Fracking is advertised as a modern and advanced method that can reduce Mexico's dependence on foreign natural gas imports, reduce the cost of petrol and domestic gas, create business opportunities, improve infrastructure, and offer much-needed employment. Many people remain unconvinced, however, because of the environmental impact fracking can have.

Fracking remains an extraction technique that has a number of environmental consequences. First of all, it involves the extraction of fossil fuel sources which are not renewable. Secondly, it uses excessive amounts of water and causes water pollution. Since the majority of water used for fracking comes from rivers and streams, this can contribute to drought in places where water is in short supply, such as in Coahuila in northeastern Mexico. Another problematic issue associated with fracking involves the use of harmful chemicals, some of which are linked to cancer and infertility in humans and loss of life among mammals and aquatic fauna. Moreover, the water used in fracking activities cannot be returned back into the environment until it is treated. Untreated post-fracking water poses a risk of contaminating groundwater. Fracking is also linked with the risk of earthquakes after large quantities of contaminated water have been injected into underground disposal wells after fracking activities are concluded. Because of the negative environmental consequences of fracking, many countries and localities have banned the use of this technique, including the states of New York, Maryland, and Vermont in the US, and Scotland and Wales in Europe (Meyers, 2015).

In Mexico, another controversy involves how fracking could affect the Indigenous peoples, their land, and their natural resources. They would be one of the most affected groups as many of the prospective fracking activities would be performed within or near Indigenous areas, especially in northern Veracruz and in the state of Puebla.[3] As I was told by Rogel del Rosal Valladaraes (p.c., March 20, 2020, Xilitla), an advocate for Indigenous human rights affiliated with Coordinadora de Organizaciones Campesinas e Indígenas de la Huasteca Potosina (COCIHP, Coordinating Committee of the Rural and Indigenous Organizations of the Huasteca Potosina), most Indigenous peoples who live outside the urban centers and rely on the local environment often associate fracking (as well as conventional methods of fossil fuel exploration) not with an opportunity for development, but rather with deception aimed at the exploitation of the Native land and its resources. This hesitancy is justified in view of a long history of external actors taking advantage of the poverty and lack of information among the Indigenous peoples. On numerous occasions Indigenous land was leased and local resources were misused without explaining the reasons or informing the local population about health or environmental hazards (Meyers, 2015). In Coahuila, Tamaulipas, and Nuevo León many farmers felt they were not consulted and were disrespected in the course of fracking activities (Morales Ramírez & Roux, 2018; Morales

---

[3] Consult maps available from CartoCrítica (2014) to see the exact location of the Indigenous groups in relation to the Round zero and Round one exploration plans.

Ramírez et al., 2018). During the experimental phase, wells were drilled without informing farmers about the details of hydraulic fracturing and its potential effects on the land, water, and health. Moreover, residents of many communities complained about problems with water supply and crops, as well as an increase in violence (Morales Ramírez & Roux, 2018). The health of many Indigenous communities has also been affected (Ramírez, 2015).

Taking into account the above-mentioned controversies and learning from the environmental devastation caused by fracking in the United States, many activists in Mexico became alarmed by the prospect of similar consequences in Mexico. Therefore, around the time when the energy reform was announced, a nationwide resistance movement against fracking started in Mexico. One of the regions where anti-fracking activities were undertaken was the Huasteca.[4]

Resistance Against Fracking in the Huasteca Potosina

The oil and gas reserves in the Huasteca Potosina are part of the Tampico Basin and this region is considered an attractive zone for fracking due to its rich water reserves. A map showing the Huasteca Potosina and the municipalities that could potentially be affected by fracking is presented in Fig. 8.1. According to this map, fracking was most likely to affect the municipalities of the region classified as *alto grado de exploración* ('high level of exploration') located in the eastern part of the region. According to the official data, in Round zero, PEMEX explored parts of only two municipalities in the state of San Luis Potosí, namely Ébano and Tamuín (CartoCrítica, 2014). However, according to Francisco Peña de Paz from El Colegio de San Luis Potosí (Valadez Rodríguez, 2018), the Mexican Secretaría de Energía(Secretariat of Energy) also authorized PEMEX to

---

[4] Other regions where anti-fracking activities took place include, e.g. the northern part of the state of Chiapas where the Zoque people expressed their opposition to fracking (Ledesma, 2021).

inspect other municipalities including Tanlajás, San Antonio, and San Vicente Tancuayalab.

The anti-fracking movement in the Huasteca Potosina began around the time when the Energy Reform was announced in 2013 and slowed only in 2018 when the new president announced a temporary ban on fracking in Mexico. The most active organization involved in informing local communities about the potential threat of fracking and their collective rights was COCIHP, which, previously had been involved in other activities aiming at the defense of Indigenous rights in the Huasteca Potosina. COCIHP was also one of approximately forty NGOs that worked together as the Alianza Mexicana contra el Fracking (Mexican Alliance Against Fracking). All of the anti-fracking events were planned in cooperation with the local authorities corresponding to the location of each event.

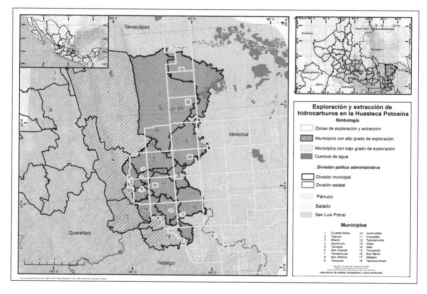

Fig. 8.1 Exploration of fossil fuels in the Huasteca Potosina. Copyright: Francisco Peña de Paz and Mario Hernández Infante, Laboratorio de Análisis Cartográfico y Socioambiental, El Colegio de San Luis

The anti-fracking movement in the Huasteca Potosina involved a number of activities including community-based participative research (i.e. studies of popular epistemology), creation of alternative reports, development of collective action, judicial activism, media-based activism, public campaigns including informative meetings, marches, and street protests, as well as artistic and creative actions such as painting. The biggest event took place on August 25, 2018, when as many as 4,200 people gathered in the village of Chimalaco in the municipality of Axtla de Terrazas (Hernández Borbolla, 2018). The core of the activities comprised open informative meetings organized in approximately 150 villages (Cravioto Lagos, p.c., June 30, 2021, phone interview) located across different municipalities of the Huasteca Potosina. The villages included both Indigenous Nahua or Tének communities, as well as rural *mestizo* communities. Some gatherings also occurred in local municipal capitals, including Xilitla in 2016. Often, in addition to the residents of the villages where meetings took place, the participants also included people from other communities who were transported free of charge to the location of the event. The activities quickly became eagerly awaited events with hundreds of participants present in each of them. The aim of these events, which generally lasted 4–6 hours, was to explain what fracking is, to raise awareness of its consequences as well as to discuss Indigenous peoples' rights in the face of the threat. The meetings included introductory sessions in which invited guests would talk and show slides about fracking, followed by work in groups and presentation of the results of group work. Throughout the events, posters with anti-fracking slogans were usually displayed. Another form of anti-fracking activities also involved street marches which took place in, among others, Tamazunchale, Tanlajás, and Xilitla.

The anti-fracking movement in the Huasteca also involved the development of collective actions and judicial activism.[5] First of all, in

---

[5] The movement arose in opposition against the violation of human rights to life, water, nutrition, health and a healthy environment. Its aim was also to protect a number of Indigenous rights: the right to land and territory, the right to be consulted, the right to information, the right to an autonomous and free choice and the right to defend themselves guaranteed by the Article 43 of the Constitution of the state of San Luis Potosí (COCIHP, 2017).

more than 100 villages (Cravioto Lagos, p.c., June 30, 2021, phone interview) general assemblies of the *ejidatarios* and community members declared their *ejidos* and communities *territorios libres del fracking* ('fracking-free zones'). What is more, the councils of 11 municipalities (Cravioto Lagos, p.c., June 30, 2021, phone interview) reached agreement and rejected fracking or other conventional fossil fuel extraction projects in their respective *Plan de Desarrollo Municipal* (The Municipal Development Plan) for the years 2018–2021. These municipalities included, among others, Ciudad Valles, Tamuín, Tanquián de Escobedo, San Antonio, Tanlajás, Tancanhuitz, Tamazunchale and Xilitla (del Rosal Valladares, p.c., March 20, 2020, Xilitla). One of the most radical decisions taken in Chimalaco in August 2018 involved increasing the existing number of community policemen to protect the residents from the Mexican army and the PEMEX staff in case they came to engage in the extraction of fossil fuels. Another activity of the anti-fracking movement in the Huasteca involved support for the creation of the *Ley General de Aguas* (General Law of Water) that would guarantee that water is not used to exploit hydrocarbons (Hernández Borbolla, 2018).

Use of Indigenous Languages in the Anti-Fracking Movement

One of the aspects of anti-fracking activities in the Huasteca involved the use of the Indigenous languages Nahuatl and Ténck. Domains of use for spoken Nahuatl and Ténck included simultaneous translations of experts' presentations (in Spanish) explaining what fracking is, a practice that normally initiated all informative meetings and workshops. Moreover, Nahuatl and Ténck were also used in discussions among participants during group work. In addition to spoken language, the two languages were also utilized in written form on banners, posters, T-shirts, and on a mural.

Several reasons were mentioned as the motivation behind using Indigenous languages in the movement. Francisco Cravioto Lagos, a consultant who took part in numerous gatherings in the region between

2014 and 2017, claims that Nahuatl and Ténet were spoken because of their practical communicative value. Use of everyday plain language allowed for clarification of difficult terminology related to fracking. In his words:

> It was mainly used to clarify terms, to ground concepts... It is a very complex technique ... and in the same way ... it is not very easy to explain it in Spanish. Well, it is not easy to explain it in Nahuatl or Ténet either, however [what] is being done is precisely to give the most basic explanation but not lacking the details about what the technique consists of and ... on many occasions the authorities and other colleagues supported us with translations.[6] (Cravioto Lagos, p.c., June 30, 2021, phone interview)

The need to come up with Native equivalents for the term 'fracking' arose early on in the resistance movement. The term was translated by Native speakers as *kitlapanas tetl* in Nahuatl and *an pok'ol tújub* in Ténet, which translates as 'breaking of the rock' in both languages. These terms later became part of the anti-fracking slogans *¡Amo kitlapanas tetl!* in Nahuatl and *¡Ibá ka pok'ow an tújub!* in Ténet, which both translate as 'No to fracking!'. These were frequently used on banners and T-shirts. The origin of the Nahuatl Native creation term for fracking is explained by a Native speaker of Nahuatl who was involved in coining the term *¡Amo kitlapanas tetl!*:

> In the region where we live, the majority speak Nahuatl, so in order to recognize the location [it was necessary that] a phrase should come out to defend the territory, our Nahuatl region, and from there the phrase that

---

6 This and the following quotes are English translations of the transcriptions of my interviews with the anti-fracking activists that were originally performed in Spanish. The original text: *Principalmente se usaba para aclarar los términos, para aterrizar conceptos... Es una técnica muy compleja pues, y de la misma forma, pues, no es muy fácil explicarlo en castellano. Pues, tampoco es fácil explicarlo en náhuatl o en tének, no obstante [lo] que se trata de hacer es precisamente dar la explicación más básica pero no carecida de cierto detalle sobre que es lo que consiste la técnica y en muchas ocasiones los autoridades y otros compañeros nos apoyaron con las traducciones.*

could be used to in defense of our territory was born: *¡Amo kitlapanas tetl!*[7] (FM, p.c., July 12, 2021, phone interview)

Del Rosal Valladares (p.c., March 20, 2020, Xilitla) remarks that the translation of the term 'fracking' to Indigenous languages is more literal and descriptive. The term itself conveys what fracking is. In the case of such languages as Spanish for example, which adopted the English term, the true meaning of the destructive technique is omitted. Del Rosal Valladares said:

> The translation of the meaning of fracking is essential . . . It is not a literal translation, let's say, it is – rather – a meaning that is directly related to the life, the present and the future of Indigenous peoples . . . It means a lot more to them – doesn't it? – and it has to do with life itself, with the sacred, in this case water, land and the pollution it would cause, and of course . . . the relationship it has with the peoples themselves, their future . . . they also like to be spoken to in their own language.[8] (del Rosal Valladares, p.c., March 20, 2020, Xilitla)

The Nahuatl version of the slogan also appeared on an anti-fracking mural in Uxtuapan, a Nahua community located in the municipality of Xilitla (see Fig. 8.2). In addition to the two official anti-fracking slogans, other creative anti-fracking messages written in Nahuatl and Ténck were coined and placed on the banners used in marches and meetings (see Fig. 8.3).

---

[7] Spanish: *En la region donde nosotros estamos en la mayor parte es náhuatl entonces para ubicarse bien que saliera una frase donde se defienda el territorio, donde es nuestra region náhuatl, y de allí nació la palabra que se podía usar como la defensa de nuestro territorio: ¡Amo kitlapanas tetl!*

[8] Spanish: *Es fundamental la traducción de significado del fracking... no es una traducción literal, digamos, es - más bien - un significado que tiene relación directa con la vida, con su presente y su futuro de los pueblos indígenas... a ellos le significa mucho más, ¿no? y tiene relación con la vida misma, con lo sagrado, en este caso el agua, tierra y la contaminación que provocaría, y por supuesto... la relación que tiene con el pueblo mismo, su futuro... además les gusta que se lo digan en su idioma.*

Fig. 8.2 Anti-fracking mural in Uxtuapan, municipality of Xilitla. Photograph by Elwira Dexter-Sobkowiak

Discussion

With regard to the use of Indigenous languages in the anti-fracking initiatives, several interesting observations can be made. In this part I aim to answer the questions of why languages were used in this type of resistance, what can be inferred from it, and what implications this phenomenon can have for the future of Nahuatl and Tének.

Although the reason for the use of Nahuatl and Tének in the anti-fracking movement that was most often mentioned in the interviews was simply to facilitate communication, there are several considerations that seem to contradict this explanation. The first counterargument is that it is now almost impossible to find speakers of Nahuatl and Tének who would not also speak Spanish. As the official language of Mexico, Spanish would be an obvious choice for a communication tool during events in which many participants

Fig. 8.3 Anti-fracking poster in Nahuatl (¡Ximanaui mo "a"! 'Defend your water!') and Spanish during a protest in Chimalaco in August 2018. Photograph: Sebastián Coronado, Facebook/everydaysanluispotosi, 2020

(including presenters) did not speak Nahuatl or Tének. On average, as many as 95% of speakers of Nahuatl and Tének in the Huasteca Potosina are bilingual in Spanish (INEGI, 2020).[9] What is more, the monolingual speakers are usually the elderly community members who tend to stay at home, so their participation in the events was very limited. Taking these facts into consideration, in theory it would be perfectly possible to use Spanish as the sole language of the anti-fracking activities.

---

[9] This percentage is calculated from the publicly available INEGI 2020 data, using the total number of speakers of Indigenous languages who also speak Spanish divided by the total number of speakers of Indigenous languages, totaled over the 18 municipalities of the Huasteca Potosina that have significant Nahuatl or Tének presence.

In our attempt to explain the reasons behind the use of Nahuatl and Tének in the movement, an important consideration involves the nature of the threat that kindled the resistance. Fracking became a strongly opposed project because of its potential consequences regarding water contamination. Water, as mentioned earlier, has a special significance for the Indigenous communities of the Huasteca not only because of its vital role in their everyday life, but also because it has a central role in the Indigenous cosmovision. This is shown, for example, by a Nahuatl oral tradition compilation project in 2015–2016 in which tales concerning the protection of water occurred especially frequently (Sobkowiak, 2016). One such story was *Iyoltsi atl* 'The heart of the water' which was written down by a secondary school student from Itztacapa in the municipality of Xilitla, who heard the tale from his grandparents. Essentially, it is a warning against external people who can come and 'stain' the water. The story, as follows, could almost be interpreted as a warning against fracking:

Iyoltsi atl

Nepa kampa nieua Itstakapa, kiampa kitokaxtijkej katli paya itstokej, pampa paya toktoya se ueyi kuauitl kampa amextok. Nochi maseualmej kitekiuiaj ni atl. Ni ueyi kuauitl itokax itstakuauitl. Panokej tonalmej ipan ni teyouali, inka kema tlantinenki atl, kiampa kijtouaj. Nopa ateno kipixtoya iyolo tlen tetl. Ni tetl tleya eltoya ipan atl, kiampa inka uaktinenki. Maseualmej kiapixtinenkej kiampa inka kiichtekisej.Se tonali inka aki kiapixki uan maseualmej tlen seyok teyouali panokej uan imoyajkej nopa atl. Kampa yajuantij itstokej kena atlamij, uan kiitakej inka aki tlajpia uajka kixtejkej iyoltsi atl. Kiuikakej kampa yajuantij itstokej uan kiampa inka kema tlami inia. Uan ama tojuantij tijpiaj tijkuitij ipan oksejko teyoualmej uan kiampa tijmaluisej toa. (José Hernández, 2016, p. 28)

The heart of the water

Itztacapa, the community where I was born, was named by its residents in honor of a huge tree that was planted next to a spring. Everybody needed water, the vital liquid. This tree bore the name of *itstakuauitl*.

They say that in ancient times the community never lacked water. The well had a heart of stone that floated on top of the water, preventing it from drying out. The villagers took much care of this treasure so it wouldn't be stolen. But one day no one looked after the well and people from another community came and made the water dirty. In their village the water had run out and they came here and when they saw that no one was looking after [the spring], they stole the heart of the water. They took it to where they live so they would never run out of water. Now we have to go to search for water in other communities and that's why we take care of our water. (translation by E. Dexter-Sobkowiak)

The threat of fracking and water contamination provoked a strong reaction among the local residents of the Huasteca because of the special place water has in the Indigenous cosmovision. They associate these environmental dangers with a threat to their cultural survival, of which local language is an integral part. The use of Nahuatl and Ténck in the social movement organized around resistance against fracking can be interpreted as a way of giving this movement a special distinction, and as means of strengthening community solidarity. In the story *Iyoltsi atl* the residents of Itztacapa became careless and stopped taking care of their water source and, as a result, people from another village could rob the stone. As the story demonstrates, solidarity among residents and care for the environment are crucial to ensure the survival of the Indigenous community. Speaking a local heritage language can therefore be also considered essential for achieving this goal.

The history of the Huasteca shows that environmental changes were indeed partly responsible for cultural decline and language loss in Nahua and Tének communities. In recent times, for example, climate change has provoked crop failures, such as coffee, which used to be a reliable source of income for many Nahua and Tének people. As a result, many individuals were forced to find employment in urban centers away from their original speech communities. Their new city life kept them away from their Indigenous traditions and their ancestral language, and they shifted to the mainstream *mestizo* culture and Spanish. The link between changes in local environment, including the disappearance of animal and plant species, and language displacement has also been noted in other Indigenous communities of Mexico and in other countries. As a result of deforestation, many of approximately 1000 languages of Papua New Guinea became threatened as the natural barriers that allowed local languages to develop in isolation were broken down (Loh & Harmon, 2014). Splintering of speech communities as a result of climate change was also reported for Sulawesi, Indonesia (Riehl, 2018).

In the case of many such dispersed language communities, the loss of their heritage languages caused identity problems. This occurred because local languages are often important components of identity and it is no different in the case of Nahuatl and Tének. It is certainly possible to be a Nahua without understanding or speaking Nahuatl, nevertheless, having a language used only with certain people is a powerful tool for connections and a sense of community. Few who have visited Nahua or Tének communities in the Huasteca would deny that those *lugareños* ('the locals') who speak the respective Indigenous language form part of a more tightly knit community and are less at odds with questions regarding their Indigenous identity. Therefore, I argue that instead of being simply a means of verbal communication, Nahuatl and Tének are a fundamental part of the Indigenous local identity.

The link between the heritage language, resistance against fracking, and the expression of local Indigenous identity was expressed directly in the anti-fracking mural in the Nahua village of Uxtuapan in the municipality of Xilitla (Fig. 8.2). The mural was created as a joint effort between the Tinta Negra collective and the Uxtuapan community members. It depicts the Aztec rain god Tlaloc protecting the local community and its resources (water springs, corn, coffee, and local fauna) against the danger of fracking. The design of the mural is, in fact, closely connected with an oral tradition of Uxtuapan (in English 'The cave of the water spring') whose residents have a special relationship with its local landscape (see e.g. the tale *Toteyouali Ostoapan* 'Our village Uxtuapan' in Sobkowiak, 2016, p. 47).

Another reason behind the use of Indigenous languages may involve the expression of solidarity in the face of external oppression and an instrument to mobilize the Indigenous population to fight for their collective rights.[10] Cravioto Lagos (p.c., June 30, 2021, phone interview) affirms that local languages are part of the Indigenous identity, becoming a resistance tool ("it is a unifying element, an element of identity and, of course, a seed of resistance").[11] Del Rosal Valladares (p.c., March 20, 2020, Xilitla) claims that the expressive value of Nahuatl and Tének is greater than that of Spanish. He claims that the Indigenous people are more impacted if the message is passed on to them in their respective ancestral language. As a consequence, they also feel more obliged to defend their land and resources:

> They translate the talk into their own language and of course people are much more convinced, very shocked about the subject. That makes them have a very strong reaction and get committed to the fight to defend the territory.[12] (del Rosal Valladares, p.c., March 20, 2020, Xilitla)

Cravioto Lagos (p.c., June 30, 2021, phone interview) perceives the ancestral languages not only as symbols of the fight against fracking, but also the fight against all big transnational neoliberal projects. He also regards the use of language as an integral part of the Indigenous mobilization in the Huasteca ("the use of those languages contributed to the process of popular mobilization").[13] An anonymous Nahuatl speaker (FM, p.c., July 12, 2021, phone interview) regards Nahuatl as an aid in the fight for the natural resources ("translate it into Nahuatl… is to defend what we have, [to defend] the nature").[14] Moreover, the presence of Nahuatl and Tének in the linguistic landscape is considered by him as something akin to a protective shield:

---

[10] The topic of use of language as a solidarity component is explored in more detail in e.g. Dołowy-Rybińska and Hornsby (2021).

[11] Spanish: *es un elemeno aglutinador, un elemento de identidad y, por supuesto, semilla de resistencia.*

[12] Spanish: *Traducen el discurso a su propio idioma y por supuesto la gente queda mucho más convencida, muy impactada demás del tema, eso los hace tener una reacción muy fuerte de compromiso con la lucha del defensa del territorio.*

[13] Spanish: *fue el alcance de nuestro contacto con estas lenguas en el proceso de movilización popular.*

[14] Spanish: *traducirlo al nahuatl… es para defender que no nos acaben lo que tenemos, lo que es la naturaleza.*

It is like a shield in which one defends that those who come from outside do not enter this place ... to destroy ... It is like a sign, so that they see that they cannot enter, [that] they cannot do harm, where one has the richness of the water...[15] (FM, p.c., July 12, 2021, phone interview)

Although the vitality of Indigenous languages in the Huasteca should not be regarded as stable or safe, their use in the anti-fracking movement certainly appears to be an expression of positive language attitudes and ideology. These, in turn, are of great importance in any potential language revitalization effort (Dołowy-Rybińska & Hornsby, 2021).[16] The use of Nahuatl and Tén, as means of oral communication, as well as increasing their visual presence on banners and on the Uxtuapan mural has, without doubt, contributed to creating positive language attitudes not only among the Indigenous peoples but also among their *mestizo* neighbors. In the case of the Uxtuapan mural, its location was chosen by the community members who decided to paint it on the wall of the community basketball court in the center of the village. This artwork quickly became the object of local pride also because it was publicized in various information portals and in social media.

Apart from the positive language attitudes that were generated in the anti-fracking movement in the Huasteca, another favorable aspect for language vitality is the creation of new domains of language use in public for Nahuatl and Tén.. This is an unexpected development since until then the two languages were mostly confined to the more intimate environments of home and the local community. Whereas the overall vitality of the two languages can be described as declining due to a decrease in the total number of speakers,[17] loss of intergenerational

---

[15] Spanish: *Es como un escudo en que uno defiende que no entren en este lugar, los que vienen así externas... para destruir... como una señalética, para que lo vean, que no pueden entrar, no pueden hacer daño, donde uno tiene la riqueza del agua...*

[16] As explained by Dołowy-Rybińska and Hornsby (2021), language attitudes operate at a conscious level, while language ideology operates at a sub-conscious level.

[17] According to data provided by INEGI, the total number of speakers of Nahuatl over three years old dropped by 4% between 2015 and 2020 (from 1,725,620 in 2015 to 1,651,958

language transmission, migration to urban centers, and shift to Spanish, an additional domain of use for spoken and written Nahuatl and Tének is a positive development.

In view of these findings, I propose that the perceived link between environmental damage and language and Indigenous culture loss should be explored and used in future language and culture revitalization projects. As seen, for example, in the Nahuatl oral tradition compilation project (Sobkowiak, 2016), revitalization initiatives are more welcomed if they make the participants more engaged with their local environment and themes they can relate to more easily, and which reflect an inherently relational perspective that is still present in local Indigenous ontology. I argue that the link between environment and language protection that can be perceived in the Huasteca can be regarded as a 'good idea' (Penfield, 2021, p. 46ff.) for a revitalization project. It may generate much-needed excitement, enthusiasm, and commitment to help the effort of reversing language loss that can be currently observed in the case of Nahuatl and Tének.

Concluding Remarks

The Indigenous residents of the Huasteca have a strong relationship with their land, environment, natural resources, and water in particular. These elements are central in their worldview and, along with the ancestral language they speak, form part of their Indigenous identity. The use of Nahuatl and Tének in the anti-fracking movement in the Huasteca demonstrates that there is a perceived relationship between environmental degradation and cultural decline including language loss. Despite the dwindling vitality of both languages, activities related to anti-fracking activism created another important domain of use for both of them. Moreover, Nahuatl and Tének were employed in the resistance not only for their purely communicative value, but also

---

in 2020 [INEGI, 2015, 2020]). The total number of speakers of Tének over three years old dropped by 3% between 2015 and 2020, from 173,765 in 2015 to 168,729 in 2020 (INEGI, 2015, 2020).

because of their greater potential for conveying emotions, strengthening Indigenous identity, building solidarity, and mobilizing against the threat of fracking. The presence of spoken Nahuatl and Ténec in public spaces during street marches and protests, as well as the increased visual presence of both languages in the linguistic landscape, contributed to improving local attitudes about these languages.

The changing language ideology relating to Indigenous languages in the Huasteca, as well as the link between the environment, language, and culture, should be explored and used in future language revitalization activities and also in broader sustainable development projects. An integral, inclusive, and sustainable development plan should involve the protection of Indigenous cultures and languages which are in a symbiotic relationship with the Huasteca's biodiversity and natural resources. Environmental protection should be planned along with cultural and linguistic revitalization, and language revitalization projects should, in turn, acknowledge the special relationship the speakers of Nahuatl and Tének have always had with their habitat. Although the situation of Nahuatl and Tének is still precarious because of the pressure from Spanish, the use of both languages in the anti-fracking movement is a phenomenon that allows us to feel more optimistic about future revitalization projects and the strengthening of the position of the two languages in the region.

Acknowledgements I would like to thank Justyna Olko and Cynthia Radding for inviting me to contribute to this volume. I also thank the participants of the UNC Chapel Hill Summer School in July of 2019 from whom I received many helpful comments and inspiration for this research. My contribution would be impossible without the insights I received from many people fighting for a better today and a better tomorrow for the Nahua and Tének people in the Huasteca. Because of the sensitivity of the topic of fracking and a possible danger mentioning individuals by name could bring upon them, many of them will need to remain anonymous. I will only list the *compañeros* who expressed their will to be mentioned. First and foremost, I would like to express my gratitude to Rogel del Rosal Valladares and Francisco Cravioto Lagos, who are both activists and experts on fracking in Mexico. I am also indebted to Paul and Benjamin who were extremely understanding and encouraging while I was working on this chapter. Part of my fieldwork in Mexico would be impossible without the financial help I received from RISE COLING, a European Union Horizon 2020 research and innovation program under the Marie Skłodowska-Curie grant agreement No. 778384, for which I am also extremely thankful.

References

Alianza Mexicana contra el Fracking. (2015). *Acción 2015 en la Huasteca Potosina.* https://nofrackingmexico.org/accion-2015-en-la-huasteca-potosina/

Alire Garcia, D. (2013, May 8). Mexico still far from tapping shale potential, minister says. *Reuters.* http://www.reuters.com/article/2013/05/08/mexico-shale-idUSL2N0DP29H20130508

Barrera, A. (2018, March 1). Mexico to offer first-ever shale blocks in Sept auction. *Reuters.* https://www.reuters.com/article/mexico-shale/update-1-mexico-to-offer-first-ever-shale-blocks-in-sept-auction-idUSL2N1QJ1XG

Bertram, R. (2019). Will fracking be banned in Mexico? *Energy Transition: The Global Energiewende.* https://energytransition.org/2019/04/will-fracking-be-banned-in-mexico/

Burnett, J. (2015, March 16). Excitement over Mexico's shale fizzles as reality sets in [Transcript]. *National Public Radio.* http://www.npr.org/sections/parallels/2015/03/16/393334733/excitement-over-mexicos-shale-play-fizzles-as-reality-sets-in

CartoCrítica. (2014, October 2). *Hidrocarburos: Ronda cero y Ronda uno.* http://cartocritica.org.mx/2014/hidrocarburos-ronda-cero-y-ronda-uno/

COCIHP. (2017). Los proyectos que amenazan a la Huasteca Potosina. *Slideshare.* https://www.slideshare.net/AIDA_Americas/los-megaproyectos-que-amenazan-la-huasteca-potosina

Cruz, J. de la (1571). *Doctrina Cristiana en la lengua Guasteca cõ la lengua castellana.* Casa de Pedro Ocharte.

Dołowy-Rybińska, N., & Hornsby, M. (2021). Attitudes and ideologies in language revitalisation. In J. Olko & J. Sallabank (Eds.), *Revitalizing endangered languages. A practical guide* (pp. 104–116). Cambridge University Press. https://doi.org/10.1017/9781108641142.008

EIA. (2013). *EIA/ARI World shale gas and shale oil resource assessment. Technically recoverable shale oil and shale gas resources: An assessment of 137 shale*

formations in 41 countries outside the United States. https://www.eia.gov/ana
lysis/studies/worldshalegas/pdf/overview.pdf

Escobar Ohmstede, A. (1997). Los pueblos indios en las Huastecas, México, 1750–1810: Formas para conservar y aumentar su territorio. *Colonial Latin American Historical Review, 6*(1), 31–68.

Escobar Ohmstede, A. (1998). *De la costa a la sierra: Las Huastecas, 1750–1900.* CIESAS, Instituto Nacional Indigenista.

Facebook/everydaysanluispotosi. (2020, August 24). https://www.facebook.com/everydaysanluispotosi/photos/bc.AbptCR4PDyr9232Vq7Jkp7XW
uV99LSkj8iRKE3bXnnWiXyu2gye1gpPx_VVTDTxBSxvYoEmAwMN
8oxTOmJd1fk-DtmpGqqbdM1GLEb7wYD1gBUtFz9bbHfGwPyXxUX
65a42-FmLgNCrH2jRS_KmIqsi-/2715363848723095/

Fountain, C. (2015). Transculturation, assimilation, and appropriation in the missionary representation of Nahuatl. In K. Zimmermann & B. Kellermeier-Rehbein (Eds.), *Colonialism and missionary linguistics* (pp. 177–198). Mouton de Gruyter.

Godoy, E. (2013, April 18). Mexico lacks water to frack for shale gas. *Inter Press Service.* http://www.ipsnews.net/2013/04/mexico-lacks-water-to-frack-for-shale-gas/

González, N. (2015, September 21). Roban 50 mdp al día a Pemex: Cifras de enero a agosto de 2015. *Excelsior.* https://www.excelsior.com.mx/nacional/2015/09/21/1046849

Grin, F. (2007). Economics and language policy. In M. Hellinger & A. Pauwels (Eds.), *Handbook of language and communication: Diversity and change* (pp. 271–297). Mouton de Gruyter.

Harbert, W. (2011) Endangered languages and economic development. In P. K. Austin & J. Sallabank (Eds.), *The Cambridge handbook of endangered languages* (pp. 403–422). Cambridge University Press. https://doi.org/10.1017/CBO9780511975981.020

Heath, S. B. (1972). *Telling tongues: Language policy in Mexico, colony to nation.* Teachers College Press.

Hernández Borbolla, M. (2018, August 30). La Huasteca, el paraíso natural mexicano en peligro por los devastadores efectos del "fracking". *RT.* https://actualidad.rt.com/actualidad/287008-huasteca-paraiso-natural-peligro-devastadores-fracking

Hooft, A. van't. (2007). *The ways of the water: A reconstruction of Huastecan Nahua society through its oral tradition.* Leiden University Press.

Hooft, A. van't. (2009). Indigenous oral traditions from the Huasteca, Mexico. *Rozenberg Quarterly. The Magazine* (pp. 63–78).

Hooft, A. van't, & Cerda Zepada, J. (2003). *Lo que relatan de antes. Kuentos tén'ek y nahuas de la Huasteca.* Ediciones del Programa de Desarrollo Cultural de la Huasteca.

INEGI. (2015). *Lenguas indígenas en México y hablantes (de 3 años y más) al 2015.* http://cuentame.inegi.org.mx/hipertexto/todas_lenguas.htm

INEGI. (2020). *Lenguas indígenas en México y hablantes (de 3 años y más), 2020.* http://cuentame.inegi.org.mx/hipertexto/todas_lenguas.htm

José Hernández, D. A. (2016). Iyoltsi atl. In E. Sobkowiak (Ed.), *Kamanaltlajtolmej Xilitlan* (pp. 28–29). Nauatlajtoli Xilitlan.

Ladefoged, P. (1992). Another view of endangered languages. *Language, 68,* 809–811.

Ledesma, F. (2021). El pueblo zoque contra bloques de hidrocaburos en el norte de Chiapas, México. *EJatlas. The Global Atlas of Environmental Justice.* https://ejatlas.org/conflict/lucha-del-pueblo-zoque-contra-bloques-de-hidrocaburos-mexico

Ley General de Derechos Lingüísticos de los Pueblos Indígenas. (2003). http://www.diputados.gob.mx/LeyesBiblio/pdf/257_200618.pdf

Loh, J., & Harmon, D. (2014). *Biocultural diversity: Threatened species, endangered languages.* WWF Netherlands.

Márquez, E. (1986). Tierra, clanes y política en la Huasteca Potosina (1797–1843). *Revista Mexicana de Sociología, 48*(1), 201–215.

Meyers, K. (2015, July 12). Fracking and poverty in Mexico. *The Borgen Project.* https://borgenproject.org/fracking-and-poverty-in-mexico/

Miroff, N., & Brooth, W. (2013, May 8). For Mexico's leader, energy reform is path to future. *The Washington Post.*

Morales Ramírez, D., Alvarado Lagunas, E., Picazzo Palencia, E., Tobías Jaramillo, R. (2018). Local perceptions of hydrocarbon exploitation operations in the northeastern region of Mexico. *Frontera Norte, 30*(60). https://doi.org/10.17428/rfn.v30i60.1139

Morales Ramírez, D., & Roux, R. (2018). Efectos esperados de la extracción de gas shale en el noreste de México: Un enfoque cualitativo. *Región y Sociedad, 30*(72). https://doi.org/10.22198/rys.2018.72.a378

Palka, J. W. (2015). The Huasteca in Mesoamerican studies. In K. A. Faust & K. N. Richter (Eds.), *The Huasteca: Culture, history, and interregional exchange* (pp. 214–224). University of Oklahoma Press.

PEMEX. (2012). *PEMEX: Exploration & production: Integrated contracts. Frequently asked questions.* http://contratos.pemex.com/en/Paginas/faq.aspx

Penfield, S. D. (2021). Planning a language revitalization project. In J. Olko & J. Sallabank (Eds.), *Revitalizing endangered languages: A practical guide*

(pp. 62–71). Cambridge University Press. https://doi.org/10.1017/978110
8641142.005

Pérez Zevallos, J. M. (1983). *La Huasteca en el siglo XVI: Fragmentación de los señoríos prehispánicos, organización social y tributo* [Escuela Nacional de Antropología e Historia, B.A. dissertation].

Pérez Zevallos, J. M. (2010). Las visitas como fuente de estudio del tributo y población de la Huasteca (siglo XVI). *Itinerarios, 12*, 41–64.

Pskowski, M. (2020). Mexico's fracking impasse. *NACLA. Reporting on the Americas since 1967*. https://nacla.org/news/2020/10/22/mexico-fracking-impasse

Quirós, S. B. de (2013). *Arte y vocabulario del idioma huasteco (1711)* (Edición crítica con anotaciones filológicas de Bernhard Hurch). Iberoamericana), Vervuert, Bonilla Artigas Editores.

Ramirez, E. (2015, July 12). Fracking: More than 900 wells are opened in six states of Mexico. *Contralínea*. https://contralinea.com.mx/english/fracking-900-wells-opened-states-mexico/

Riehl, A. (2018, November 26). The impact of climate change on language loss. *The Conversation*. https://theconversation.com/the-impact-of-climate-change-on-language-loss-105475

Sallabank, J. (2013). *Attitudes to endangered languages: Identities and policies*. Cambridge University Press. https://doi.org/10.1017/CBO9781139344166

Smith, D. A., Herlihy, P. H., Kelly, J. H., & Ramos Viera, A. (2009). The certification and privatization of Indigenous lands in Mexico. *Journal of Latin American Geography, 8*(2), 175–207. https://doi.org/10.1353/lag.0.0060

Sobkowiak, E. (Ed.) (2016). *Kamanaltlajtolmej Xilitlan. Narraciones en náhuatl de Xilitla*. Nauatlajtoli Xilitlan.

Speake, J. (Ed.) (2009). *The Oxford dictionary of proverbs* (5th ed.). Oxford University Press.

Tapia Zenteno, C. de (1753). *Arte novissima de lengua mexicana*. Viuda de D. Joseph Bernardo de Hogal.

Tapia Zenteno, C. de (1767). *Noticia de la lengua huasteca*. Imprenta de la Biblioteca Mexicana.

Trejo Arenas, R. M. (2015). *Yo soy Dhipak. Naná' pel in Dhipak*. Secretaría de Cultura, CONACULTA, FONCA, PECDA-SLP.

Valadez Rodríguez, A. (2018, August 6). Pemex alista fracking en la Huasteca Potosina sin consultar a pobladores. *La Jornada*. https://www.jornada.com.mx/2018/08/06/estados/025n1est

Villers Negroponte, D. (2014). Mexico's energy reforms become law. *Brookings*.

9

The Interrelation Between Language, History, and Traditional Ecological Knowledge Within the Nahuat-Pipil Context of El Salvador

Ebany Dohle

Introduction

This chapter looks at the interrelation between Traditional Ecological Knowledge (TEK), language, and historical events within the context of the Nahuat-Pipil language of El Salvador. It deals with what some refer to as Indigenous Knowledge (IK), Traditional Knowledge (TK), or Traditional Ecological Knowledge (TEK), domains that position knowledge within broader contexts and social systems. Consider the following (Brockman et al., 1997, p. n.p.):

> Traditional Ecological Knowledge is a body of knowledge built up by a group of people through generations of living in close contact with nature. Traditional Knowledge is cumulative and dynamic. It builds upon the historic experiences of a people and adapts to social, economic, environmental, spiritual and political change. The quantity and quality of Traditional Knowledge differs among community

E. Dohle (✉)
School of Oriental and African Studies, University of London, London, UK
e-mail: ebanyd_gcfn@outlook.com

J. Olko and C. Radding (eds.), *Living with Nature, Cherishing Language*,
https://doi.org/10.1007/978-3-031-38739-5_9

members according to their gender, age, social standing, profession and intellectual capabilities. While those concerned about biological diversity will be most interested in knowledge about the environment, this information must be understood in a manner which encompasses knowledge about the cultural, economic, political and spiritual relationships with the land.

Despite its name, TEK encompasses much more than traditional knowledge about the environment. It is adaptive, cumulative, and dynamic and is deeply rooted in the different facets of the social systems of its holders. It cannot be understood or applied without looking at the broader social, economic, historical, and political dimensions (Stevenson, 1996, p. 281), and thus it is not possible to understand how TEK is perceived, conceptualized, and embedded within cognitive processes by simply looking at how it is represented in language. Nevertheless, analyzing language can provide insights into the types of knowledge which are prioritized and deemed essential within a larger cultural context. If a plant is significant enough to be named, for example, and features prominently in daily life and local folklore, then it is deemed to be part of TEK.

The approaches and theories applied in this chapter are based on interactions with Indigenous people in western El Salvador from the towns of Santo Domingo de Guzmán, Nahuizalco, and Cuisnahuat where Nahuat-Pipil is most widely spoken. The research questions are a response to their specific request to conduct research on TEK, as 'We are losing this knowledge, and without it we cannot call ourselves Indigenous' (T. Pedro, personal communication, July 2012). This request led me to examine contemporary constructions of cultural identity in relation to the environment in accordance with a Nahuat-Pipil worldview as expressed in the Nahuat-Pipil language. Thus, the first premise of this chapter is that possession of TEK, is a key component of Indigenous identity for the Nahuat-Pipil of El Salvador. The interaction between TEK and Nahuat-Pipil experiences are examined by exploring the outside factors that have contributed to the collective accrual of traditional knowledge and its integration into Nahuat-Pipil culture, language, and construction of identity. These factors include historical events such as migration in Mesoamerica, the cultural, linguistic, and biological diversity of Central America and El Salvador specifically, as well as the major socio-political events that have affected speakers of Nahuat-Pipil.

Having established a baseline for understanding the motivations behind Indigenous interest in TEK, it is possible to then focus on the Nahuat-Pipil linguistic repertoire and how TEK is encoded within it. Thus, the investigation turns to the question of the ethnobiological categorization and classification of plants, how this is achieved by speakers of Nahuat-Pipil, and whether cognitive categorization strategies are reflected in the language itself. It is hypothesized that some cognitive categorization practices are reflected within the lexicon of the language, however this is not to say that all categorization strategies are lexically marked, as is demonstrated by instances of covert categorization.

The investigation then examines folk nomenclature of plants by presenting their internal linguistic composition. The investigation of plant names is used to further inform documentation efforts of Nahuat-Pipil by adding new focalized materials to the existing range of resources. The naming strategies used within Nahuat-Pipil are compared to Berlin's (1992) theories on nomenclature as it has been found that ethnobiological naming conventions in Nahuat-Pipil are in line with Berlin's theories (1972, 1973, 1992).

The theoretical framework and methods employed to collect data for this body of research are interdisciplinary and draw largely upon ethnobotany, anthropology, the collection of oral histories, and sociolinguistics, in addition to my core background as a linguist and language documenter. The concrete methods used for data collection were informed by anthropological (Davies, 2008; Hill, 2006) and ethnobiological approaches (Berlin, 1992; Martin, 2007; Puri, 2014a, 2014b, 2014c). These included the use of ethnographic observation, semi-structured and topic-led interviews as well as experimental methods (free listing, pile sorting, and forest and hill walks).

An underlying aspect of my investigations of semantic categories in Nahuat-Pipil is the idea that language and culture are intrinsically tied to each other (Sharifian, 2017). This is a notion that is long-studied and interest in the relationship between language and culture can be traced back as far as the eighteenth century. William von Humboldt (1767–1835), Franz Boas (1858–1942), Edward Sapir (1884–1939), and Benjamin Whorf (1897–1941) are prominent scholars who investigated the relationship between language, culture, and thought.[1] Within the Nahuat-Pipil context and this study, I view language and culture as two inter-related yet separate concepts in which evidence of the close relationship between humans and nature can be found. I analyze how language is used to refer to and name elements of nature. In turn, there is evidence of the way that nature can inspire metaphors that describe human concepts e.g. in Nahuat-Pipil *nushuchiu*, literal translation 'my flower' is used to refer to a female romantic partner. From a wider cultural perspective, we can see how trees and plants can take on specific characteristics and are prominent in local folklore and origin stories, as well as the complementary use of plants and herbs in spiritual practices. While there is evidence to support the importance of nature within Nahuat-Pipil culture, however, there are other aspects which need to be considered. Historical and political events, for example, have also contributed to the formation of Nahuat-Pipil worldview.

By seeking to listen to and understand the requests of the language and speech community, this chapter thus aims to investigate how TEK informs the construction of sociocultural identity through language use, and how TEK itself is cognitively, culturally, and linguistically encoded in Nahuat-Pipil. This approach raises a question of reflexivity for researchers: to what extent do they serve the needs of the community, and how can they conduct research that is beneficial? It is my belief that this approach of seeking a common ground between the motives of the 'researcher' and the 'community', which may be potentially but not necessarily different, more effectively includes Indigenous viewpoints and perspectives into academic research.

---

[1] There are many complexities and nuances in the study of the relationship between language and culture which will not be discussed in this chapter, however, consider reading an overview of the history of the study of language and culture in the US (Duranti, 2003) and the innovations developed by the subdiscipline of Cultural Linguistics (Sharifian, 2017) for more insights.

Brief Historical Context

Linguistic

In texts on Uto-Aztecan languages (Campbell, 1985; Karttunen, 1983), the term 'Nahua' is used to refer to all languages, dialects, and variants of the Nahuat subgroup, a group to which the Nahuat-Pipil language belongs. Internationally among academic circles, the Nahuat-Pipil language spoken in El Salvador is known as 'Pipil' and its corresponding ISO Code is [ppl]. Language speakers in El Salvador, however, reject the name 'Pipil' because they view it as a derrogatory term; they opt instead to use the name 'Nawat' or 'Nahuat'. The naming of the Nahuat-Pipil language is a complex matter given some of the socio-political relations between the Nahuat-Pipil Indigenous community in El Salvador and the Nahuatl communities in Mexico as well as its international presence and representation in academia (Arauz, 1960; Campbell, 1985; Hernandez Gonzalez, 2011; King, 2011; Lara-Martinez & McCallister, 2012; Schultze-Jena, 2014). However, in an attempt to decrease the confusion and remain faithful to the language name used by speakers themselves, I have opted to use a hybrid of the various naming conventions and refer to the [ppl] language as Nahuat-Pipil.

At the time of the first contact with Spanish colonial forces in 1525, El Salvador was home to at least three distinct Indigenous languages and cultures, the Nahuat-Pipil, the Cacaopera or Cacahuiles, and the Lenca or Potones (Lemus, 2011), all of which continue to exist in El Salvador. Of these three, Nahuat-Pipil has the most speakers, however, it is classified as being 'critically endangered' (Moseley, 2010) or 'nearly extinct' (Lewis et al., 2016), meaning the language is no longer being transmitted to younger generations and is therefore at risk of being lost. It is difficult to arrive at solid statistics on the numbers of speakers of Nahuat-Pipil in El Salvador, as there is much mistrust toward outsiders and researchers given the historical and ongoing violence (Dohle, 2020). The violent and turbulent history of El Salvador has given rise to an environment where it is difficult, if not dangerous, to conduct research and attain accurate and up-to-date demographics. This is exemplified by the observations of two scholars from the late twentieth century (Ching & Tilley, 1998, p. 123):

> Until recent years, historical research was difficult or impossible in El Salvador, a country that has presented daunting conditions for even the most diligent of researchers. Government archives were strictly off-limits until the late 1980s, travel was dangerous, and during the Civil War of 1980–1992 local suspicions of researchers were high.

Despite the passage of over 20 years since this statement was made, conditions have not improved. Violence has increased greatly, making travel around the country difficult, archives are largely inaccessible as natural disasters have destroyed the buildings in which they were housed, and the little funding for archive accessibility is often poorly managed and misplaced. Furthermore, negative attitudes toward the language and culture prevail, and I have observed a distrust from within Indigenous communities toward outsiders and researchers who may be perceived to benefit from power imbalances.

In terms of numbers, it is estimated that in 2008 there were approximately 200 speakers of Nahuat-Pipil (Censo Nacional, 2008); however, by 2013 numbers had increased to 300 (Censo Nacional, 2013), possibly due to increased instances of self-reporting. From my own field experience, having visited the sites where Nahuat-Pipil is spoken in El Salvador (Santo Domingo de Guzmán, Cuisnahuat, and Nahuizalco), I would estimate that the number of speakers could have been as high as 500 (2013), however, the recent COVID-19 pandemic paired with limited access to health services and transportation between rural and urban centers, has meant that the population of Nahuat-Pipil speakers has been particularly hard hit. It is currently estimated that there are between 100 and 200 native speakers of Nahuat-Pipil, or 'nahuahablantes', in El Salvador (W. Hernandez, personal communication December 2021, M. Ramirez, personal communication, December 2021),[2] although this number does not take into account new speakers of the language, or 'neohablantes'.

---

[2] These estimates have been provided by M. Ramirez, director of the *Casa de la Cultura*, a government office which oversees cultural and Indigenous affairs in Santo Domingo de

In the Nahuat-Pipil context, the terms 'neohablante' and 'nahuahablante' were coined by Anastacia Lopez, a Nahuat-Pipil speaker, during one of our discussions on what it means to be Indigenous. *Nahuahablante* is used to refer to the native speaker and *neohablante* refers to new speakers of Nahuat-Pipil. These terms are intended to highlight the differences between native speakers and new speakers who have in recent years begun visiting the older speakers in Santo Domingo de Guzmán. Anastacia Lopez does not have any formal training in language documentation, nor does she have any knowledge of the literature that has been written on language endangerment and speaker profiles. However, through her experience of working as a Nahuat-Pipil language teacher, in addition to her involvement in revitalization activities and as a member of the *nahuahablante* language community itself, she has coined the above terms to explain some of the linguistic trends taking place in the Salvadoran context. What follows is a binary description of speakers as perceived and described by native speakers of Nahuat-Pipil raised within a wider Nahuat-Pipil speaking community. This view, or classification of speakers does not consider the full complexity of the spectrum of different types of speakers. There is no reference, for example, to latent or semi-speakers who were raised within a Nahuat-Pipil speaking environment, have full comprehension, yet are not active speakers of the language. I suspect that the purpose of classifying speakers into a binary of *nahuahablantes* versus *neohablantes* is less about language proficiency and speaks more to the need to distinguish between perceived insiders and outsiders.

For the most part, *nahuahablantes* are generally of the age 50+ and all are bilingual Spanish speakers (Lemus, 2011). These speakers live in more rural settings and are often laborers. They usually have low levels of formal education. In contrast, the *neohablante* is younger, often in higher education, and tend to live in urban centers such as the capital city. Their occupations vary, but they have similar levels of formal education. All new speakers have obtained a university degree or are in the process of obtaining it.

Guzmán, and W. Hernandez, a Salvadoran Nahuat-Pipil researcher and activist who has been active for over 20 years.

Despite these differences, there is one commonality between the two groups: language. Both profiles are fluently bilingual in Spanish and Nahuat-Pipil. Although the order and manner of language acquisition may be distinct, both groups have a deep-seated interest and love for the Nahuat-Pipil language. For the *nahuahablante*, Nahuat-Pipil was acquired first, Spanish second. The acquisition of Spanish usually took place around the age of 7 upon entry into formal primary education, but in some instances, it took place when the person was in their early 20s. The reverse is the case for the *neohablante* whose first language is Spanish. They learn Nahuat-Pipil as young adults. The manner of acquisition is also very distinct. Whereas the *nahuahablante* has learned organically via immersion with caregivers and other available speakers, the *neohablante* has chosen to learn and is thus self-taught, usually by whatever means is available. Fortunately, thanks to the work of previous researchers, Nahuat-Pipil does have online resources[3] which facilitate self-study of the language (Alej, 2017; Hernandez Gonzalez, 2011; King, 2011, 2013, 2018), in addition to the documentation and descriptions of the language that have been carried out by previous linguists (Arauz, 1960; Campbell, 1985; Hernandez Gonzalez, 2011; King, 2011; Lara-Martinez & McCallister, 2012; Schultze-Jena, 2014). Furthermore, social media sites such as Facebook, YouTube, Instagram, and Twitter offer opportunities for interaction in the target language, which is of great benefit for new speakers wishing to practice their developing language skills.[4]

Given the different backgrounds of the two types of speakers described, it is not unusual to hypothesize that the two will have different outlooks on what it means to be Indigenous. For a *nahuahablante*, material poverty, traditional knowledge, and direct links to ancestry are all integral to the concept of Indigeneity (Dohle, 2020). There is a notion within the community of *nahuahablantes* that you cannot be Indigenous if the link between generations has been cut. In other words, the oral transmission of traditional knowledge, language,

---

[3] www.tushik.org (Accessed: 15/02/2018); https://www.youtube.com/channel/UCbYqsaNZAzdRq94OfRO4odQ/feed (Accessed: 15/02/2018).

[4] https://www.facebook.com/pg/NawatElSalvador (Accessed: 15/02/2018); https://www.facebook.com/Tzunhejekat (Accessed: 15/02/2018).

culture, behavior and shared experience is an integral aspect of Indigeneity,[5] something *neohablantes* cannot learn and absorb into their own identities from books alone.

It is important to acknowledge that there is a broader continuum of speakers who cannot be easily classified as being a nahuahablante or a neohablante, but who nevertheless are carriers of Indigenous social and cultural values and belief systems. As was previously mentioned, it is likely that the need to create distinctions between native speakers and new speakers has nothing to do with judging language proficiency, but rather has more to do with the needs arising from within the Nahuat-Pipil speaker community to highlight differences between the insider and the outsider, or the Indigenous and non-Indigenous speaker. Considering this, to better understand the factors that contribute to the construction of Indigenous identity, it is important to look at the collective history of speakers of Nahuat-Pipil and how they came to reside in El Salvador.

Migration

According to classic ethnohistoric research, Nahua populations first came to settle in the geographical area that is now El Salvador in Central America somewhere between 1000 and 1100 AD, after 300 years of constant migration in the southern regions of Mexico near the modern states of Veracruz and Chiapas (Arauz, 1960), although some scholars place Nahua migration at around the year 900 AD (Escalante Arce, 2004). Arauz (1960) argued that during this time contact with Mayan languages in addition to Nahua technological advancements in agriculture accelerated the process of language evolution and contributed to its separation from classical Nahuatl.

---

[5] It is important to note that there are other types of speakers that this article will not touch upon such as dormant or semi-speakers. This is not because the author does not deem such speakers to lack importance, rather because it was not possible to conduct research with these types of speakers for a variety of reasons. It is my belief that to gain a more complete understanding of the linguistic situation and notions of indigenous identity, other types of speakers should also be consulted.

More recent findings have pointed to the presence of Nahuatl speakers in the southern Maya region, which includes Guatemala, El Salvador's northern neighbor, meaning that Nahua influences in the region were likely to be present much earlier than previously thought (Hill, 2001; Macri & Looper, 2003). In terms of the evolution of the language, however, the time its speakers spent in migration is more important than the discrepancy in the dates. I would also propose that changing geographies and environment would have had an impact on the language and that the discovery of new plants and animals would lead to the early creation of neologisms.

Language and Biodiversity

Central America is home to the Mesoamerican Biodiversity Hotspot which extends from Southern Mexico to the Panama canal (CEPF, 2016). It is ranked second among the 25 richest and most threatened biodiversity hotspots of the world (Olivet & Asquith, 2005), outranked only by the Andean mountains. North and South America were initially two separate landmasses, thus the flora and fauna of the two continents evolved separately from each other. The narrow strip of land which now connects the two is highly diverse, representing the convergence of two biogeographic regions. The geography of Central America includes highlands (mountains and volcanoes) and lowlands (swamps, coasts, grasslands, and valleys) which simultaneously provide the perfect conditions for isolation and speciation as well as movement and migration (CEPF, 2016).

If we consider the figures, approximately 17% of the roughly 17,000 plant species found in the Mesoamerican Biodiversity Hotspot are endemic species unique to this landmass as is illustrated further in Table 9.1. This means that knowledge of the uses, treatment, and possible cultivation of the nearly 3000 endemic species will most likely be held by the people living in this area. In relation to the languages spoken in Central America, it is likely that such knowledge is encoded within the communication systems of distinct language groups. Furthermore, plant naming conventions within a language can provide

Table 9.1 Diversity and endemism of mesoamerican flora and fauna (CEPF, 2016)

	Taxonomic group	Nr. species	Nr. endemic species	Endemism %
1	Plants	17,000	2941	17.3
2	Mammals	440	66	15.0
3	Birds	1113	208	18.7
4	Reptiles	692	240	34.7
5	Amphibians	555	358	64.5
6	Freshwater fish	509	340	66.8

further insights into the mobility and migration patterns of a given population (Berlin, 1972). For example, by looking at the prevalence of certain plant names across different languages and cultures in the Americas, it is possible to see how the Nahua cultures expanded across Mesoamerica reaching as far south as Peru on the South American continent (Escalante Arce, 2004).

Mesoamerica has a range of geographical terrains: from the arid deserts of Northern Mexico to the rich and fertile soils of the volcanic valleys of El Salvador. Mexican terrain where the Nahua reside is arid and desert-like, whereas Salvadoran terrain is mostly composed of rich humid rainforests and volcanoes. Temperature-wise, central and Northern Mexico are 10 degrees Celsius cooler than El Salvador. These differences of terrain, altitude, temperature, and humidity mean that the flora and fauna are also different.

Considering these differences in terrain, it is worthwhile to consider how movement across different geographical realities would affect human behavior, language, and culture. As communities of speakers migrate south and their physical environment changes, the language used to talk about and describe their natural surroundings must be adapted to reflect the new geographical reality (Berlin, 1972, 1973). Therefore, an object may no longer be 'X' it may be 'X-like' or 'Xish'. Furthermore, changes in the natural environment can also result in changes to the way it and its subspecies are conceptualized.

Biological and Linguistic Diversity

In line with the claim that there is a positive correlation between biological and cultural diversity (Gorenflo et al., 2012; Nettle & Romaine, 2000; Stepp, 2015), Mesoamerica is also a hotspot of linguistic diversity. Furthering the discussion around the link between language, culture, and biodiversity, is the view that there is not only a correlation but rather an interdependence between the three concepts (Terralingua, 2017). The link between humans and nature is seen to have evolved over time, and collective knowledge spanning generations offer a unique insight into the mutually beneficial relationship between people and nature in some contexts. The physical aspects of geographical terrain that allow for a wealth of biodiversity are the same aspects that encourage the development of a high density of linguistic diversity.

Through improved availability of the geographic distribution of language and biodiversity data, it is possible to see that there is in fact a positive relation between the two (Gorenflo et al., 2012). Of the estimated 5000–7000 languages spoken around the world, 3202 are found in 35 of the 50 biodiversity hotspots. The biodiversity hotspots with particularly high linguistic diversity include the East Melanesia Islands, the Guinea Forest of West Africa, Indo-Burma, Mesoamerica, and Wallacea, with each acting as a home for more than 250 languages. In terms of language endangerment, 50–90% are endangered, 85% have less than 100,000 speakers, 52.5% have less than 10,000 speakers and roughly 4000 languages have not been adequately described (Nettle & Romaine, 2000).

The Living Planet Index demonstrates trends in the populations of several thousand vertebrate species worldwide and demonstrates an overall decline in biodiversity of 30% in the last four decades (Loh et al., 2005). The Index of Linguistic diversity shows the same rate of decline over the same period worldwide; however, at a regional level, it demonstrates linguistic loss of as high as 60% in the Americas (Harmon & Loh, 2010; Loh, 2017). In the top 5 biodiversity areas there are 1622 languages, and 2166 languages in biodiversity hotspots are endemic to individual regions (Gorenflo et al., 2012).

Major Political and Social Events

Although El Salvador is situated in the center of the Mesoamerican Biodiversity Hotspot, an area that has nearly 300 languages, it is the widespread belief of the Salvadoran population that it is itself completely homogenous (Lara Martínez, 2006). During my fieldwork in 2016, I observed that those who were not directly involved with or worked on Indigenous issues were unaware of the fact that separate cultures and languages continue to co-exist within the country's political borders (Dohle, 2020). Although the Nahuat-Pipil language is still spoken and has in recent years gained more visibility in the media (2014a, 2014b; Miranda, 2016), the lack of awareness of the cultural diversity of El Salvador can be found at different levels, from the personal to the institutional, and this is largely a result of the political and social history of the country.

It is estimated that before the genocide of 1932, or La Matanza, there were approximately 300,000 people who identified as and lived as Indigenous people, 25,000 of whom were speakers of Nahuat-Pipil (Gould & Lauria-Santiago, 2014). Scholars disagree on the exact figures, however, it is estimated that somewhere between 5000 and 45,000 Indigenous people of different ethnicities were killed during the Indigenous uprising and subsequent genocide of 1932 (Delugan, 2013; Gould & Lauria-Santiago, 2014; Hernandez, 2017; Lara Martinez, 2006). This genocide arose due to land disputes, a result of an agricultural reform which sought to privatize the communal lands which were used by the Indigenous people to subsist (A. Lauria-Santiago, 1999a, 1999b; Seligson, 1995) The privatization of communal lands meant that Indigenous people no longer had access to land on which to grow and cultivate their crops, and needed to resort to their knowledge of the natural world to source food, medicine, and building materials in order to survive. Moreover, shortly after the genocide, the government imposed a cultural and educational reform that prohibited the use of any markers of Indigeneity such as language, dress, or the use and practice of Indigenous belief systems within public space (Gould & Lauria-Santiago, 2014). In turn, the educational reform sought to replace existing knowledge systems with ones that were more aligned with the ideologies present within the governing bodies. As a result, survivors of the genocide were faced with the decision to either flee or assimilate and erase their connection to their Indigenous identity.

The Civil War (1980–1992) further damaged the delicate balance between the oppressors and the oppressed, and a further 8000 lives were claimed during the 12 years of conflict (Gould & Lauria-Santiago, 2014). It officially began in January 1981, when the FMLN launched their military offensive marking the beginning of the end of the military regime which had ruled the country for most of the twentieth century (Lindo-Fuentes et al., 2012). I would argue, however, that tension had been brewing since 1932.

In 1968, another educational reform was initiated partly as a result of La Matanza resulted in horizontal mixing between social groups and a freer flow of information and ideas (Lindo-Fuentes et al., 2012). Tensions began to build in the following decade as large numbers of youth from Indigenous settlements such as Santo Domingo de Guzmán, Nahuizalco and Cuisnahuat began joining the political party which was to become the leftist Frente Farabundo Martí para la Liberación Nacional (FMLN)[6] (Gould & Lauria-Santiago, 2014). In the late 1970s, guerrilla fighters began to assassinate top government officials and in response, those who promoted national dialogue on education reform were targeted by pro-government paramilitaries, death squads who acted in opposition to the FMLN and their allies. These increasing tensions and assassinations of countless peasants, teachers, office workers, intellectuals, politicians, ministers, and civilian bystanders, in addition to the deaths of guerrilla fighters and professional soldiers, marked the start of the Civil War.

A direct result of the violence of El Salvador's history has been the conditioning of the Salvadoran population to believe that a homogenous cultural identity exists throughout the entirety of the political territory (Delugan, 2013; Tilley, 2002). The perception is that this homogenous culture is composed of mestizos and ladinos only, and

---

[6] Translated as the Farabundo Martí Front for National Liberation (Gould & Lauria-Santiago, 2014).

disregards the existence of Indigenous groups (Lara Martinez, 2006). Although demeaning attitudes toward Indigenous people already existed before 1932, the view of homogeneity was reinforced when the government decided to 'restore order' in 1932 by taking military action against Indigenous groups of farmers. The result is that to this day the majority of the population in rural and urban settings believe that Indigenous groups were completely wiped out during this time, a view reinforced by the indifferent attitudes that government policies adopted toward Indigenous groups after La Matanza (Lara Martinez, 2006, p. 9).

I have witnessed a significant change since 2013, in which awareness-raising efforts have been successful, and members of the public are slowly being sensitized to Indigenous issues and the cultural diversity of the country. This observation is based on my own engagement in awareness-raising activities, some of which have resulted in concrete action such as the integration of Nahuat-Pipil into the national curriculum by the Ministry of Education (MINED) (2019; Sibrián, 2019). Further efforts are also being made by young activists with a legal background to ensure the state provides legal recognition of the Indigenous names of towns and villages with a large presence of native language speakers.

Today, the largest factor which limits access to any rural community is the high crime rate in El Salvador, which is primarily caused by criminal activity and territorial disputes between two rival gangs: MS-13 and Barrio 18. From 2009 to 2012, gang violence, including shootings, kidnapping and armed robbery, and the trade of drug, sex, and weapons, were most prevalent in rural areas (Valencia, 2015) such as the state of Sonsonate, the location of my primary field-site. Since 2013, due to the government's change of strategy toward organized crime, there has been a shift in the manifestation of gang violence. In protest of new governmental policies such as the 'mano dura'[7] (Hume, 2007), gangs and their members have focused their activities on urban areas. By the end of 2015, they were actively attacking all city, government, and police offices (Labrador & Rauda Zablah, 2015;

---

[7] 'Mano dura', or the 'iron fist' refers to the Salvadoran government's strict policies and approaches to tackling gang-related crime and violence.

Valencia, 2015) resulting in the spread of panic across many populations and social groups as death tolls surpassed those of the Civil War era (1980–1992) (ContraPunto, 2015; Watts, 2015).

The violent history of El Salvador, along with the crime and violence it faces today, has meant that Indigenous communities have lived with social marginalization, reinforced by violent attitudes and demeaning policies towards their language, knowledge and culture for many generations. These attitudes and policies have affected Indigenous communities' access to inclusive spaces, health care systems, education, and opportunities to enter the labor market. As a result of this, communities who still identify as Indigenous have needed to be self-reliant in order to survive. Those who have maintained their traditional languages have also been able to maintain the traditional knowledge encoded in the language, knowledge that has been integral to their survival. The following section will examine how TEK is encoded in language.

The Encoding of TEK in Nahuat-Pipil

When it comes to TEK and its position within the context of endangered languages, it is observed that the parts of the lexicon dealing with TEK are the most threatened (Si, 2011). This heavily affects Nahuat-Pipil as well. For example, when children are pushed to enter the Spanish-medium education system in El Salvador they are taught knowledge systems that are greatly influenced by Western ideologies and are exposed to Western-based attitudes towards learning and priorities of what should be learned. As a result, a Nahuat-Pipil child's exposure to the oral transmission of Nahuat-Pipil TEK is greatly minimized. This is also seen when young adults migrate to larger cities, or abroad to Mexico, the United States, Canada, and beyond in search of work. As their environment changes, along with the type of work and the language used to conduct this work, they are likely to lose much of their pre-existing knowledge as it might be considered less relevant or even backward in their new context (Gould & Lauria-Santiago, 2014; Little, 2008). Therefore, in the interest of the preservation of traditional knowledge and minority and endangered languages, it is of vital importance to create accessible materials to document TEK, when possible, for the benefit of future generations.

Here we consider how TEK is encoded in language by looking at categorization and classification, as well as folk nomenclature. To better understand these, it is important to define the concepts:

- *Categorization* refers to the human mind's need to lump information together in order to create order and be able to engage with the world in an efficient manner. It is important to note that the act of categorizing is innate, however, the trends observed within what is categorized are often cultural and learned.
- *Classification* refers to the hierarchical organization of knowledge, and within this chapter, this is applied to ethnobotanical knowledge.
- *Folk nomenclature* refers to the naming systems applied to a group of objects in the physical world as developed within a specific culture. This often develops as a result of a natural need to categorize, classify, and name objects of significance within their physical surrounding. This might refer to functional, cultural, or spiritual importance.
- When we speak about *Ethnobiology* we are referring to the systems of terminology relating to the biological world as employed by individuals who are not trained according to Western scientific conventions.

The major impetus for cognitively oriented ethnobiological research can be traced back to Harold Conklin's influential doctoral dissertation based on research in the Philippines (Conklin, 1954, 1972, 1980). It was the first ethnographically and botanically sophisticated description of a full ethnobotanical system of classification for a nonliterate society (Berlin, 1992, p. 4), and his research stressed the importance of discovering Indigenous categories for plants and their conceptual relationships to one another as complete, self-contained systems.

The work presented here is based on Brent Berlin's (1992, 1993) models on ethnobiological categorization, classification, and folk nomenclature, although it also draws on the work of Conklin (1954). Berlin's (1992) work is based on 20 years of ethnobiological fieldwork and research, and it is seminal for the topics of taxonomy, classification, and nomenclature in ethnobiology. Berlin argues for the human mind's innate need to categorize, which creates the necessary conditions to understand the natural system of the world around us and classify plants and animals accordingly. He emphasizes how the mind acts as an objective computer, creating categories on the basis of objects' morphological similarities and differences (Berlin, 1973, p. 260).

Categorization and Classification

The human need to deal with any input or information received in order to make sense of the world gives rise to the grouping of information into categories, thereby allowing for the creation of order in an otherwise complex and data-rich environment. Without this ability individuals would be unable to handle the complexity of daily interactions with the world (Aitchison, 2004; Markman, 1989). Therefore categorization is seen to be a fundamental cognitive mechanism (Polzenhagen & Xia, 2015) that simplifies the individual's experience of the environment by grouping information and thereby reducing the load on memory and facilitating the efficient storage and retrieval of information from the cognitive system (Jacob, 2004, p. 518).

The process of categorization involves dividing information into groups, and members of each group can share similarities and features (Jacob, 2004). A set of features associated with a given category can be both context-dependent and context-independent (Barsalou, 1987, p. n.p.). Context-dependent features are only relevant within a given context, such as in relative statements of truth. For example, 15 °C may be regarded as cold in the summer but warm in the winter. Context-independent features on the other hand provide information on a category that is relevant across various contexts, e.g. in the way that 'fire' connotes light, heat, and energy.

Some view categorization as being intrinsically cultural, a view represented by Glushko et al. (2008, p. 129):

Categorization research focuses on the acquisition and use of categories shared by a culture and associated with language - what we call 'cultural categorization'. Cultural categories exist for objects, events, mental states, properties, relations, and other components of experience. Typically, these categories are acquired through normal exposure to caregivers and culture with little explicit instruction.

Considering the above quote, semantic categories are not explicitly transmitted and taught, rather they seem to be absorbed organically based on the shared context, behavior of speakers, and treatment of language. We can therefore view semantic categorization as contextually and culturally dependent. Lexical categorization on the other hand refers to the way that categories are marked within language, usually by way of a classification system. The relationship between the two concepts is such that context, or semantic meaning, can dictate the way that information is lexically marked or categorized by means of classification.

In ethnobiology, the hierarchical grouping of information is applied to the natural world. Berlin (1992, p. 2) argues that:

The observed structural and substantive typological regularities found among systems of ethnobiological classification of traditional peoples from many different parts of the world can be best explained in terms of human being's similar perceptual and largely unconscious appreciation of the natural affinities among groupings of plants and animals in their environment - groupings that are recognised and named quite independently of their actual or potential usefulness or symbolic significance to humans.

Berlin draws a line between those who view reality as a social and cultural construct unhindered by the parameters of the physical world, and those who recognize the seemingly innate ability of humans to recognize 'natural affinities' among the living entities of our biodiverse planet: plants and animals. These affinities result in taxonomical grouping by means of classification, which is reflected in language. In the case of the natural world, taxonomy refers to the hierarchical categories which are applied to plants and animals, whereas classification is how categories of plants, animals, objects, or information in general are grouped, based on features such as size, shape, or color. Within Ethnobiology, classification is the means by which humans navigate, utilize, and survive in the natural world, as 'before humans can begin to utilise the biological resources of the local environment, they must first be classified' (Berlin, 1992, p. 5). Thus, '[c]lassification… is an absolute and minimal requirement of being or staying alive' (Simpson, 1961, p. 3).

Based on the above assumptions, Berlin's Principles of Nomenclature and Principles of Categorization reflect the presumed natural system observed by humans. This, and a discussion on how individual variation and evolutionary processes are factors to be considered within classification, are presented in the following section.

Basic Principles of Classification

Berlin applied his analysis and theoretical framework developed for the study of color terms (Berlin & Kay, 1969) to folk taxonomies of living organisms seeking to identify universal patterns of categorization and lexical encoding. He noted that cross-culturally, ethnolinguistic folk classification can be organized into 5–6 categories arranged hierarchically in taxonomical groups, much like those identified by Western scientific methods for identification. Berlin's folk classifiers do not correspond exactly to scientific categories, but the following tables provide an approximation of the correspondence between the two (Tables 9.2 and 9.3):

Linnaean categories, one of the scientific systems of botanical hierarchies, include seven obligatory ranks, in addition to other ranks which are optional and are used as needed. These ranks classify all living things, making folk taxonomies very shallow by comparison (Hunn & Brown, 2011, p. 326). Berlin (1992) identifies six universal ranks, of which generic taxa is identified as being highly salient meaning they are the first terms to be encountered in ethnobiological investigations because they stand out in the landscape.

Table 9.2 Scientific classification of plants

Scientific division	Example
Kingdom	*Plantae*
Class	–
Order	*Fagales*
Family	*Fagaceae*
Genus	*Quercus*, Oak
Species	*Q. ilex*
Varietal	*Q. ilex* subsp *gracilis*

Table 9.3 Folk classification of plants (Berlin, 1992)

Berlin's ranks	Example
Unique Beginner	Plant
Life form	Tree, bush
Intermediate (optional)	Covert e.g. flowering
Intermediate (optional)	Covert e.g. evergreen
Folk Genera (basic level)	Oak
Folk Specifics	Red oak
Varietal Taxa	Red dwarf oak

A taxa, or an item, of a given rank may be found at different taxonomic levels, defined by the number of nodes between a taxon and a unique beginner. This is a result of ambiguity or 'fuzziness' in a category's boundaries. An example of this 'fuzziness' can be seen in the distinction between tree and bush, as there are bushes that may be very tree-like and vice versa.

The most basic rank as defined by Berlin (1992) is the *Folk Generic*. Generic taxa are typically understood as sets of living organisms that reproduce after their own kind. They are structurally basic, meaning they are the foundation on which the elaborated taxonomic hierarchy is built. Examples of generic taxa include, 'squash' or 'berries' in English, however, in some contexts, ambiguous or unaffiliated generic taxa may be considered different from other plants. In other words, they cannot be described as 'a kind' of something else because they carry so much cultural significance or because they are very distinct. An example of this would be 'corn' for Indigenous people of Mesoamerica, and 'cactus' for English speakers.

Life Form taxa are 'polytypic', meaning they always include at least two named taxa (usually more than two). *Folk generic* taxa may be either polytypic or monotypic e.g. the generic category coyote only includes a single species Canis latrans, as opposed to a polytypic folk generic taxa which includes two or more folk specific taxa e.g. White Oak and Black Oak.

Folk specific taxa are generally named through the use of secondary lexemes (e.g. White Oak, Cutthroat trout), however, some are named with primary lexemes as is often the case with protypical folk specifics. Generally, in traditional local systems, very large polytypic generics are cultivated plants and domestic animals.

At the *Varietal* level we see structural replication of the folk specific. *Folk varietal* names are characteristically binomial or even multinomial unless abbreviated e.g. Eastern diamond-backed snake vs. Western diamond-backed snake. Usually, this level is monotypic unless there are cultivars of high value. Polytypic cases are relatively rare.

Categorization and Classification Examples

Through categorization tasks and general observation and conversation, I have identified what I believe to be semantic and lexical categories in Nahuat-Pipil. Within Berlin's (1992) model for plant classification, all categories would be found at the Generic level, and individual taxa within each category are subgeneric taxa. No distinction is made between fruits and vegetables in Nahuat-Pipil, rather the categorization of fruits and vegetables is done in relation to their interaction with the human body (Dohle, 2020). This is an important aspect to remember as category names feature in the nomenclature of items within the category. This has the benefit of efficiently communicating to the listener how they should expect to identify and handle the item. Features such as size are measured according to how the human would have to hold said fruit, i.e. will it be held in the palm of my hand or the tips of my fingers? Does it need to be torn, sliced, or parted to be consumed? Interestingly, references to these different categories are occasionally accompanied by a respective hand motion. Table 9.4 presents an overview of the identified categories.

This list of categories is by no means a finished list, rather it represents an overview of the categories within which plant names were placed throughout the data collection process. It was compiled using my own primary research (Dohle, 2020) as well as by referencing previous descriptions of Nahuat-Pipil (Campbell, 1985; Hernandez Gonzalez, 2011) and Salvadoran botany (Choussy, 1975, 1976, 1977, 1978). Almost all the categories presented are productive; they are still being used to create new words whenever a new fruit is encountered as is the case with *muyulala* or *lalashukuk*, two terms for 'mandarin' which were spontaneously created and agreed on by speakers during one of my elicitation sessions. Rows 1–9 and 12–15 represent the Generic taxa in the system of folk taxonomies; rows 10–11 represent Life Form taxa. Finally, rows 16 and 17 represent covert categories that are undergoing a process of lexical shift. The term puputukat, for example, means 'reeds that smell good', and it is applied to all herbs which are not classified as *kilit*. In contrast to other categories however, taxa within the category puputukat do not have Nahuat-Pipil names, and thus they are not subject to lexical nominal classification. Finally, the category of citric fruits is one that appears to be expanding as neologisms are created for taxa within this category. The prototype is *lala*, 'orange', and it is possible that the nominal classifier for this category might also be *lala*, but it remains to be seen and depends on the lexical form of future neologisms for items within this category.

The data presented here and in the following sections was compiled and collected via active discussions and interactions with native Nahuat-Pipil speakers. I worked with 73 participants (45 females; 28 males) in total, and of these 52 (33 females; 19 males) took part in the plant identification and pile sorting tasks. All participants were willing and comfortable to speak Nahuat-Pipil in my presence and were all between the ages of 52–90+.[8] The goal was to include as many participants as possible who varied in age and gender. Ideally

---

[8] In the case of the individual who was 90+, their exact age was unknown, but known to be at least 90.

participants of different social statuses and professions would also be represented (Berlin, 1992, p. 204), however, considering within this context the limits of intergenerational transmission, the perception of Indigeneity as a social status, and the gendering of profession or skilled work), it was not possible to vary these factors much.

A range of methods were used to collect different types of data. Ethnographic observation was used throughout the data collection process as a means to gather social contextual data for TEK. Initial elicitation sessions focused on individual free listing of plants, fruits and vegetables (Martin, 2007; Puri, 2014a, 2014b, 2014c). In addition, semi-structured interviews and discussions as well as experimental methods such as elicitation of plant names, plant identification, and pile sorting were utilized. As my research is mostly focused on consultants' biological knowledge, interaction with the natural world was maximized. This was done by taking consultants on forest and hill walks, doing individual and group-based tasks, and asking direct and indirect questions about consultants' knowledge of agricultural practices, staple foods, as well as prominent local flora (Martin, 2007; Puri, 2014a, 2014b, 2014c).

Folk Nomenclature

When it comes to nomenclature, Berlin (1992) argues that universals are found in ethnobiological nomenclature i.e. the patterns that underlie the naming of plants and animals in systems of ethnobiological classification. Such nomenclature is said to represent a natural system of naming that reveals much about the way people conceptualize living things in their environment (Berlin, 1992, p. 26). To analyze nomenclature, we look at the linguistic patterns found in the emic naming of plants by looking at the internal lexemic composition of folk nomenclature. Generally, lexemes of nomenclature are productive, or transparent, allowing meaning to be embedded in names. Proto-Aztecan nomenclature for generic and life form taxa, for example, often have descriptive or onomatopoeic names for birds, (Hunn & Brown, 2011, p. 322). Proto-Aztecan examples include *kakalo*, 'crow' or 'animal that says /kaka/', and *tukulo*, 'owl' or 'animal of the night which says /tuku/'. Nomenclature for specific and varietal forms is often accompanied by an adjective which describes color or size e.g. 'red sparrow' or Nahuat-Pipil *ayutzin*, 'small turtle'. Berlin argues that the above two systems can be found in all languages, and while many have confirmed these claims, some skeptics (Baker, 2007; Dwyer, 2005; Si, 2011) maintain that far more languages need to be investigated in detail before these naming conventions and levels of categorization can be treated as universal.

In terms of ethnobotanical folk nomenclature, Berlin (1992, p. 26) claims that the formal linguistic structure of plant names is similar in all languages. He also claims that salient morphological and behavioral features of plant and animal species are often encoded directly in the ethnobiological names used to refer to these species. He suggests that this purposeful assignment of names to plants and animals may have adaptive significance because such terms will be less difficult to learn, easier to remember as well as to utilize, thus reducing the cognitive effort required of people of non-literate traditions who must control large ethnobiological vocabularies. Furthermore, the transparency and productivity of lexemes allow meaning and knowledge to be embedded in plant nomenclature, meaning the name of a plant can communicate important information such as how to identify the plant and how to interact with it. Finally, it is important to note that a comparative analysis of the mapping of names demonstrates that while a name indicates the existence of a particular taxon, the absence of a name does not imply the absence of a category.

Naming Conventions and Examples

In this section we can look at the naming conventions of folk nomenclature in Nahuat-Pipil by looking at the internal linguistic composition of names and considering how they can be analyzed differently to account for semantic meaning. Concrete results include a 300-item wordlist of local names in Nahuat-Pipil for local plants. A note on data presentation: All Nahuat-Pipil plant names are accompanied by their corresponding Salvadoran Spanish names (SV) and, when possible, their corresponding English names (EN). Most plant names have been identified by their scientific name presented in italics and prefaced with 'bot.', and they are also presented when available.

As was previously mentioned, some participants were unable to identify many of the samples provided during the plant identification task. Taking the mentioned factors into consideration, the plants that consultants had the most problems with were those which are not native to the Americas and would have been introduced in the years following contact with the Spanish colonizers. Such plants include onions, rue, and carrots and were often found in the consultant's gardens and houses. While in most cases the consultants were able to correctly identify these plants in Spanish, they were not able to produce their corresponding names in Nahuat. Not surprisingly, none of the consultants had any issues identifying and naming plants native to Central America in Nahuat-Pipil, such as tomatoes, avocados, maize, beans, and so on. Comparison to plant terminology found in previous documentation of Nahuat-Pipil (Campbell, 1985), showed that terms were already missing from the lexicon when Lyle Campbell carried out fieldwork in the 1970s. Comparison to early 1920s botanical guides (Carpenter Standley, 1922; Choussy, 1975, 1976, 1977, 1978) did not bring up any Nahuat-Pipil terms for non-endemic plants.

Morphosyntactic Structure

The processes of breaking down the internal composition of botanical nomenclature clearly demonstrate that many of these terms include reference to the root category to which they belong. Research on a different variant of Nahuatl, Cuetzalan Nahuatl spoken in Mexico, demonstrates a similar strategy for the creation of color terms (Castillo Hernández, 2000).

As an agglutinating and polysynthetic language, the lexical items of Nahuat-Pipil are a composite of root words or concepts paired with morphemes to signpost lexical items' relationships with each other. As a result, it can sometimes be difficult to uncover the etymology of a given plant name. Thus, the following presents approximations of etymologies based on the nominal structure of plant names by analyzing in detail the internal morpho-syntactic structure of naming conventions of plant names. These are ultimately indicative of the noun phrase structure. The internal nominal structure incorporates three principal strategies: N, NN, and NAdj. These are outlined in the following subsections.

N

Nomenclature of plants classified as Generic within Berlin's (1992) Folk classification paradigm are often primary names, as outlined previously. Primary names can be simple and composed of single nouns (e.g. fish), productive and composed of compound nouns (e.g. catfish), or unproductive and composed of Adj + N (e.g. silverfish). In Nahuat-Pipil, simple primary names usually refer to both the superordinate category and the prototype of the category, which is in line with Berlin's (1992) views on protypicality. Examples include: tzaput, shukut, and et.

Given the agglutinative nature of Nahuat-Pipil, however, it can be difficult to analyze the lexical composition of plant names. It is not clear where one form ends and another begins. For example, as seen in Table 9.4, eshut has been analyzed as one semantic category but lexically it could be analyzed as either a single noun (N) or a compound noun (NN). Therefore, eshut could also be analyzed as a productive primary name, however more research would need to be carried out to see if it behaves in a productive manner, or if it has become fossilized as a simple primary name. I would suggest that generally, the shorter the name, the less likely it is to be composed of a compound. For example, et 'beans', and 'water', are single concept words.

Nn

I found that the use of compound nouns in ethnobotanical naming conventions is most common when the referent of the term is less likely

Table 9.4 Nahuat-Pipil plant categories (Dohle, 2020)

	Plant category	Category description	Prototype	Examples
1.	Akat/Uwat	Plants with partitioned, tube like stems such as bamboo, cane and 'bara de castilla'	Akat 'Sugar cane' or 'grass'	Akat 'sugar cane, small'; Elutzakat 'tall grass'; Istahuat 'fine grass used for building'; Tekumajakat; Tzakat 'marijuana' or 'grass'
2.	Ayut Ayuj	Squashes, pumpkins and edible creeper plants that grow along the ground, members of the Cucurbiteae family. Generally, have thick outer skin, a soft, fleshy interior and come in all shapes and sizes. Its seeds are highly sought after and used in cooking	Ayut 'squash' or 'pumpkin'	Ayut 'squash' or 'pumpkin' Witzayut 'chayote'
3.	Chil	All chillies, members of the Capsicum family. Used to add flavor to cooked food, though not necessarily spicy	Chil 'red chilli pepper'	Chil Chukulat, Chil Tzupelek, Chil Tata, Chil Tepet, Chil tekpin 'different types of chilies'
4.	Eshut	All young beans which are edible without being cooked		Eshut 'green been'
5.	Et	All bean-like fruits which grow in 'pods'	Et 'red bean'	Istaket 'white bean'; Perunhet 'lentil'
6.	Kamut Kamuj	All starchy and edible roots	Kamut—sweet potato, yuca	Chiltikamuj—red sweet potato; Kamuj—yucca; Shuchikamuj—carrot
7.	Kilit	Small, edible, and leafy with a woody stem. Objects which belong to this category are 'torn' from their woody stem	Kilit 'chipilin' Crotallaria vitellina	Masakilit 'lorocco' Fernaldia pandurata; Ishkilinit 'tamarind' Tamarindus indica
8.	Kushit	Fruit encased in pods like beans, however, the inner flesh is fleshy. Like a combination of a banana and a peapod. Outer skin is thin and can be peeled easily	Kushit 'pepeto'	Kushit 'paterna' Inga paterno Harms; Kushit 'pepeto negro' Inga leptoloba Schecht; Kushit 'pepetillo' Inga espuria Humb. and Bonpl; Kushit 'cojin de finca' Inga preussi Harms

	Plant category	Category description	Prototype	Examples
9.	*Kuyul*	All small, hard ornamental fruits or nuts. They usually have some sort of economic value or spiritual function. Highly valued	*Kuyul*	*Kuyul ne mistun* 'coyol de gato' Solanum hirtum *Kulyulmatzaj*, 'pineapple',
10.	*Kwawit*	Tree	*Kwa-*	*Kwaulut* 'caulote' Guasuma umifolia *Kwashilut* 'kwajilote' Parmentiera edulis
11.	*Mekat*	Vines		*Mekat pal corral* 'Petastoma patelliferum' *Mekat* 'vieja'—n/a *Mekayo* 'hammock'
12.	*Tzaput*	All palm-sized sweet fruits with a hard skin and smooth shiny seed. Need to be consumed using a 'splitting' or 'slicing' motion in order to qualify	*Tzaput* 'sapote'	*Tzaput istak* 'custard apple' Annona squamasa *Muyutzaput* 'nispero' Maniłkara achvar *Atzaput* 'pear'
13.	*Tumat*	n/a	*Tumat* 'tomato'	*Witztumat* 'huistomate' Solanum donnell
14.	*Shukut*	All small fruits which are sour and can be held by the tips of the fingers	*Shukut* 'jocote' Spondias purpurea	*Shukut jobo—jocote verano*, Spondias mombin *Shukut—jocote invierno*, Spondias lutea *Lalashukuk*—mandarin
15.	*Pajti*	Medicinal plants	*Pajti* 'medicine'	*Siwapajti*—'ciguapate' Pluchea odorata *Tempajti* 'Tempate' Jatropha curcas
16.	*Puputukat*	Herbs which add flavor to food (coriander, mint, spring onion, sometimes include carrots and celery, etc.)		Coriander Mint Spring onion
17.	Citric fruits	Covert category: there is no term for this category, possibly because citric fruits are not native to the area	*Lala* 'orange'	*Muyulala* 'mandarin'

to resemble the prototype of the category it belongs to. This strategy is always agglutinative and combines key features of the fruits or categories in question. When combined, the listener can determine in an intuitive manner what the object being referred to might resemble. For example, *kuyulmatza* combines the concept *kuyul* 'small hard fruit', with *matzaj* 'prickly but not spiny fruit'. This is the same strategy used in English when compounding names for fruits e.g. crab-apple or eggplant. This is an example of what Berlin (1992) refers to as 'secondary names', which are often employed to refer to taxa that are folk specific.

Alternatively, these could also be analyzed as primary productive names, or folk-specific or folk varietal names that are monothetic. Generally, to be monothetic, names must be defined in relation to a simple feature contrast such as color or size. In Nahuat-Pipil, the size of a fruit is usually indicated by referring to a small or large animal. Something small, for example, would be indicated in relation to *michin* 'fish', or *muyut* 'fly' e.g. *muyulala* 'mandarin', and *muyutzaput* 'nispero' (bot. *Manilkara achvar*).

NAdj

Without getting into a discussion about whether adjectives exist in Nahuat-Pipil, for this specific section, an adjective is defined as terms with the morpheme ending /-tik/. These terms can sometimes be interpreted as nouns meaning 'something coloured X' (Campbell, 1985). The combination of a noun or root word with an adjective is in some cases, agglutinative and other cases paired as a compound. Thus, *et istak* can also be called *istaket*. In both of these cases, there is emphasis on the color of the bean *istak* 'white'. These names are secondary names within Berlin's (1992) principles of nomenclature, as they only appear in contrastive sets, the most common contrastive sets being made between *istak* 'white', and *chiltik* 'red' as in *et istak* and *et chiltik, tzaput istak, tzaput chiltik*.

On occasion, however, the NAdj compound has become fossilized as is the case for *tzanajtultik* 'stem'. It would appear that in this example, the color term is no longer an active component or a focal point of the term. This needs to be investigated further to see if distinctions are made between different types of stems within a plant. In general, folk names with this linguistic construction are secondary names and are classified as folk specific taxa.

Non-Agglutinative Noun Phrases

Non-agglutinative noun phrases are also employed in ethnobotanical naming strategies, particularly when speakers are not confident in their responses. For example, *tzaput pal mico* 'category.prep.monkey', is a descriptive noun phrase (NPrepN) as is *kuyul ne mistun* 'category.art.cat' (NArtN). When this strategy is employed, the superordinate classifier is always used, as seen in this case with *tzaput* and *kuyul*. Noun phrases are more commonly employed to name plants which were identified but not frequently seen. This is to say that these plants did not feature in the everyday lives of the speakers and thus had to be described rather than named. Within Berlin's Folk Classification paradigm, these names refer to varietal taxa, however he does not refer to this naming strategy within his principles of nomenclature.

Within Berlin's (1992) principles of nomenclature, N names are simple primary names assigned to the Generic level and prototype of that level, NAdj are secondary names used for taxa at subgeneric level such as folk-specific levels as they usually only occur in contrastive sets, and NN names are productive primary names. Finally, the use of noun phrases to refer to varietal taxa and the inclusion of superordinate classifiers within the naming convention is evidence that these classifiers are still productive.

Conclusion

In this chapter, we have explored the interactions between language, migration, environment, and historical events which affected speakers of Nahuat-Pipil. We have discussed how the disruption of the transmission of traditional knowledge and the historical trauma endured has resulted from the historical violence against Indigenous people in El Salvador. Nevertheless, it is observed that their marginalization has also resulted in the preservation of TEK as it is this knowledge which has contributed to their survival. The ability to name and identify plants is essential to being able to live off the land, and certain aspects of language such as lexical categorization and folk nomenclature are essential strategies for quickly grouping and communicating information about the plants which are of cultural and practical significance. We see evidence of how this is used in Nahuat-Pipil in the way that lexical categorization indicates how items are to be handled to be consumed. All members of the *tzaput* category must be held in the palm of the hand compared to members of the *shukut* family which are held with the tips of the fingers. This is aligned with Berlin's (1992) view that behavioral and morphological features are encoded directly into names. Linguistic analysis of folk nomenclature also demonstrates that the more prominent and important items have simpler names e.g. *et* 'beans', a staple in Nahuat-Pipil diet, compared to *perunhet* 'lentils' a lesser available legume. We can consider how the purposeful assignment of names to plants and animals may be of functional significance. Such terms will be less difficult to learn, easier to remember, and utilize, and thus reduce the cognitive effort required of people of non-literate traditions who must control large ethnobiological vocabularies.

The Nahuat-Pipil have a shared background knowledge that is unique to El Salvador. Aspects of this knowledge base are found among people who do not identify as Indigenous, however, this constructed world or belief system is much richer and more 'real' within the Indigenous communities and Nahuat-Pipil speakers themselves because of the language they share. The descriptions in this chapter of TEK as it is encoded within the Nahuat-Pipil language offer examples of knowledge that are still relevant and productive within the language-speaking community. As the Nahuat-Pipil speaking population dwindles and access to land and the natural world becomes more restricted, this type of knowledge begins to disappear, however, this does not mean that it is any less relevant or significant, especially given the current environmental crisis. I believe there is still much to be learned from those who have lived in such close proximity to nature for many generations, and who despite facing adversity have continued to survive as a result of this close relationship.

Bibliography

Aitchison, J. (2004). The Rhinoceros's problem: The need to categorize. In C. J. Kay & J. J. Smith (Eds.), *Categorization in the history of english* (pp. 1–17). John Benjamins Publishing.

Alej. (2017, October 14). *Náhuat El Salvador* [Video sharing platform]. *YouTube.* https://www.youtube.com/channel/UCbYqsaNZAzdRq94Of RO4odQ

Arauz, P. (1960). *El Pipil de la Region de los Itzalcos.* Departamento Editorial del Ministerio de Cultura.

Baker, B. (2007). Ethnobiological classification and the environment in Northern Australia. In A. Schalley & D. Khlentzos (Eds.), *Mental states.* John Benjamins Publishing.

Barsalou, L. W. (1987). The instability of graded structure: Implications for the nature of concepts. In U. Neisser (Ed.), *Concepts and conceptual development: The ecological and intellectual factors in categorization* (pp. 101–140). Cambridge University Press.

Berlin, B. (1972). Speculations on the growth of ethnobotanical nomenclature. *Language and Society, 1,* 51–86.

Berlin, B. (1973). Folk systematics in relation to biological classification and nomenclature. *Anual Review of Ecology and Systematics, 4,* 259–271.

Berlin, B. (1992). *Ethnobiological classification: Principles of categorization of plants and animals in traditional societies.* Princeton University Press.

Berlin, B., & Kay, P. (1969). *Basic color terms: Their universality and evolution.* University of California Press.

Brockman, A., Masuzumi, B., & Augustine, S. (1997). *When all peoples have the same story, humans will cease to exist: Protecting and conserving traditional knowledge: A report to the biodiversity convention.* Dene Cultural Institute. http://nativemaps.org/?p=1401

Campbell, L. (1985). *The Pipil language of El Salvador.* Mouton Publishers.

Carpenter Standley, P. (1922). *Expedition history: Botanical explorations in Central America*. Smithsonian Institution.

Castillo Hernández, M. A. (2000). *El mundo del color en Cuetzalan: Un estudio etnocientífico en una comunidad nahua*. Instituto Nacional de Antropología e Historia.

Censo Nacional. (2008). *VI Censo de Población y de Vivienda: Cifras oficiales de población, vivienda y hogar*. Direccion General de Estadistica y Censos.

Censo Nacional. (2013). *VII Censo de Población y VI de Vivienda*. Direccion General de Estadistica y Censos.

CEPF. (2016). Critical ecosystem partnership fund: Protecting nature's hotspots for people and prosperity. *Mesoamerica*. http://www.cepf.net/resources/hot spots/North-and-Central-America/Pages/Mesoamerica.aspx

Ching, E., & Tilley, V. (1998). Indians, the military and the rebellion of 1932 in El Salvador. *Journal of Latin American Studies, 30*(1), 121–156.

Choussy, F. (1975). *Flora Salvadoreña: Tomo I* (2nd ed., Vol. 1). Editorial Universitaria.

Choussy, F. (1976). *Flora Salvadoreña: Tomo II* (2nd ed., Vol. 2). Editorial Universitaria.

Choussy, F. (1977). *Flora Salvadoreña: Tomo III* (2nd ed., Vol. 2). Editorial Universitaria.

Choussy, F. (1978). *Flora Salvadoreña: Tomo IV* (2nd ed., Vol. 2). Editorial Universitaria.

Conklin, H. C. (1954). *The relation of Hanunóo culture to the plant world* [Ph.D.]. Yale University.

Conklin, H. C. (1972). *Folk classification: A topically arranged bibliography of contemporary and background references through 1971*. [New Haven] Department of Anthropology, Yale University. http://archive.org/details/fol kclassificati0000conk

Conklin, H. C. (1980). *Folk classification: A topically arranged bibliography of contemporary and background references through 1971: Revvised reprinting with author index*. Deptartment of Anthropology Yale University.

ContraPunto. (2015, May 18). *Cifra de asesinatos supera a los de la guerra civil salvadoreña* [Newspaper]. Contrapunto.com.sv. http://www.contrapunto. com.sv/sociedad/violencia/cifra-de-asesinatos-supera-a-los-de-la-guerra-civil-salvadorena

Davies, C. A. (2008). *Reflexive Ethnography: A guide to researching selves and others* (2nd ed.). Routledge.

Delugan, R. M. (2013). Commemorating from the margins of the nation: El Salvador 1932, indigeneity, and transnational belonging. *Comemorando a*

Partir Das Margens Da Nação: El Salvador 1932, Indigenismo and Pertença Transnacional, *86*(4), 965–994.

Dohle, E. (2020). *Nahuat-Pipil: The encoding of ecological knowledge in semantic and lexical categorization systems* [Ph.D.]. SOAS University of London.

Duranti, A. (2003). Language as culture in U.S. anthropology: Three paradigms. *Current Anthropology*, *44*(3), 323–347.

Dwyer, P. (2005). Ethnoclassification, ethnobiology and the imagination. *Journal De La Societe Des Oceanists*, *120–125*, 12–25.

Escalante Arce, P. (2004). *Los Tlaxcalas en Centro América* (2nd ed., Vol. 11). CONCULTURA, DPI.

Glushko, R. J., Maglio, P., Matlock, T., & Barsalou, L. W. (2008). Categorization in the wild. *Trends in Cognitive Sciences*, *12*(4), 129–135.

Gorenflo, L. J., Romaine, S., Mittermeier, R. A., & Walker-Painemilla. (2012). Co-occurrence of linguistic and biological diversity in biodiverse hotspots and high biodiversity wilderness areas. *PNAS*, *109*(21), 8032–8037.

Gould, J. L., & Lauria-Santiago, A. (2014). *1932: Rebelión en la Oscuridad*. Museo de la Palabra y la Imagen.

Harmon, D., & Loh, J. (2010). The index of linguistic diversity: A new quantitative measure of trends in the status of the world's languages. *Language Documentation*, *4*, 55.

Hernandez Gonzalez, W. (2011). *Nawat Mujmusta*. NawaCoLex.

Hernandez, W. (2017, August 24). *Buenos Resultados con Poco Dinero*. Escuela de campo: documentación, investigación participativa, eseñanza y revitalización de idiomas en peligro, San Miguel Xaltipan, Tlaxcala, Mexico.

Hill, J. H. (2001). Proto-Uto-Aztecan: A community of cultivators in Central Mexico? *American Anthropologist*, *103*(4), 913–934. https://doi.org/10.1525/aa.2001.103.4.913

Hill, J. H. (2006). The ethnograpy of language and language documentation. In J. Gippert, N. P. Himmelmann, & U. Mosel (Eds.), *Essentials of language documentation* (pp. 113–128). Mouton de Gruyter.

Hume, M. (2007). Mano Dura: El Salvador responds to gangs. *Development in Practice*, *17*(6), 739–751.

Hunn, E. S., & Brown, C. H. (2011). Linguistic ethnobiology. In E. N. Anderson, D. M. Pearsall, E. S. Hunn, & N. J. Turner (Eds.), *Ethnobiology*. Wiley-Blackwell.

Jacob, E. K. (2004). Classification and categorization: A difference that makes a difference. *Library Trends*, *52*(3), 515–540.

Karttunen, F. (1983). *An analytical dictionary of Nahuatl*. University of Texas.

King, A. R. (2011). *Timumachtikan! Curso de lengua náhuat para principantes adultos*. NawaCoLex.

King, A. R. (2013). *NawaCoLex project*. Alan R. King. http://alanrking.info/ncl.php

King, A. R. (2018). TUSHIK | Lenguas del centro de Centroamérica [Blog]. *TUSHIK*. http://tushik.org/

Labrador, G., & Rauda Zablah, N. (2015, July 29). Pandillas logran sostener pulso con el gobierno por el transporte público. *elfaro.net*. http://www.elfaro.net/es/201507/noticias/17232/Pandillas-logran-sostener-pulso-con-el-gobierno-por-el-transporte-público.htm

Lara Martinez, B. C. (2006). *La Población Indígena de Santo Domingo de Guzmán: Cambio y continuidad sociocultural* (Vol. 3). CONCULTURA.

Lara-Martinez, R., & McCallister, R. (2012). *El Legado Náhuat-Pipil de María de Baratta*. Fundación AccessArte.

Lauria-Santiago, A. (1999a). *An Agrarian Republic: Commercial agriculture and the politics of peasant communities in El Salvador, 1823–1914*. University of Pittsburgh Press. https://doi.org/10.2307/j.ctt9qh7vd

Lauria-Santiago, A. (1999b). Land, community, and revolt in late-nineteenth-century Indian Izalco, El Salvador. *The Hispanic American Historical Review, 79*(3), 495–534.

Lemus, J. E. (2011). Una aproximación a la definición del Indígena salvadoreño. *Científica: Revista de Investigaciones de la Universidad de Don Bosco, 12*.

Lemus, J. E. (2014a, February 11). ¿Para qué mantener vivo el náhuat? [Newspaper]. *elfaro.net*. http://www.elfaro.net/es/201402/el_agora/14722/

Lemus, J. E. (2014b, June 23). Los pueblos salvadoreños indígenas siempre han existido [Newspaper]. *Elfaro.Net*. http://www.elfaro.net/es/201406/el_agora/15560/Los-pueblos-salvadoreños-indígenas-siempre-han-existido.htm

Lewis, M. P., Simons, G. F., & Fennig, C. D. (2016). *Ethnologue: languages of the world* (17th ed.). Ethnologue. http://www.ethnologue.com/country/SV/languages

Lindo-Fuentes, H., Ching, E., & Johnson, L. L. (2012). *Modernizing minds in El Salvador: Education reform and the cold war, 1960–1980*. University of New Mexico Press. http://ebookcentral.proquest.com/lib/soas-ebooks/detail.action?docID=1118955

Little, W. E. (2008). A visual political economy of Maya representations in Guatemala, 1931–1944. *Ethnohistory, 55*(4), 633–663.

Loh, J. (2017). *Indicators of the status of, and trends in, global biological, linguistic and biocultural diversity* [Ph.D., University of Kent]. https://kar.kent.ac.uk/61424/

Loh, J., Green, R. E., Ricketts, T., Lamoreux, J., Jenkins, M., Kapos, V., & Randers, J. (2005). The living planet index: Using species population time series to track trends in biodiversity. *Philosophical Transactions of the Royal Society B: Biological Sciences, 360*(1454), 289–295. https://doi.org/10.1098/rstb.2004.1584

Macri, M., & Looper, M. (2003). Nahua in ancient Mesoamerica: Evidence from Maya inscriptions. *Ancient Mesoamerica, 14*, 285–297. https://doi.org/10.1017/S0956536103142046

Markman, E. M. (1989). *Categorization and naming in children: Problems of induction.* MIT Press.

Martin, G. J. (2007). *Ethnobotany: A methods manual.* Earthscan.

MINED. (2019). *¡Titaketzakan Nawat!* [Government of El Salvador Website]. Ministerio de Educación. https://www.mined.gob.sv/nawat-titaketzakan.html

Miranda, J. (2016, May 1). Paula López, voz del Río de Espinas. *La Zebra.* https://lazebra.net/2016/05/01/jazz-miranda-paula-lopez-voz-del-rio-de-espinas-cronica/

Moseley, C. (2010). *Atlas of the world's languages in danger* (3rd ed.) [Atlas]. UNESCO Publishing. http://www.unesco.org/culture/languages-atlas/index.php

Nettle, D., & Romaine, S. (2000). *Vanishing Voices: The extinction of the world's languages.* Oxford University Press.

Olivet, C. R., & Asquith, N. (2005). *Mesoamerica hotspot: Northern Mesoamerica briefing book* (Improving linkages between CEPF and World Bank operations, Latin America forum). Critical Ecosystems Partnership Fund. http://www.cepf.net/Documents/final.mesoamerica.northernmesoamerica.briefingbook.pdf

Polzenhagen, F., & Xia, X. (2015). Prototypes in language and culture. In *The Routledge handbook of language and culture* (pp. 253–268). Routledge.

Puri, R. K. (2014a, May 10). *Ethnobotany I: Cultural domain analysis.* Plants. Animals. Words. University of Kent.

Puri, R. K. (2014b, May 10). *Ethnobotany II: Voucher Specimens.* Plants. Animals. Words. University of Kent.

Puri, R. K. (2014c, May 10). *Tree Trail.* Plants. Animals. Words. University of Kent.

Schultze-Jena, L. (2014). *Mitos en la lengua materna de los Pipiles de Izalco en El Salvador: Gramática* (R. Lara-Martinez, Trans.; First). Editorial Universidad Don Bosco.

Seligson, M. A. (1995). Thirty years of transformation in the agrarian structure of El Salvador, 1961–1991. *Latin American Research Review, 30*(3), 43–74.

Sharifian, F. (2017). *Cultural linguistics: Cultural conceptualisations and language.* John Benjamins Publishing.

Si, A. (2011). Biology in language documentation. *Language Documentation and Conservation, 5,* 169–186.

Sibrián, W. (2019, November 12). 70 docentes ya pueden enseñar el idioma náhuat en las escuelas de El Salvador. *La Prensa Gráfica.* https://www.laprensagrafica.com/elsalvador/70-docentes-ya-pueden-ensenar-el-idioma-nahuat-en-las-escuelas-de-El-Salvador--20181211-0232.html

Simpson, G. G. (1961). *Principles of animal taxonomy.* Columbia University Press.

Stepp, R. (2015, January 7). Global biocultural diversity. *The Decolonial Atlas.* https://decolonialatlas.wordpress.com/2015/01/07/global-biocultural-diversity/

Stevenson, M. G. (1996). Indigenous knowledge in environmental assessments. *Arctic, 49*(3), 278–291.

Terralingua. (2017). *Biocultural diversity.* http://terralingua.org/biocultural-diversity/

Tilley, V. Q. (2002). New help or new hegemony? The transnational indigenous peoples' movement and 'Being Indian' in El Salvador. *Journal of Latin American Studies, 34*(03). https://doi.org/10.1017/S0022216X0200651X

Valencia, R. (2015, September 3). La Tregua redefinió el mapa de asesinatos de El Salvador. *El Faro.* https://public.tableau.com/views/Manodurismoversus usTregua/ManodurismoversusTregua?:embed=y&:toolbar=no&:display_count=no;:showVizHome=no;

Watts, J. (2015, August 22). One murder every hour: How El Salvador became the homicide capital of the world. *The Guardian.* http://www.theguardian.com/world/2015/aug/22/el-salvador-worlds-most-homicidal-place

10

Cenotes and Placemaking in the Maya World: Biocultural Landscapes as Archival Spaces

Khristin N. Montes, Dylan J. Clark, Patricia A. McAnany, and Adolfo Iván Batún Alpuche

Yo, cenote

¡Cuídame!
Porque soy un lugar grandioso,
sé que a muchos les gusto,
y a otros tantos asusto,
poseo historias grandiosas,
que algunos desean conocer,
grandes piedras tengo,
que me gustaría proteger.

K. N. Montes (✉)
Regis University, Denver, CO, USA
e-mail: klandrymontes@gmail.com

D. J. Clark
North Carolina Office of State Archaeology, Asheville, NC, USA

P. A. McAnany
University of North Carolina at Chapel Hill, Chapel Hill, NC, USA

A. I. B. Alpuche
Universidad de Oriente, Valladolid, Yucatán, México

© The Author(s) 2024
J. Olko and C. Radding (eds.), *Living with Nature, Cherishing Language*,
https://doi.org/10.1007/978-3-031-38739-5_10

Introduction

The poem above reflects the sentiments of a middle school student at the Escuela Secundaria Tecnica Num. 69 in the Maya community of Xocén, Yucatan, Mexico. In English, the poem translates closely as follows (and in keeping with the original rhyme scheme):

I, Cenote

Take care of me!
Because I am an impressive place,
I know that many like me,
And some others I scare,
I possess grandiose stories,
That some wish to know,
I have great stones,
That I would like to protect.

The student's poem is written from the first-person perspective of a Yucatecan *cenote*, a natural freshwater sinkhole common in the karst geology of the otherwise dry, Yucatan Peninsula. Known as *ts'ono'oto'ob* (singular: *ts'ono't*) in the Indigenous Yucatec Mayan language of Maaya t'aan, cenotes form when the surface limestone erodes and dissolves, exposing the groundwater beneath that has, over time, carved out channels within the layers of limestone bedrock and sediment that make up the peninsula (Fig. 10.1).

Thousands of these solution sinkholes dot the Yucatan Peninsula, and they come in a variety of types, ranging from closed cenotes that are entirely subterranean to open cenotes that have the appearance of surface ponds. Between these are a variety of other types that are all connected to the subterranean aquifer system (Anda Alanís, 2010; Batún Alpuche et al., 2021; Landry-Montes et al., 2020; López-Maldonado & Berkes, 2017). Prior to the drilling of modern wells, cenotes provided the main access to natural sources of freshwater, as there are very few surface rivers and lakes in the region (Beddows et al., 2007). Cenotes have long been important and powerful places in

Fig. 10.1 Students gathering knowledge while photographing plants and animals in the Cenote Yax Ek' (Green Star in the Yucatec Mayan language) in the community of Kaua, Yucatan. Photograph by Khristin N. Montes

the Maya world. The reliance on groundwater here means that Yucatecan people have had to develop multiple strategies for collecting and storing freshwater. Cenotes are essential *biocultural* resources, simultaneously carrying biological, ecological, cultural, and historical significance for Indigenous people who have lived in this tropical savanna biome for millennia.

In ancestral Maya times before the Spanish invasion, as well as more recently, cenotes were understood as central places and animated features of sacred geographies (Bassie-Sweet, 2008; Crumley, 1999). Associated with agricultural fertility and cyclical renewal, they were often considered portals where rain and maize deities, along with other supernatural forces, dwelt and could be reached by humans (Brady &

Ashmore, 1999; Brady & Prufer, 2005; Hernandez & Vail, 2013; Luzzader-Beach et al., 2016, p. 428; Moyes & Brady, 2012). Cenotes retain a visual language specific to Yucatec Maya communities that is shared with this ancestral past.

Specific cenotes, such as the Sacred Cenote in the Maya city of Chichén Itzá (late sixth–early eleventh centuries), were sites of religious offerings and foci of rituals essential for the well-being of the community as is apparent from the vast array of highly valued prestige objects dredged from that cenote between 1904 and 1961. Among these include gold and jade offerings associated with, and dedicated to, the rain god Cháak, a deity intimately linked to cenotes (Coggins, 1984). This suggests that Chichén Itzá's largest cenote remained a place to make offerings and petition for non-human intercession beginning in at least the sixth century, through the Postclassic period (thirteenth–early sixteenth centuries), and into Colonial times—well after the city's apogee.

Returning to the student's poem, we see how the animacy and agency of cenotes continue to be appreciated and expressed in contemporary Yucatecan communities. Certainly, a rich tradition of oral histories and legends surrounds cenotes today, many of which are preserved in memory and communicated by village elders, including *jmeen,* or healers. Maaya t'aan (Yucatec Mayan) is the primary language of many *jmeeno'ob* (the plural term for jmeen). Unfortunately, both Maaya t'aan and cenotes in Yucatan face sustained threats from the effects of colonization and globalization. With less than 500,000 speakers, Maaya t'aan and much traditional knowledge are in danger of being lost as intergenerational transmission declines (López-Maldonado & Berkes, 2017, p. 10). Yucatec Mayan is rarely spoken in the home today, even in communities where the majority of the population are direct descendants of Indigenous Maya (Lizama Quijano et al., 2011; Quijano & Sosa, 2014). Most adolescents and young adults learn Spanish as a primary and "official" language, taking up English as a way to expand opportunities within the global tourist market—the economic vehicle that supports Yucatan through the state's internationally celebrated archeological sites, beaches, and proximity to resorts in Cancun and the Maya Riviera (Re Cruz, 2003). In the case of cenotes, pollution resulting from increases in population and untreated wastewater, trash-dumping, intensive farming, animal husbandry, tourism development, and other industries like thermoelectric power and tortillerias, all adversely affect the delicate environments of cenotes and the aquifer (López-Maldonado & Berkes, 2017). Furthermore, cenotes' importance and traditional cultural value in Maya history are not standard components of state or national primary and secondary school curricula.

As anthropologists, archeologists, and art historians who have worked and lived in Maya communities for many years, we recognize the ongoing threats to Maya biocultural heritage and language, as well as the potentially positive impact that young people could have at the local level in resource conservation, provided they have greater opportunities for exploration and access to information. Like many of our collaborators, we agree that education is an integral component of any effort to overcome heritage distancing, bolster the transmission of traditional knowledge, and restore control over tangible and intangible biocultural resources to Indigenous people (López-Maldonado & Berkes, 2017, p. 12; McAnany & Parks, 2012).

In 2018, InHerit: Indigenous Heritage Passed to Present at the University of North Carolina at Chapel Hill partnered with the Universidad de Oriente (UNO)—a community university in eastern Yucatan with a majority Maya student population—and nine middle schools in Maya communities in Yucatan, Mexico, to undertake the *Cultural Heritage, Ecology, and Conservation of Yucatec Cenotes* project. It is also referred to as *Patrimonio Cultural, Ecología y Conservación de Cenotes Yucatecos* (PACECCY) in Spanish. With financial and material support from the National Geographic Society and other institutional partners, we collaboratively created middle school-level experiential education resources focused on the environmental and cultural preservation of cenotes—with language survival included as part of cultural preservation. Three broad and interrelated, cenote-centered themes provided entry points and a flexible framework around which educator workshops and lesson plans were designed. These themes coalesced through our connections and collaborations with students, teachers, and researchers from both Mexico and the U.S. These included: (1) oral history and folklore of cenotes, (2) science and safety of cenotes, and (3) the archeology and heritage of cenotes. Over the course of the project, our team implemented experiential education activities in the nine middle schools related to each of these themes. To enhance the sustainability of the program, we also authored an open source textbook (Batún Alpuche et al., 2021; available at https://in-herit.org/resources-2/resources-for-teachers/) that organizes learning activities and related resources on cenotes into specific units designed to be used by secondary school teachers in Yucatán for their classes. In November 2022, hard copies of the textbook were published and disseminated to the participating middle schools, another 459 middle schools across the state of Yucatan, libraries in the state capital of Merida including the city's Central Library, the main library at the Universidad Autónoma de Yucatán, the state hemerotheque, and the state archive. More detail on textbook themes, activities, and preliminary results can be found in Landry-Montes and colleagues (2020).

In many ways, the Cultural Heritage, Ecology, and Conservation of Yucatec Cenotes project focused on dialogs of Maya placemaking. Keith Basso (1996) has explored placemaking in *Wisdom Sits in Places: Landscape and Language Among the Western Apache* as the dynamic act of making meaning in the world, a method of world-building that requires multiple, yet culturally agreed-upon methods for remembering the past in ways that inform and influence the present. The concept suggests that particular sites, or environments, can elicit powerful cues for this kind of remembering and world-building. *What happened here? Who was involved? What was it like?* These are the kinds of questions that involve place in the function of creating realities in the present. Remembering is intimately linked to imagination and futurities. New realities are formed from the specific set of verbal and visual accounts that a place embodies (Basso, 1996; Casey, 1976, 1987). As Basso (1996) states,

> What is remembered about a particular place—including, prominently, verbal and visual accounts of what has transpired there—guides and constrains how it will be imagined...Essentially, then, instances of

placemaking consist in an adventitious fleshing out of historical material that culminates in a posited state of affairs, a particular universe of objects and events—in short, a place-world—wherein portions of the past are brought into being. (pp. 5–6)

For the Apache, this kind of living in a place and engaging with memory through a landscape's features, laden as they are with mnemonic capabilities, ultimately serves as a context through which traditional wisdoms are taught and preserved across generations. In the Maya world, certain places and certain kinds of placemaking carry similar agency.

In the Yucatan Peninsula, cenotes play important, active roles in Indigenous placemaking and can be further conceptualized as a type of living archive. The student's poem introducing this chapter, for example, invokes an image of the cenote as an animate being, one who holds and guards stories. From a Western perspective, archives are places where records have been selected for permanent or long-term preservation on grounds of their enduring cultural, historical, or evidentiary value. They are normally unpublished and almost always unique. One of their most important functions includes the preservation of primary sources, including oral, written, and visual accounts (these descriptions of archives and their functions in the West are adapted from the Society of American Archivists. "A Glossary of Archival and Records Terminology", http://www2.archivists.org/glossary/terms/a/archives).

Archives are also situated places—locations where primary sources are collected, stored, and shared. Archives are often associated with powerful institutions like local and national governments. Much like archives, cenotes are dynamically situated places capable of collecting, preserving, protecting, conveying, and informing intergenerational memory and knowledge. In this sense, they have the unique capacity to function as centers for promoting and developing sustainable measures for knowledge collection, language revitalization, and maintenance. In the ancestral past, many cenotes in the Yucatan Peninsula—including the Sacred Cenote at Chichén Itzá already discussed, as well as Cenote Ch'en Mul at the nearby Maya city of Mayapán (1220–1440

CE)—were important to the overall cosmological significance and social memory of those centers. Offerings to cenotes or, rather, offerings to the sacred forces that cenotes embodied and provided access to, materialized stories connected to them. For the ancestral Maya, these stories were about primordial times, places, and their protagonists. Objects deposited within cenotes as offerings represent these stories and point to the considerable longevity of cenotes to function as knowledge collectors and disseminators—much in the same vein as living archives.

Cenotes were also, historically, cosmological anchors or "axis mundi" in city layouts and planning. There are many examples of the deep connections between watery caves, mountains, and pyramids that were widely shared across the entire *Mesoamerican* world (the term given to the broad Indigenous cultural area now comprised by the nations of Mexico, Guatemala, Belize, Honduras, and El Salvador) (Carrasco et al., 2002; Pasztory, 1997; Vail & Hernández, 2010). At Mayapán, for example, the major pyramid was built just over the cenote, Ch'en Mul. The pyramid rises nine levels above the cenote, in reference to the levels of the underworld, and was constructed as a "radial" pyramid featuring four staircases oriented to the cardinal directions (Masson & Peraza Lope, 2014). Like a world tree or creation mountain rising to the heavens, as referenced in the later K'iche' Maya creation story the Popol Vuh (Tedlock, 1996), the pyramid-cenote complex was an embodiment of the centered cosmos. The pyramid and cenote anchored vertical and horizontal directionality and provided a fulcrum around which all life moved. Anchoring these powerful directions and linking the world itself to a watery, life-giving space, the pyramid-cenote complex was also a symbol of agricultural fertility (for further discussion on Mayapán and a general overview of the pyramid-cenote complex see Landry, 2018).

Today, cenotes and caves across the Yucatan Peninsula maintain many connections to the built environments they are part of and are still situated and centering places in hundreds of Maya communities. For the majority of these Maya towns, the centrality of cenotes likely began out of necessity, as cenotes have always provided Maya towns with access to groundwater for drinking, bathing, and irrigation. When Spanish colonial grids were imposed on Maya mission towns, such grids carefully worked around cenote locations, as happened at Tahcabo in eastern Yucatan. Today, cenotes are widely used for bathing, municipal plumbing systems, irrigation, and to a lesser extent for consumption. Many have also been re-branded as tourist destinations. Like their Western archival counterparts, cenotes embody important and priceless cultural and historical value.

In contemporary Yucatan, cenotes as biocultural places are also intimately related to the Indigenous language, Maaya t'aan, as well as to primary source materials including local oral histories, shared folklore, and Maya codices—illustrated books of prophecy and fate created by Maya scribes in the fifteenth century. Cenotes are particularly strong placemaking devices, which serve as memory-keepers, and contemporary world-making features on the Yucatecan landscape. In short, cenotes are places that elicit both storytelling and documenting, in the sense that they simultaneously preserve and create Maya *mythhistories*.

Art historian Elizabeth Boone (1999) coined the term *mythhistory* to describe a common form of Mesoamerican narrative tradition that blends elements typically characterized by the Western literary tradition as belonging in distinct categories, namely "myth" and "history." In Native American literature, there is less of a dichotomy between accounts of the past that describe events, places, people, and activities that were witnessed or experienced first-hand and tied to specific calendrical dates and locations versus those associated with primordial time and celestial, spiritual, or supernatural realms. Whether spoken, performed, or written, Mesoamerican mythhistories are inclusive of primordial and historical times and places that all contribute to true accounts of the past. We agree that this concept more accurately expresses how Indigenous languages communicate histories and the peoples' relationships to diverse landscapes across Mesoamerica.

During the implementation phase of the PACECCY project, middle-school students collected mythhistories from knowledge holders in their communities. Stories were usually related to them by community elders, including their grandparents and, in some cases, local *jmeen*, or healers, who often serve as de facto community historians and keepers of traditional knowledge (Fig. 10.2).

Fig. 10.2 Oral history interview conducted in the town of Xocén, Yucatán, Mexico. Middle school students interview the community's *jmeen* about the town's cenote. Photography by Khristin N. Montes

Stories related to cenotes were also shared by students through project surveys taken both before and after the implementation of class activities. These were recorded in Spanish as the students' primary language (although several students are also fluent or passive speakers of Maaya t'aan) (see Landry-Montes et al., 2021). Students collecting mythhistories from elders used oral history backpacks developed specifically for the project. Later donated to the participating schools, the backpacks contained tools for collecting community stories, including voice recorders, notebooks, pens, and flashcards with cenote-focused vocabulary and oral history prompt questions in both Spanish and Maaya t'aan. The bilingual oral history and vocabulary cards gave students the option to interview community elders in Maaya t'aan and practice translating from one language to the other.

The students' answers to survey questions, as well as the content of the oral mythhistories they collected contained certain reoccurring themes. Foremost and to this day, a cenote is associated with a specific deity, spirit guardians, or *dueños* (a pairing that corresponds to precontact times based upon the content of the extant Maya codices). Many of these dueños are thought to guard precious treasure and offerings within the cenote's depths—a point that students across communities recounted. Second, cenotes are sanctuaries for, and linked with, powerful animals and Maya ancestors. It is said that certain cenotes can themselves embody zoomorphic and anthropomorphic qualities particularly those of snakes and feathered serpents. Though some aspects of contemporary Maya stories have changed from pre-Spanish contexts, the survival of feathered serpent ideologies, in particular, reflects a clear cultural continuity with the ancestral past. In Yucatan, the feathered serpent was a hero-deity who was intimately linked to the religion and politics of several of its major Indigenous cities (including Chichén Itzá and Mayapán). He was called Kukulcan in the Yucatec-speaking world and known as Queztalcoatl in Nahuatl-speaking regions further north (both translate to "feathered serpent") (for further background on this figure see Carrasco, 1982; Landa & Tozzer, 1941; Masson & Lope, 2014; Miller & Taube, 1993; Nicholson, 2001; Ringle, 2004; Ringle et al., 1998).

The connections between cenotes, mythhistories, and animacy continue to be important to Maya communities. More important, perhaps, is the notion among Maya youth that cenotes should be cared for today because of these connections. It was explicitly stated in student survey feedback, for example, that cenotes were—and are—sacred and animate resources once cared for by their ancestors. It was noted that they have been handed down to Maya youth as patrimony to be protected now and in the future. Notably, however, these same connections and the belief that cenotes are living and influential places, can make cenotes dangerously powerful locations—many contemporary Maya stories about cenotes and their ecosystems include some element of peril. For example, there is a continuing notion that beautiful, but malevolent, female spirits live within certain cenotes and come out to lure men to their deaths at night (particularly inebriated men). Lastly, both oral mythhistory interviews and student surveys reveal continuance of the popularly held idea that cenotes were places of human sacrifice in the ancestral Maya past. We recount some of these collected survey answers and oral mythhistories below.

One particularly popular mythhistory shared throughout Yucatan concerns a giant feathered serpent (La Tzucán) that inhabits a cenote. A common version of the story that we heard, shared across smaller Maya communities in Yucatan, as well as in larger cities including Valladolid and Merida is as follows:

> It is said that some cenotes have guardians who care for, protect and live within them. They can be animals or spirits, but they all have special powers. A story that is shared in various Mayan communities throughout Yucatan is about "La Tzucán," a large snake that lives in the cenotes. In Yucatan there are all kinds of snakes, but La Tzucán is unique. He is very big, much bigger than any snake you have ever seen, and he has feathers. The history of this snake began thousands of years ago when there was a terrible drought in the land of the Mayab. Cháak, the God of rain, searched and searched for water, but the earth and the skies were dry. There was no water in the cenotes, caves or rivers, so he rode away on a winged animal in search of water.

> He traveled for many, many days and finally got tired and to regain his strength he sat down on a thick log. And just as he sat down, the trunk moved! Because it was the body of a great snake! The snake was hungry and ate the animal that Cháak was riding, Cháak was so angry with the snake that he said—now it will be your duty to travel with me in search of water— The snake wanted to run away, but the god stopped it and after a fierce fight finally subdued it, and then Cháak with his power made the snake grow large, feathered wings and so it took flight, with Cháak on his back, looking for water. They traveled until they reached the sea and there Cháak filled hundreds of pots with water.

> The snake wanted to stay in the sea, because there was so much water that it could swim in, but Cháak tricked him by telling him that as soon as they filled the cenotes and caves with the water they collected, there he would have as much water as he wanted, and he would be the guardian of that water. Cháak also promised him that he could return to the sea after having seen all

the cenotes, that is, when he was about to die. But Cháak is smart because he knows that the snake will always lose its skin, it seems to die, but it will grow new skin year after year. And that is how the great Tzucán never died, and it is said that the plumed serpent, Tzucán, guards the water of the cenotes to this day. (Batún Alpuche et al., 2020, p. 42)

The feathered serpent guardian from the story above was described to us by many students in surveys and oral mythistories and seems to be an especially important figure in the town of Calotmul. There, students commonly recounted tales of a feathered serpent emerging during certain times of the year:

During Semana Santo (Holy Week) a giant snake with wings emerges from the cenote and flies around its rim to protect it.

A giant serpent lives in our [Calotmul's] cenote. The cenote is very dangerous because of the serpent. It emerges during Semana Santa and flies around the outside of the cenote. It is the guardian of the cenote.

In the towns of Xocén and Kaua, three local middle-school students collected the following stories about other dueños of cenotes, the treasures protected by the dueños within the cenotes, and the concept of these treasures as offerings to cenotes.

They say about the history of my town, that all cenotes have a lord. A long time ago in ancient times, two girls went to take water out of the cenote. They just had a week of being married and living with their new families. There, at the cenote, they were talking, and they asked one another to show their wedding rings, but when they took the rings off of their fingers, they dropped them by accident. When they tried to take the rings out of the water, both of them were sucked into the cenote by the evil air living in the cenote.

According to the history of my town, they say that all cenotes have their own lord, and it is said that this cenote's lord is a woman who gets out of the cenote during the night to take drunk men inside the cenote…

It is said that in the cenote during a special day, in ancient times, you may see the shadow of a treasure chest full of jewels and money. People say that all these jewels and money belong to the princesses who were given as sacrificial offerings to the ancient gods.

Cenotes are clearly elements of living, and lived, landscapes, as is attested to by the stories above. They also braid together and preserve rich Maya histories, environmental ecology, and the sense that such patrimony must be protected. Students from Kaua, Cuncunul, and Tahcabo shared the following stories with us about their connections to cenotes as special ecosystems—important for their own lives as well as those of their ancestors. The students' stories reflect sentiments that as inheritors of the cenotes, it is now their job to care for them.

> For me, the cenotes are most important because they were left to us by our ancestors, the ancient Maya. They protected the cenotes and cared for their plants and animals. It is important that we care for them now.
>
> Cenotes are a natural gift left to us. They are the patrimony of humanity. They are especially important for their flora and fauna.
>
> Cenotes are important for their natural, fresh water and they serve as a special ecosystem for different animals. We need to protect them because our grandparents and ancestors left them to us.
>
> For me, cenotes are significant because their biodiversity is part of our patrimony. For many years, cenotes have helped the people in the community of Tahcabo live. We inherited the cenotes from our ancestors. Cenotes are beautiful places. It is there that we can forget our troubles and there is no noise, just the song of the birds.
>
> Cenotes are important to me because we can consider them part of our patrimony left to us by our ancestors. One of our histories about the cenote in our town is that a meteorite fell here in our town of Kaua and formed a large hole. Water started to fill there and then a star fell in the hole and turned the cenote green (this is why the cenote in our town is called Yax Ek' or Green Cenote).

Ultimately, telling stories about cenotes links language, people, and place to one another. This practice of storytelling about cenotes has a deep history in Yucatan. Stories featuring supernatural animals, like la Tzucán and Xtabay, correspond directly to the most common categories of the oral tradition documented by ethnographers and shared widely across the peninsula, namely supernatural animals, anthropomorphic beings, and witchcraft, all linked to caves and cenotes, which are both dangerous and guarded (Cervantes & Augusto, 2010, p. 44). Cenotes have long been key elements in visual as well as oral language systems. They appear, for example, in the iconographic and epigraphic contexts of the precontact Maya codices introduced earlier in this chapter (Hernandez & Vail, 2013; Vail, 2000). Although only four pre-Hispanic codices exist today (not one is currently archived in the Maya region), there certainly were hundreds, if not thousands, of pictographic documents curated in Maya communities throughout the Peninsula in the early sixteenth century. Unfortunately, the majority were destroyed by Spanish conquistadors or missionaries in the Colonial Period (Landa & Gates, 2015). Of the four that remain, at least two—the Dresden Codex and Madrid Codex (so-named for the European cities in which they are archived today)—were likely authored by scribes within communities located in the Yucatec zone of cenotes (Chuchiak, 2004). These books utilize standardized forms of glyphs and visual imagery that communicate Maya creation stories, chart and prognosticate ritual and agricultural events related to the Maya cyclical calendars, and document local celestial and terrestrial sacred landscapes (Knowlton, 2010; Vail, 2000; Vail & Aveni, 2008; Vail & Hernandez, 2013). The Madrid Codex, in particular, features several examples of cenotes, which are paired with important patron deities such as the rain god Cháak and refer either to underworld spaces or the central/fifth direction in Maya cosmography. Notably, Cháak is a deity that is still called upon by local jmeen during *Ch'a Cháak*, or rain-calling, ceremonies in Yucatan wherein altars are set up and oriented to the four directions with a special position on the altar saved for the center direction (Hernandez & Vail, 2013; Salvador Flores & Kantun Balam, 1997).

Like contemporary Ch'a Cháak rituals, cenotes from the ancestral Maya codices are closely associated with specific flora and fauna of Yucatan. Additionally, the ancestral Maya relationships among patron deities, cenotes, and animals from Yucatan closely mirror the contemporary oral histories recorded by students participating in the PACECCY project. In Fig. 10.3, Cháak is depicted as a patron deity of a cenote from the Madrid Codex. The cenote's rim is formed by the

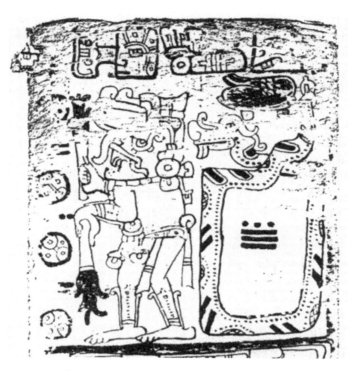

Fig. 10.3 Image of the rain deity, Cháak, depicted as the patron deity to a zoomorphic cenote from the Madrid Codex, page 5a, top register. Image altered after the combined Léon de Rosny, "Codex Cortesianus", (Rosny 1883) Libraires de la Société d'Ethnographie, Paris, 1883 & Brasseur de Bourbourg, "Manuscrit Troano", Imprimerie Impériale, Paris, 1869–1870.[1] The full codex can be found on FAMSI: http://www.famsi.org/mayawriting/codices/madrid.html

body of a serpent. This visual representation of a cenote and serpent in Madrid is strikingly similar to the accounts given by students in Calotmul in which the feathered serpent emerges from their cenote and flies in circles around it—effectively becoming the circumference of the cenote itself through this action.

---

[1] The Codex Cortesianus and Manuscrit Troano represent two pieces of the Madrid Codex (they were reunited from two separate collections).

Given that Maya codices are considered to be priceless examples of world heritage and embody a great wealth of information about placemaking, environment, and Indigenous language of Yucatan, it is a remarkable and tragic consequence of colonialism that most of the Maya students with whom we worked did not know of the codices' existence prior to the project. This is one result of heritage distancing or the process whereby the dominant culture worked to systematically de-legitimize Indigenous cultural practices and knowledge systems (especially religions), separate Indigenous people from their lands and sovereignty, and ultimately attempt (unsuccessfully) to erase them and reduce their history to a footnote (McAnany, 2016, pp. 3–35, 2020, p. 322). Of course, the structural violence that has coalesced through colonialism over several centuries has taken many forms, including the nationalization of heritage as patrimony and the re-casting of the precontact, autochthonous civilizations of Mexico as progenitors of the *mestizaje*, a mixed race comprised of the equal blending of two great civilizations (Bonfil Batalla, 1994; Castañeda, 2009). While Indigenous people in Mesoamerica continue to be marginalized socially, economically, and structurally within the system, they are increasingly estranged from a sense of shared identity based on strong connections to the tangible and intangible expressions of cultural heritage like language, ancestral traditions, archeological sites, and artifacts (McAnany, 2020, p. 322). This is arguably reinforced not only in Mexico, but throughout the Americas in the educational curricula that eschew Indigenous knowledge systems and languages while presenting colonialist social hierarchies, landscapes, and resource extraction as innate or inevitable.

As we observed this kind of heritage distancing in operation first-hand during our project, one goal became the introduction of the codices into school curricula, including how they reference and valorize the visual cultural and linguistic environment of Yucatan. Since the collaborating teachers were also largely unfamiliar with the codices, significant resources were directed toward capacity-building in order to ensure that students could learn about their ancestors' book making and literary tradition. This capacity-building is still an ongoing component of the project. The experiential learning activities included students authoring their own personal codices using the visual vocabulary and structure of the ancestral books (Fig. 10.4).

The study of cenotes as biocultural heritage provides a context for the validation of community wisdom, with a strong capacity to link

Fig. 10.4 Students in the town of Xocén, Yucatán create their own codices. Photograph by Yaremi Tuz May

Indigenous knowledge systems (such as language) to environmental conservation and sustainability. Cenotes are capable of acting in these ways because of their powerful roles as placemaking devices and living archives. They have the power to center and preserve human stories, while simultaneously linking people to ecological environments and to one another, even across generations. Storytelling is an important Indigenous research methodology, as Linda Tuhiwai Smith (1999) points out:

> Each individual story is powerful, but the point about the stories is not that they simply tell a story or tell a story simply. These new stories contribute to a collective story in which the Indigenous person has a place. For many Indigenous writers stories are ways of passing down the beliefs and values of a culture in the hope that the new generations will treasure them and pass the story down further. The story and the storyteller both serve to connect the past with the future, one generation with the other, the land with the people and the people with the story. (pp. 144–145)

Stories and storytelling ultimately provide bridges between people and people, and between people and places. Unfortunately, as López-Maldonado and Berkes (2017) discuss, one of the key issues limiting cenote conservation is the lack of cultural valorization of cenotes or, at least, the weak support for transgenerational bridging of knowledge/values about them. They suggest that one solution is to provide education about cenotes at the local level (the community level) and support emotional engagement people have with these special places. Our own project followed this recommendation and sought to create access to new educational opportunities, build capacity with teachers, and encourage a local "revalorization" of cenotes as archival and placemaking devices, linking together intergenerational knowledge shared through stories *in* place. Oral mythhistories recorded during this project and descriptions of cenotes recounted in student surveys strengthen intergenerational knowledge transmission and preserve narratives about cenotes as irreplaceable examples of cultural heritage. These reflections articulate a certain "iconography of cenotes" made tangible in the lived environment and visualized in cenote imagery, such as that presented in the Postclassic codices.

Cenotes are precious, unique places. They are contexts that inspire connection and deep emotion, and anchor entire communities—they have long done so. Cenotes are also spaces in desperate need of care, in the sense of ecological conservation and in terms of maintaining and preserving their priceless links to biocultural Maya heritage. For those who can hear them, cenotes themselves seem to tell us these things,

> Take care of me,
>> I possess grandiose stories,
>> That I would like to protect.

Acknowledgements Our project was made possible by funding from the National Geographic Society, generous donations from EarthEcho International and OpenROV, and the dedication of all of our team partners including InHerit and UNO faculty, staff, researchers and students; our middle school partner directors, teachers, and students; and all of our invited speakers, researchers, designers and artists. We are particularly grateful to the community members of Kaua, Yalcobá, Cuncunul, Tikuch, Calotmul, Tahcabo, Tixhualactún, Xocen, and Hunukú for supporting their children throughout our project. During July 1–12, 2019, members of our project including Patricia A. McAnany, Iván Batún Alpuche, Dylan Clark, and Khristin Montes were invited to present key themes from our project at the University of North Carolina Chapel Hill as part of COLING, an international organization that promotes and develops sustainable measures for language revitalization and maintenance. More recently, in the spring of 2021, we participated in a lighting roundtable discussion at the Native American Art Studies Association (NAASA) conference at which we were encouraged to think about whether or not cenotes could be conceptualized as community archives. We thank the participants of both of these meetings for their encouragement and feedback in bringing these ideas together for this chapter.

Bibliography

Anda Alanís, G. (2010). En los profundos dominios de los dioses: Arqueología subacuática en Yucatán. In L. F. Souza (Ed.), *Los antiguos reinos del juagar* (pp. 133–145). Secretaria de Educación del Gobierno de Yucatán.

Bassie-Sweet, K. (2008). *Maya sacred geography and the creator deities*. University of Oklahoma Press.

Basso, K. (1996). *Wisdom sits in places: Landscape and language among the Western Apache*. University of New Mexico Press.

Batún Alpuche, I., Clark, D. J., Landry-Montes, K., & McAnany, P. A. (2020). *Ciencia y saberes de cenotes Yucatecos*. Universidad de Oriente Press.

Batún Alpuche, I., Clark, D., Montes, K. N. L., & McAnany, P. A. (2021). *Ciencia y saberes de cenotes yucatecos*. Universidad de Oriente. https://in-herit.org/resources-2/resources-for-teachers/

Beddows, P., Blanchon, P., Escobar, E., & Torres-Talamante, O. (2007). Los cenotes de la Península de Yucatán. *Arqueología Mexicana, 14*(83), 32–35.

Bonfil Batalla, G. (1994). *México profundo: una civilización negada*. Grijalbo.

Boone, E. (1999). *Stories in red and black: Pictorial histories of the Aztec and Mixtec*. University of Texas Press.

Bourbourg, B. (1869–1870). *Manuscrit Troano*. Imprimerie Impériale.

Brady, J. E., & Ashmore, W. (1999). Mountains, caves, water: Ideational landscapes of the ancient Maya. In W. Ashmore & A. B. Knapp (Eds.), *Archaeologies of landscape, contemporary perspectives* (pp. 124–148). Blackwell Publishers.

Brady, J. E., & Prufer, K. M. (Eds.). (2005). *In the maw of the Earth monster: Mesoamerican ritual cave use*. University of Texas Press.

Carrasco, D. (1982). *Queztalcoatl and the irony of empire: Myths and prophesies in the Aztec tradition*. University of Chicago Press.

Carrasco, D., Jones, L., & Sessions, S. (Eds.). (2002). *Mesoamerica's Classic heritage: From Teotihuacan to the Aztecs*. University Press of Colorado.

Casey, E. (1976). *Imagining: A phenomenological study*. Indiana University Press.

Casey, E. (1987). *Remembering: A phenomenological study*. Indiana University Press.

Castañeda, Q. E. (2009). Aesthetics and ambivalence of Maya modernity: The ethnography of Maya art. In J. Kowalski (Ed.), *Crafting Maya identity: Contemporary wood sculptures from the Puuc region of Yucatán, Mexico* (pp. 133–152). Northern Illinois University Press.

Cervantes, E., & Augusto, C. (2010). La mitología en Yucatán. In F. F. Repetto (Ed.), *Estampas etnográficas de Yucatán* (pp. 43–75). Ediciones Universidad Autónoma de Yucatán.

Chuchiak, J. F. (2004). Papal bulls, extirpators, and the Madrid codex: The content and probable provenience of the M. 56 patch. In G. Vail & A. Aveni (Eds.), *The Madrid codex: New approaches to understanding an ancient Maya manuscript* (pp. 57–88). University Press of Colorado.

Clark, D. J., & Anderson, D. S. (2015). Past is present: The production and consumption of archaeological legacies in Mexico. *Archaeological Papers of the American Anthropological Association, 25*, 1–18.

Coggins, C. C. (1984). Introduction: The Cenote of Sacrifice Catalog. In C. C. Coggins & O. C. Shane (Eds.), *Cenote of sacrifice: Maya treasures from the sacred well at Chichén Itzá* (Vol. III, pp. 22–29). University of Texas Press.

Crumley, C. L. (1999). Sacred landscapes: Constructed and conceptualized. In W. Ashmore & A. B. Knapp (Eds.), *Archaeologies of landscape, contemporary perspectives* (pp. 269–276). Blackwell Publishers.

Hernandez, C., & Vail, G. (2013). The role of caves and cenotes in Late Postclassic Maya ritual and worldview. *Acta Americana, 18*, 13–45.

Knowlton, T. W. (2010). *Maya creation myths: Words and worlds in the Chilam Balam*. University Press of Colorado.

de Landa, D., & Tozzer, A. M. (1941). *Landa's Relaciones de las cosas de Yucatán*.

Landa, D. de, & Gates, W. (2015). *Yucatan before and after the conquest*. Dover Publications.

Landry, K. (2018). *The sacred landscape of Mayapán, a Postclassic Maya center*. University of Illinois at Chicago. https://hdl.handle.net/10027/23011

Landry-Montes, K. N., McAnany, P. A., Clark, D. J., & Alpuche, I. B. (2020). Karst water resource management and sustainable educational practices in nine Yucatec Maya communities. In L. Land, C. Kromhout, & M. Byle (Eds.), *Proceedings of the Sixteenth Multidisciplinary Conference on Sinkholes and the Engineering and Environmental Impacts of Karst. NCKRI Symposium 8* (pp. 18–29). National Cave and Karst Research Institute.

Landry-Montes, K. N., McAnany, P. A., Clark, D. J., & Alpuche, I. B. (2021). Decolonizing the classroom and centering the biocultural heritage of cenotes in Yucatán, Mexico. *The Mayanist, 3*(1), 19–38.

López-Maldonado, Y., & Berkes, F. (2017). Restoring the environment, revitalizing the culture: Cenote conservation in Yucatán, Mexico. *Ecology and Society, 22*(4), 7–20. https://doi.org/10.5751/ES-09648-220407

Luzzadder-Beach, S., Beach, T., Hutson, S., & Krause, S. (2016). Sky-earth, 8 Lake-sea: Climate and water in Maya history and landscape. *Antiquity, 90*(350), 426–442.

Masson, M., & Lope, C. P. (2014). *Kukulcan's realm, urban life at Ancient Mayapan*. University of Colorado Press.

McAnany, P. A. (2016). *Maya cultural heritage: How archeologists and Indigenous communities engage the past*. Rowman and Littlefield.

McAnany, P. A. (2020). Imagining a Maya archaeology that is anthropological and attuned to Indigenous cultural heritage. *Heritage, 3*, 318–330.

McAnany, P. A., & Parks, S. (2012). Casualties of heritage distancing: Children, Ch'orti' Indigeneity, and the Copán archaeoscape. *Current Anthropology, 53*(1), 80–107.

Miller, M., & Taube, K. (1993). *The Gods and symbols of ancient Mexico and the Maya: An illustrated dictionary of Mesoamerican religion*. Thames & Hudson.

Moyes, H. (Ed.). (n.d.). *Darkness: A global perspective on the ritual use of caves*. University Press of Colorado.

Moyes, H., & Brady, J. E. (2012). Caves as sacred space in Mesoamerica. In J. E. Brady (Ed.), *Sacred darkness: A global perspective on the ritual use of caves* (pp. 151–170). University of Colorado Press.

Nicholson, H. B. (2001). *Topiltzin Quetzalcoatl, the once and future lord of the Toltecs*. University Press Colorado.

Pasztory, E. (1997). *Teotihuacan: An experiment in living*. University of Oklahoma Press.

Quijano, J. L., & Sosa, B. P. (2014). Decir que somos quien somos. In *Compendio de resultados de encuestas a población maya de la Península de Yucatán (2004–2012)*. Raíz de Sol.

Quijano, J. L., Sosa, P. B., & Robleda, G. S. (2011). *Un mundo que se desaparece. Estudio sobre la región maya peninsular*. CIESAS, CDI.

Re Cruz, A. (2003). Milpa as an ideological weapon: Tourism and Maya migration to Cancún. *Ethnohistory, 50*(3), 489–502.

Ringle, W. M. (2004). On the political organization of Chichen Itza. *Ancient Mesoamerica, 15*(2), 167–218.

Ringle, W. M., Negrón, T. G., & Bey, G. J. (1998). The return of Quetzalcoatl: Evidence for the spread of a world religion during the Epiclassic period. *Ancient Mesoamerica, 9*(2), 183–232.

Rosny, Léon de. (1883). *Codex Cortesianus*, Libraires de la Société d'Ethnographie.

Salvador Flores, J., & Kantún Balam, J. (1997). Importance of plants in the Ch'a Chaak Maya ritual in the peninsula of Yucatan. *Journal of Ethnobiology, 17*(1), 97–108.

Smith, L. T. (1999). *Decolonizing methodologies: Research and indigenous peoples.* Zed Books.

Tedlock, D. (1996). *Popol Vuh: The Mayan book of the dawn of life* (Revised). Simon and Schuster.

Vail, G. (2000). Pre-Hispanic Maya religion: Conceptions of divinity in the Postclassic Maya codices. *Ancient Mesoamerica, 11*, 123–147.

Vail, G., & Aveni, A. (2008). El Codice Madrid, un viejo documento revela nuevos secretos. *Arqueología Mexicana, XVI*, 75–81.

Vail, G., & Hernandez, C. (Eds.). (2010). Introduction to Part III: Intellectual interchange between the Northern Maya lowlands and highland Mexico in the Late Postclassic Period. In *Astronomers, scribes and priests: Washington* (pp. 263–278). Dumbarton Oaks Research Library and Collection.

Vail, G., & Hernandez, C. (2013). *Re-creating primordial time: Foundation rituals and mythology in the Postclassic Period*. University Press of Colorado.

11

Nakua nukuu ini Ñuu Savi: Nakua jíno, nakua ka'on de nakua sa'on ja jee Koo Yoso

Omar Aguilar Sánchez

Kijeo

Koo Yoso kuu in yaa ja kanùu nu Ñuu Savi ode kivi nda ndoso. *Quetzalcoatl* náani yaa ya ityi Ñuu Ko'oyo, suka jíninda nda ntyivi a ka'an nda tu'un *nahuatl*. Ini Ñuu Savi jekun ini nda ja su *Quetzalcoatl* kuu Koo Yoso. Yaa ya de kanuu ne'e tyi yaa ya kuu ja ni jekoo nda ñuu luli Ñuu Savi.[1] Ityi tyata nda ntyivi ja nki Yata Ndute (*Europe*) ndùùni nka va'a nda ja jee yaa ya, tyi nkatyi nda ja tatyi (*devil*) kuu. Ja vii kuu, ntyivi savi nduna xnaa Koo Yoso, tyi ode kivi mita katyi nda ntyivi

---

Omar kuu in se'e Ñuu Yute Suji (Santo Tomás Ocotepec), nu Ñuu Savi. Satiñu ji COLING, The Americas Research Network ji ve'e sku'a Universidad Autónoma Comunal de Oaxaca. Sava tiñu yaa, ja kuenta tu'un savi, suni nkee nu in tutu kuenta CIESAS kuiya un'un tuvi oko uu kuiya.

---

[1] Nu Ñuu Ko'oyo.

O. Aguilar Sánchez (✉)
Universidad Autónoma Comunal de Oaxaca, Oaxaca, Mexico
e-mail: oas.kooyoso.03@gmail.com

ña´anu tu nenda Koo Yoso de kuun savi-tatyi. De suni katyi nda ja
kuatyi yaa ya kuu ja nasa´a nda "viko kasiki" ode kivi mita nu Ñuu
Yute Suji, nu Ñuu Savi.[2]

Ku ka'ano ino ndo ja ta ka'on ja sá'o nasa ni nkuu kivi yata, suni ku
ka'ano ini ndoo ja ni nkundao nda tu'un, nakua jíni nda nda ntyivi inka
ñuu. De mita, tu kuno nasa ni nkuu kivi yata suni kuni kuka'ano nakua
ni nkuu ta nki'in nda ntyivi Yata Ndute. Tani katyi Jansen ji Pérez (2011,
p. 210).

Tu kuno ka'on in uu tu'un ja kuenta Ñuu Savi ji ntyivi savi kivi mita,
su ni kúni kuno nasa ni nkuu kivi yata, kúni skua'o nakua ni nsa'o ityi
tyata ja ni nkiji tee xtila, de suni na kua ni nasamo ja ni nkundao ta ni
nkiji nda tee xtila. Suni skua'o ñii ñu'u, tutu, nakua sa'o kivi mita, tu'un
kuenta nasa jekoo ña'a yivi, de suni nasa jekoo ñuu moo. Suni skua'o
nasa ka'an nda ntyivi jika ndavi, nasa ka'an nda ta ni tanda'a nda ntyivi,
de suni nasa sa'a nda nu tyi ñu'u nda, nu sa ii nda. Suni kuni skua'o
tu'un moo, tyi nu tu'un mo jiso ja jino ñaa yivi, nasa ku ka'ano ino ja
ntyivi koo. Tu'un Savi[3] kuu tu'un moo, de su tu ka'on tu'un moo ma
xna'o nakua kuu ndo, tani nsa'a nda ntyivi ña'anu.

Koo Yoso Ñuu Savi

Koo Yoso kuu in "yaa", de katyi nda ja taka in koo kaa de suni né'e tumi
ti, katyi tu'un nda ntyivi savi. Tu ka´on ja jee koo de jeku ino ja in kiti
kuu, de matu ja "yoso" de su nikuu de vasu sunka nka´an nda ja mita
jino ja kuu "tumi". Katyi nda ja koo ya de nde túmi ti, de matu ja nsu
naní tumi kuu, tyi tumi ja luu satuni kaa, kusa ju de xiñu'u nda tumi
ya. De nuku ka´on "yoso" da? Ja ya de nu kuiya 1593 nkatyi in sutu ja
ni naní Antonio de Alvarado ja "yodzo" kuu nda tumi luu ja nasa´a nda
xíni, tumi ka´anu kuenta in saa kuu. Suka ninkuu kivi yata, de kivi mita
nduka kátyo "yodzo", ka´on "yoso" ji "tumi", de su tuu skua´o nda tu´un

---

[2] Kuta'avi ni nu nda ntyivi Ñuu Yute Suji, de su ni nu nda ña'a xte'e Justyna Olko ji Cynthia
Radding.

[3] Tu'un Savi katyi nda kuu in tu'un ja ne'e *tono*.

jana´a de kuno ja ini kuu "yodzo" ji "yoso" (Jansen ji Pérez Jiménez, 2009). Sukua Koo Yodzo ja na'a kuu Koo Yoso kivi mita (Aguilar, 2020).

Nasa jíno a tu'un Koo Yoso? Ja xina ñu'u ta ni jini sa ja jee *Koo Yoso* de nana ña'anu sa Francisca Reyes Jiménez[4] nakani ña ja yaa ya de ndée ti ini miiní, de ta saamá ti nu ndée ti de niji kuu, tyi kuu savi de yi tatyi. Ya ku ja nda ntyivi jíni nda ta saamá ti nu ndée ti, de ta yàa ti de kua'a ti de skunu ti nda ñutu. De suka kuu ja nda ntyivi jíni nda ni ítyi nki de ni ityi kua'an *Koo Yoso*. De suni katyi nda ja su suka kuu nu ke'en ñuu ini Ñuu Savi, tanu ku Ñuu Ndeya, Nuyoo, Yucuhiti, de nu nda ñuu ya ka'an nda Koo Savi (Monaghan, 1995; Perez, 2008). Ya ku ja nakani nda ntyivi ña'anu ja kuu Koo Yoso ja Koo Savi, de su nduu yaa ya íini kuu nduu.

Nda ntyivi ja skua'a nda "ñii ñu'u" (*codex*) tanu ku Jansen ji Pérez Jiménez de katyi nda ja Koo Yoso ku tikatyá, de kanuu yaa ya tyi suni kuu in suji (Jansen, 2012, p. 35). Koo Yoso de suni vaji nu ñii ñu'u ja nani Yuta Tno'o, sivi de kuu suni Ìin Tatyi, de suni kuu de "Savi-Tatyi" (tuni in) (see Fig. 11.1).

Nu ñii ñu'u ya vaji nakua nsa'a *Yaa Ìn Tatyi*, nasa jekoo ndi'i nda ñuu luli Ñuu Savi, suni nasa ni xte'e nasa sa'a nda tiñu nda ntyivi de suni nasa nakuatu nda. A kuatyi yaa de *tée* ja nani Florescano (2004, pp. 216–222) kátyi ja nu ñuu nda ntyivi ja ka'an nda tu'un *nahua* Koo Yoso kuu *Ehecatl*, tyi yaa ya kua jekoo ntyivi *ña'a yivi*. Ja kuatyi yaa kuu katyo ja *Koo Yoso* ku in yaa ja kanuu ndi'i ñuu ña'a yiyi.

Tu ka'o ja jee ntyivi Ñuu Savi de yaa ya ku "yaa Tatyi" (*Wind God*). De mita u'vi kuu tu ka'on ja jee yaa Tatyi, tyi síi ja jeku ino tyi nikuu de "yaa Tatyi" nda ntyivi xtila nkenda nda Ñuu Savi de nasa'a nda "*tatyi xee*" (*evil*), tyi va su sua nka'a nda ta ni nkenda nda ntyivi ja vaji Yata Ndute (Castilla)[5] de nka'a nda ja tatyi ku. Ja ya'a de jíno ja kanùu ne'e ka'on tu'un savi tava jekuu ino ja Koo Yoso ji Savi-Tatyi inu kuu. De suni ja kuatyi Koo Yoso iyo in tu'un nu Ñuu Yute Suji a katyi nuku sa'a nda kuiya kuiya viko tee kasiki, a kuenta in tu'un a katyi nukú ñu'u nda sa'ama ndi'i tuni teku nda tee kasiki, *chilolos*.

---

[4] Kuiya ja nka'a ña ji saa de iyo ña kumi xiko kumi kuiya.

[5] Ityi tyata de Yata Ndute ka'an nda nu kuu Castilla (España), nu Europa, nu nkiji nda ntyivi xtila, de mita tyi nduka tyi Yata Ndute kuu *norte* (Estados Unidos).

Tumi xiñu

Nu kivi oko uja, yoo in, kuiya un'un tuvi xa'on uu, ni jee ní ñuu luli ja ka'an nda Itu Tasu, nu Ñuu Yute Suji. Ya de ni natu'un ni ji in tée ña'anu ja nani Benito Cristino Cruz Sánchez.[6] De suan nkatyi de ja kuatyi tumi Koo Yoso:

> Ajá, ndee [Koo Yoso] tumi ti de sa ndaka nu teku kuu tumiti ka'an nda ii. De iyo tu'un ja ñuka nakenda nda tee kasiki a kuu sa'ama nda kasiki, katyitu nda.

Vii kuu ja ni nka'a tee ña'anu Benito, tyi nkatyi de ja Koo Yoso ndee tumi ndi tuni teku. De ja kuatyi tu'un ya de jíno nuku nda tee jana'a nkayu nda *yoso* nu *ñii ñu'u* taka yùù tumi ndi'i tuni teku. Tani kaa nu tuni uu (see Fig. 11.2).

Nuu ñii ñu'u Tonindeye ja ni nkavi Jansen ji Pérez Jiménez (2007a, p. 213) yóso in tee ja nánide Kumi Kuiñi, de katyi nda ja tee yaa kuu *Quetzalcoatl*, de yaa ya jíso xíini tumi ndi'i tuni (*Quetzalapanecayotl*). Tani kaa tuni uni (see Fig. 11.3).

Kanuu kuno tu'un yaa tyi katyinda ja kivi yata de nda to'o ñu'u nda sa'ama nu tée nda tumi. Suni katyi nda ja tumi ya'a de ñu'un nda sutu Yaa Nkandii. De nda tumi saa kuii, ja ka'an nda *Quetzal*, de su katyi nda ja ñu'un nda tu kuika ne'e ntyivi (Filloy, 2019, pp. 20–21).

Ja ya de jíno nu kuu Koo Yoso jiso xíni a ne'e tumi ndi'i tuni teku. A kuenta ya'a suni jino nuku nu Ñuu Yute Suji de Ñuu Nkuiñi katyi nda tu ni'o in tumi kuenta Koo Yoso de kuika ne'e ko'o ndo. Ta katyo ja kuika ne'e koo nsuu kuu ja kuenta xu'un tyi ja kuenta koo va'a nu de suni kune va'a nu niñi ji kiti tava kaa nu. Ja tu'un yaa de suni ka'an nda ityi Ñuu Ndeya (Witter, 2011, p. 109).

Ja ya de jíno nuku ndee nda tumi xíini nda ntyivi jana'a, ntyivi ja ni yosnuu nu nda ñuu, ntyivi ja ni ndee nda nu yuu, yaa ji ndoso ja yoso nu ñii ñu'u. Tani tutu ja nani *Codex Mendoza* ji *Matricula de Tributos*. Ja ndu tutu ya de katyi nasa tumi ni jika nda ntyivi Ñuu Ko'oyo nu nda

---

[6] Kuiya ja nka'a de ji saa ni iyo de uni xiko uja kuiya.

ntyivi savi Ñuu Yoso Koo ji Ñuu Ndinu. Ntyivi Coixtlahuaca ni je´e nda de uu tuvi tumi kuenta sàà kuii de nda ntyivi Ndinu ni je´enda de in tuvi (Codex Mendoza, folio 43r and 45r).

Viko Kasiki ji Koo Yoso

Suni natu´u ni ji tee ña´anu Benito[7] a je´e Koo Yoso, de nkatyi de ja nda teku nda sa´ama ja ñu´u nda tee kasiki Ñuu Yute Suji vaji nu nda tumi ja nde yika ji xíini Koo Yoso. A kuenta nda tee kasiki, jínisa, de suni nakani nda ja nda tee kasiki tu kua´a nda ve´e ve´e de jitaje´e nda de nda ntyivi xive´e táva nda nuni tata, tyikin, ndutyi de nda tee kasiki ya de nasa íínda nda ya. De tee ña´anu Benito suni nkatyi:

> Ajá, saa íi nda tee kasiki, sede jitaje´e nda sede jiko nda ta nuu nu iñi a kuu ndoo nuni tata, ja nda tyinkin. Sua nsa´a nda ni xina de mita tyi nduka nasa´a nda sua.

Ja je´e ya de kuu katyi ni ja "viko tee kasiki" kuu in viko ja kanùu, ja kuu ñuu ni Ñuu Yute Suji. Kuu in viko ja luu ne´e kaa de suni kua´a ne´e ntyivi jee kasiki de jitaje´e a kuatyi kua kunde´e. Kuiya 2020 de nùu nda 3519 nda *chilolos*. In viko ka´anu nu naketa´a kuane´e ntyivi. Viko ya ka´an nda tu´un xtila *Carnaval*, de tu´un *carnaval*, a kuenta sivi de jíka ne´e vaji sivi, vaji ode Yata Ndute (Hernández-Díaz & Angeles Carreño, 2005). A kuenta viko tee kasiki ja sa´o *Ñuu Yute Suji* de nsuu viko vaji Yata Ndute kuu tyi in viko kuenta ntyivi moo kuu, viko ya sa´a nda ntyivi suji ode kivi ana´a, sa´a nda tava kije´e nda kaki nda itu.

Kivi mita nda tee kasiki kasiki nda uni kivi. Ja nduu xina ñu´u kivi ndi uxi uni ñuu luli Ñuu Yute Suji de je´e nda ta ve´e ta ve´e ntyivi ñuu luli nda de kataje´e nda, ta in kua´an nda de jitaje´e nda ji yáa violin, yáa savi. Sìì satuni sa´a nda ntyivi Ñuu Yute Suji. De suni ta kua´a nda de inka ñuu, nda ntyivi sìì jíni nda de iyo ve´e nu tava nda nuni tata, tava nda nuu tee kasiki tava nasa íí nda. Iyo ta je´e nda inka ñuu, tani Nundaco, Yucuhiti, Ñuu Kuiñi, San Miguel Progreso ja Ñuu Nkaa. Ñuu Nundaco de sìì ne´e jíni nda ntyivi ya tuni nkenda nda tee kasiki tyi katyi nda

---

ntyivi ña'anu ja xina ja Ñuu Yute Suji ni kua'an nda ntyivi ja ni jekoo
ñuu Nundaco. Kuu sii ini nda tyi nda tee kasiki nasa ii nda nuni tatan
tava skee ne'e nda kuiya ñuka. A kuatyi yaa de je'e nda nu tee kasiki kaa
nda ji ko'o nda. De ja uni kivi de ndi'i nda tee kasiki ñuu luli nuu nda
ma Ñuu.

Ja kuenta sa'ama, kátyi nda ja kivi yata sa'ama ja ñu'u nda velo tava
kasiki nda kuu sa'ama kátyi, *mascada*, ña'aná ñii ji ixi yu'u, de suni ñu'u
nda xíini ja nde nanikanu *listón* de ya nditakaa xini nda de. Kivi mita de
nsama sa'ama ja ñu'u nda tee kasiki, mita de kuu naani ña'aná, núni
pañitu nuu nda, ñu'u nda tyarru (ta ñu'u nda ntyivi ko'oyo) ji sa'ama ja
sii teku, de yoso soo koton siki ndade, yi'i nda nija de jiso nda tikanaxi
(tuni kumi ji tuni un'un, see Figs. 11.4 and 11.5).

Tu nde'o de ja sa'a nda tee kasiki vaji o de kivi yata, katyo ja Koo Yoso
ku ja jekoo viko yaa. Tava ntyivi savi tyiñu'u nda ñu'un nda de va ne'e
koo itu nda de koo tava kaji nda. Sukua nda tee kasiki jínu nda ma'añu
nda ñu'u ntyivi ñoo de jita je'e nda tava kaxi nda tikanaxi ndade.

Ityi jee

Ityi tyata de *viko tee kasiki* kuatyi maa nda tée kasiki nda, niku de ndu
kasiki nda ña'a tyi ndu nde va'a ntyivi. Nda ña'a ja kúni nda kasiki
nda kusa kuni nuku'u nda sa'ama tée tava kuu kasiki nda. Kivi mita de
nku kua ne'e ña'a ja kasiki nda, de suni kivi mita de je'e nda nasa'a
nda sa'ama ña'a, tyi ñu'u nda xoo, ñu'u nda sa'ama ja ñu'u nda ña'a
tyi kua sama *tiempo* mita (tuni iñu, see Fig. 11.6). De ja je'e ya kuu ja
mita de nda ña'a nduu kuatyi ndee nda ini ve'e nda tyi nsama *tiempo*,
tyi nda ña je'e nda ve'e skua'a de ni'i nda tiñu, de suni mita nda ña katyi
nda tu kúni nda tanda'a nda a nasa se'e kúni nda koo nda. Va'a kuu ya
tyi sama *tiempo* de nda ña'a yosnuu nu nda ntyivi ñoo, tyi mita nda ña'a
suni nee nda tiñu ini ñuu. Ja kuatyi ya'a, ja kasiki nda ña'a jii tee, de ma
ka'onka viko tee kasiki tyi kuatyi *viko kasiki* ka'on. Kasiki kuu jitaje'e
de jitaje'e kuu ja kuu sii ino, sukua ku viko, de ja kuatyi ya kuatyi ka'on
viko kasiki.

Nu Ñuu Savi nduna tiñu iyo tava ni'on xu'un de suni nduna ve'e skua'a iyo tava nda ndi'i se'e ñuu skua'a in *carrera*. Nda tyivi ja ke'e nda nani nda *radicados*, ntyivi ja kua'a jika, kua'a satiñu tava ni'i xu'u, kua'a nda nuu inka ñuu, ta ku Ndinu, Ñuu Nduva, Ñuu Ko'yo ji Norti (Estados Unidos). De nda se'e nda *radicados* nduka ka'an nda tu'un savi. De suni ma kunda nda ndi'i ja iyo ñuu moo. De a luu kuu ta kije'e viko kasiki tyi nda kivi yaa nda radicados ndikoko nda kíi nda nu ñuu ja nkaku nda tava kasiki nda viko kasiki. Viko kasiki kuu in kivi ja síí ne'e nda se'e Ñuu Yute Suji, tyi jita je'e nda de kuu síí ini nda ja iyo nda ji nda ta'an nda ji se'e nda de ka'a nda sa'an savi. De nda se'e nda *radicados* jíni nda ñuu nda, nasa kuteku ñuu ya'a, ninu váji nda, de suni jíni nda nda ta'an nda de suka de jíni nda a se'e Ñuu Yute Suji kuu nda.

In, uu tu'un de ko'on

Koo Yoso kuu yaa ja kanùu nu Ñuu Savi, ode kivi nda ndoso ode kivi mita. Nuu Ñuu Ko'oyo ka'an nda *Quetzalcoatl*. Ityi nu kee nkandi, nu ka'an nda tu'un *Maya* de sivi yaa ya kuu *Kukulkan*. De Ñuu Savi ka'on Koo Yoso, de suni Koo Savi. Katyi nda ja tu ni'on in tumi de kuika ne'e koo. Suni katyi nda ja ndituni teku ku tumi ti. Ja ya de, katyi nda ja teku Koo Yoso nakenda nda sa'ama tee kasiki. Ja ya de ku katyo ja Ñuu Yute Suji kuu in ñuu jana'a, ja jekoo ode kivi nda ndoso, tyi ntyivi suji nduna xna nasa nsa'a nda ntyi ña'nu de sukua nasa'a nda ode kivi mita, ta kuu viko kasiki. De ja ya de, síí neo koo a kuu ndo se'e Ñuu Yute Suji, nu Ñuu Savi, de ka'on sa'an savi.

Memory and Cultural Continuity of the Ñuu Savi People: Ancestral Knowledge, Language and Rituals Around *Koo Yoso* Deity

Omar Aguilar Sánchez[8]

Abstract

This chapter aims to reintegrate the cultural memory of *Koo Yoso* deity in *Ñuu Savi* territory and to show its meanings from antiquity to the present. *Koo Yoso* is the Mixtec Quetzalcoatl, one of the most important Ñuu Savi deities in the Mesoamerican spiritual world. As I argue in this paper, the Feathered Serpent continues to play an important role today in the well-being of Ñuu Savi, despite the efforts by Spanish friars to suppress Mesoamerican religions in colonial times. This deity is commemorated every year by the Mixtec community of Ocotepec (Oaxaca, Mexico), in the—erroneously known as—carnival festival. This festival has taken on new meanings through the experiences of transnational migration. The language of the rain is the main way to communicate in Ocotepec, a container of ancestral knowledge, and its analysis is crucial to understand landscape, nature, rituals and cultural values of *Ñuu Savi* cultural memory. For this reason, this study is supported by cultural continuity through the Sa'an Savi-Mixtec language, showing how the Feathered Serpent is a key for the communities' identity and their relationship with nature and the landscapes they have created, from the past to the present.

---

[8] Omar Aguilar Sánchez holds a PhD. in Archaeology from Leiden University, the Netherlands, and he is a researcher belonging to the Ñuu Savi People.This article was encouraged and supported by his collaboration with the COLING project "Minority Languages, Major Opportunities. Collaborative Research, Community Engagement and Innovative Educational Tools", H2020-MSCA-RISE-2017 number 778384. Currently, Aguilar is a fellow of The Americas Research Network and a profesor of the Universidad Autónoma Comunal de Oaxaca.

Introduction[9]

The Feathered Serpent is one of the most well-known and important deities in the Mesoamerican world, known as Quetzalcoatl in the Nahuatl language. Among the *Ñuu Savi* "People of the Rain",[10] this deity, known as *Koo Yoso*, or "Feathered Serpent", has been fundamental in the history, origin and foundation of local communities. *Koo Yoso* was demonized by the Spaniards during the colonial era; however, *Koo Yoso* remains part of *Ñuu Savi*, manifests itself in a *savi-tatyi* "rain-wind" way and is remembered at the festival *viko kasiki* "feast-play". The aim of this chapter is the reintegration of cultural memory about *Koo Yoso* into the People of the Rain from immemorial to recent times.

It is necessary to be clear that cultural continuity doesn't mean a mirror between the past and the present. We are conscious that:

> the concept of cultural continuity does not mean that we have to suppose an anachronistic fossilization of society, but, quite the contrary, implies a dynamic diachronic relationship of the present with the past, in which there are bound to be important changes and in which at the same time important traditional elements and structures may be preserved. In fact, the present-day traditions and concepts become a crucial point of departure for a better identification and understanding of the themes and motifs in ancient images and texts. There is a dynamic relationship between past and present, which is captured in the term *cultural memory*. (Jansen & Pérez, 2011, p. 210)

To reintegrate cultural memory means to be conscious about our present and past, the recognition and knowledge of the painful colonial process and the disjunction that it created around our cultural heritage. It means to study our historical–cultural heritage as a whole from a decolonial framework, an integral study of precolonial artifacts and

---

[9] I want to thank all the inhabitants of the Santo Tomás Ocotepec municipality and the Ñuu Savi People. Special thanks to Justyna Olko and Cynthia Radding for your invitation to participate in this volume.

[10] The People of the Rain is one of the 68 Indigenous Peoples of southern Mexico, also known as Mixtec People.

settlements, pictorial manuscripts, colonial maps and documents and the living heritage of rituals, oral traditions and festivities among the Ñuu Savi communities on the basis of cultural continuity, being aware that any continuity also implies changes. This study has to be fundamentally linked by the *Sa'an Savi* language analysis and a strong participation of the *ntyivi savi* (rain person, Mixtec) (Aguilar, 2019, p. 335). Why the language? Not just because in our case the Mixtec language is at least three thousand years old, but it is the container of a particular relationship between the speakers, the speakers and nature, the speakers and gods. In one sentence, the rain language is the proof of a unique language to understand the world. It contains ancestral knowledge about nature, the spiritual world and landscape. Furthermore, to speak the Mixtec language today, in a globalized world, is an act of resistance, just as our ancestors did.

The Mixtec Quetzalcoatl

Koo Yoso is the Feathered Serpent. Today "feather" is commonly known as *tumi*, but the feathers of this deity do not refer to the common feather. The explanation for the different words can be found in the linguistic analysis of colonial and precolonial sources. Today *yoso* means "flat, on or over, *metate*", depending on the tone and context.[11] However, Alvarado's vocabulary of 1593 defined *yodzo* as "big feather" and associated it with precious feathers. For example, Alvarado wrote *yodzo yoco* "penacho, plumaje" and *yodziñandi saha yodzo* "atar plumas ricas haciendo plumaje" (Jansen & Pérez, 2009). In Ñuu Savi pictography, the valley is depicted as a feather board, a phonetic game where the association of *yodzo* as a valley and beautiful feathers is clear. In terms of comparative linguistic analysis, the *yodzo* of the sixteenth century corresponds to the *yoso* in *Ocotepec* nowadays, since the phoneme/dz/became/s/in the current variant. Thus, I can argue that *Koo Yoso* or *Koo Yodzo* is the "Feathered Serpent" (Aguilar, 2020).

---

[11] The Mixtec language or Sa'an Savi is a tonal language.

What do we know about *Koo Yoso* deity? The first time I heard about *Koo Yoso* was from my grandmother, Francisca Reyes Jiménez.[12] She told me that *Koo Yoso* lives in the lagoons. When *Koo Yoso* changes abode it brings with it a lot of rain and wind, known as *savi-tatyi* "rain-wind". In its wake, *Koo Yoso* throws trees and the direction in which they fall indicates where they went. This is well-known in Mixtec communities, such as Chalcatongo de Hidalgo, Santiago Nuyoo and Santa María Yucuhiti (Monaghan, 1995; Pérez, 2008). In all these versions *Koo* Yoso is named as *Koo Savi* "Rain Serpent". *Koo Yoso* and *Koo Savi*, both have a close relationship with rain and wind, therefore they are the same entity associated with natural phenomena.

Jansen (2012, p. 35), based on a decipherment by Pérez Jiménez, argues that *Coo Dzavui* (*Koo Savi*) "Serpent of the Rain" is a metaphor for the "swirl" as a creator and an important *nahual*.[13] *Koo Savi* is the protagonist of the creation narrative in the *Codex Yuta Tnoho* (Vindobonensis), where it appears as *Yaa Ìn Tatyi* "9 Wind God" (Fig. 11.1). In this scene, the relationship of the wind to rain is clear, which we can interpret as the manifestation of *Savi-Tatyi* "Rain-Wind".

Yaa Ìn Tatyi "9 Wind God" in the Codex Yuta Tnoho founded the Mixtec communities and himself is the manifestation of the civilization and religious principles. For this reason, Florescano (2004, pp. 216–222) suggests his correspondence with the Nahua God *Ehecatl*, understood as the God of the cosmos and human creation. Thus, *Koo Yoso* is the Mixtec manifestation of a pan-Mesoamerican spiritual entity.

Friars associated this deity with the devil, which may explain why in the current *Sa'an Savi*, the term *tatyi* refers to both wind and the devil, according to the tone and context of speech. In the same way, knowing the essence of the Mesoamerican Gods, both *Koo Savi* and *Savi-Tatyi* are manifestations of the same deity, *Koo Yoso*. Its spiritual power endures in the cultural memory of Ocotepec, as observed in the stories I heard from my grandmother, where *yoso* (precious feather) is the remnant of a

---

[12] In 2018, she was 84 years old.
[13] "Animal companion" or "alter ego in nature".

Fig. 11.1 Koo Yoso carrying the water of the sky (from the Coast) to Mixtec Highlands. App "Códices Mixtecos" (2019)[14]

concept that has remained through the centuries. It is also interesting to highlight that in Ocotepec a ritual to *Koo Yoso* is recreated every year, the *Viko Tee Kasiki* "festivity of the man who plays".

Precious Feathers

On 27 January 2017, I went to the community of *Itundi* (Lázaro Cárdenas) in Santo Tomás Ocotepec where I talked with Mr. Benito Cristino Crúz Sánchez.[15] I asked him about the *Koo Yoso* feathers, and he told me the following.

---

[14] https://play.google.com/store/apps/details?id=com.codice.celina.codicesmixtecos.
[15] He was 67 years old in 2017.

Ajá, ndee [Koo Yoso] tumi ti de sa ndaka nu teku kuu tumiti ka'an nda ii. De iyo tu'un ja ñuka nakenda nda tee kasiki a kuu sa'ama nda kasiki, katyitunda.

Yes, *Koo Yoso* has feathers; the elders say it has feathers of all colors. And there is knowledge that chilolos were inspired by these colors to make their clothes. The elders say that.

Mr. Benito's statement that *Koo Yoso* has feathers of different colors is very important when it is associated with significances of precolonial and colonial material culture that show us how Mesoamerican concepts have prevailed even to the present. In iconographic terms, the glyph that represents the valley in the codices is depicted with feathers of different colors (Fig. 11.2).

In addition to the above, Lord 4 Jaguar *Koo Yoso*—identified as the historic Quetzalcoatl by Jansen and Pérez Jiménez (2007a, p. 213) in the

Fig. 11.2 Valley of the Rain God. In this toponymic glyph, we can observe the head of the Rain God over the valley constituted by feathers of different colors. Codex Tonindeye Reverse (Nuttall), page 48-III. © The Trustees of the British Museum[16]

---

[16] I want to thank The Santo Domingo Centre of Excellence for Latin American Research (SDCELAR), at the British Museum, for the rights of reproduction of the images 11.2 and 11.3.

Codex Tonindeye (p. 75)—carries a headdress with feathers of different colors (*quetzalapanecayotl*) as characteristic attribute (Fig. 11.3).

Why are these data, the colors and the feathers important? For the symbolic meanings of feathers in Mesoamerican spiritual world. The feathers had a symbolic, religious and ornamental use, related to the nobility and gods. Even the specific feathers were associated with to particular Gods. For example, the eagle's feathers were attributed to the Sun God and the Feathers Quetzal were a symbol of precious, richness and fertility (Filloy, 2019, pp. 20–21).

It seems clear that this specific headdress associated with Quetzalcoatl, *Koo Yoso* in the Mixtec people, has multicolored feathers,

Fig. 11.3 Lord 4 Jaguar "Koo Yoso" in the Codex Tonindeye Reverse, page 75. © The Trustees of the British Museum

characteristic that was mentioned by Mr. Benito for this deity.[17] Furthermore, the feathers signify richness, fertility and the quality of precious, ideas that prevail to the present. In Santo Tomás Ocotepec and Santa Maria Cuquila, I have heard from several residents that if you got one feather of the *Koo Yoso*, you would be very fortunate (*tu ni'i nu in tiumi Koo Yoso de kuu kuika ne'e nu*). *Rich* does not refer to money or accumulation of wealth, as in the Western worldview; rather it refers to a fortunate person, like the narrative of *Ñuu Ndeya*-Chalcatongo, in another Mixtec community, where they said that the feathers of *Koo Savi* are "little, multicolor and shine like the sun" (Witter, 2011, p. 109).

Knowing the meaning of precious feathers, we can imagine the social status of the person who wore a headdress. Rulers, nobility and warriors were represented in murals, sculpture and codices through Mesoamerica wearing a huge variety of feather headdresses with different symbolic entanglements. It is no coincidence that feathers are one of the most valuable objects requested by the Triple Alliance from the tribute provinces, as we observe in the Codex Mendoza and the *Matrícula de Tributos* for the Mixtec area. For example, the Coixtlahuaca province—in the Mixtec Lowlands—paid the Triple Alianza "eight hundred rich green long feathers that they call Quetzale" and *Tlachquiauhco* province—in the Mixtec Highlands—four hundred feathers (Codex Mendoza, folio 43r and 45r).

Viko Kasiki and Koo Yoso

Returning to my conversation with Mr. Benito, he related the colors of *Koo Yoso* to the clothes of *tee kasiki* "men who play", *chilolos* or "masked" characters of the "Carnival" of Ocotepec. From my own experience and from what other people shared with me about this festivity, I told Mr. Benito that I had seen on a few occasions that when the chilolos went

---

[17] It is worth reflecting whether the images, sculptures that should have had a color coating and buildings associated with Quetzalcoatl or the Feathered Serpent should have been multicolored. Thus, it would be worth doing more specific studies and not only to associate the long feathers "blue-green" of the quetzal as characteristic of the "Feathered Serpent" but considering its range of colors.

to the houses, the people of the houses would take out their *nuni tata* "corn to sow", their beans, so that the chilolos could bless these seeds. He confirmed my observation and added:

Ajá, saa ii nda tee kasiki, sede jitaje'e nda sede jiko nda ta nuu nu iñi a kuu ndoo nuni tata, ja nda tyinkin. Sua nsa'a nda ni xina de mita tyi nduka nasa'a nda sua.

"Yes, they bless it, and then they dance around the corn seeds to be sown. They did it before, but they don't do it that much today".

The carnival of Santo Tomás Ocotepec is the most important festivity for its inhabitants, as seen in its colorful esthetics and the full participation of the community. According to the 2020 census, Ocotepec has a total of 4066 inhabitants.[18] In that same year, *Viko kasiki* brought together 3519 masked participants, not counting the organizers and spectators. Thus, we can observe the participation of more than 80 percent of this municipality throughout the organization and performance of the festival. The carnival, as a celebration before the Christian Lent, came with the introduction of the Catholic religion to the Americas. But after the conquest, Mesoamerican cultures took the carnival date to perform dances and rituals with different origins and meanings than those of the Christian tradition. Many dances—such as that of the devils—are performed as a satire to colonial rule, where they are personified and mocked with characters that are portrayed with European characteristics (Hernández & Angeles, 2005). Other communities, such as Santo Tomás Ocotepec, took this date to perform rituals linked to the agricultural cycle, erroneously called carnival, because they performed this ritual as part of the carnival dates. Thus, without pretending to describe in great detail the carnival—which goes beyond the objective of this article—here I will only outline the most significant parts for the themes of this book related to language and living with nature.

The *viko kasiki* festivity of Ocotepec lasts 3 days. In the first two days the chilolos—of each of the thirteen communities that integrate

---

[18] http://www.microrregiones.gob.mx/catloc/LocdeMun.aspx?tipo=clave&campo=loc&ent=20&mun=532, this has been consulted in November 25, 2019.

the municipality—go household by household, eating, drinking and dancing to the rhythm of traditional or recorded *chilenas*.[19] A few people still take out their selected seeds that will be used for planting and the masked *chilolos* bless them. For its magnitude and colorful performance, the *viko kasiki* is known in the surrounding municipalities, such as Nundaco, where the people no longer play, but they appreciate it very much when the *chilolos* of Ocotepec visit them. For example, Ms. Martina Avila—who was 84 years old in 2017—shared with me that she is thankful when the masked men of *Ñuu Ka'anu* "Big Town"[20] arrive, because they can bless her seeds. In that way, she will have a better harvest during the next agricultural cycle. In general, in all houses, these dancers are very welcome, where they receive refreshing meals and drinks. On the last day of the carnival, all the *chilolos* from all the communities meet at the center of Ocotepec.

A few decades ago, the *chilolos* used to wear a suit of *manta*[21] or silk in rich colors, silk scarf, a hat with different colored slats, leather mask with a long beard—that represents elders—and *huaraches*—Mexican leather shoes. However, with the passage of time, the community is replacing the traditional elements with others that look more like the Mexican style. Today, the chilolos' costumes consist of masks of different materials,[22] bandanas, *charro (Mexican hat)*, satin shorts and shirts of different colors, long socks, *gaban* or *sarape* "Mexican poncho", copper sleigh bells and boots (Figs. 11.4 and 11.5).

Viko Kasiki shows us the symbolic and religious connection between *Koo Yoso*, the agricultural cycle, and chilolos; they are an essence of the *Koo Yoso* personification because they go in a line to cross fields and arrive at the houses where they dance to awaken the fertility of the mother Earth with the sounds of the copper sleigh bells and blessing the seeds.

---

[19] The chilenas is the common name in Spanish for the local genre of music. In *Sa'an Savi* language we call *yaa* "music".

[20] Nundaco and Ocotepec share a territorial history since colonial times. The community of Nundaco point out that the families who founded their community were from Ocotepec. Then, *Ñuu Ka'anu* refers to Ocotepec.

[21] *Manta* is made of cotton.

[22] The masks can be made of leather, plastic or textile and represent different characters involved into the politics and Mexican wrestlers.

Fig. 11.4 Traditional viko kasiki costume at Morelos, Ocotepec, in 1963. Photograph of the "Bienes Comunales" office at Santo Tomás Ocotepec

Fig. 11.5 Chilolos of Santo Tomás Ocotepec. Photograph by the author, 2019

We do this ritual before planting and the beginning of the rainy season in Ocotepec.

New Values Around Koo Yoso and Viko Kasiki

In *Sa'an Savi* of Ocotepec, the carnival used to be called *viko tee kasiki,* which literally means "party of the men who play". This describes the festival in the past. For a long time, the women did not participate in the dance. Today, however, women's participation is fundamental for the magnificence of our festivity. In the past, women were criticized if they participated in the dances, for that reason, women had to dress like men to participate. Fortunately, in the last fifteen years, many women have decided to enjoy the festivity by highlighting their own gender. They have innovated and stylized the "traditional and masculine" costume to reaffirm their gender in front of the community. For example, in (Fig. 11.6), *masked women* or *chilolas* participate in the *Viko Kasiki*. They wear the traditional *xikin* (*huipil or blouse*), palm hat, cotton shawl and skirt made with leaves of mature maize. Probably the dress they have chosen emphasizes the importance of the harvest. Women's participation in the carnival, at the same time, is the result of changing gender roles in the Mixtec communities. Today women have more public seats in academia, business and politics in the local context. At home, in nuclear families, the women decide if they want to get married and to whom and how many children they want to have.

The main aspect of this celebration today is *kasiki* "to play" and to play is to dance and to dance is to enjoy. So, today we should call this festivity just as *viko kasiki* "the festivity of those who play". Yet, the playful enjoyment of *kasiki* brings movement, life, and fertility to the fields and it renews the agricultural cycle among the *Ñuu Savi*.

Viko kasiki festivity has the capacity to reintegrate the community in times where massive migration is a common rule in Mixtec communities due to lack of labor opportunities and universities to study a profession. It represents a huge problem for cultural continuity, because traditional knowledge is no longer shared across generations. Migrants represent more than half of the participants of *Viko Kasiki*

Fig. 11.6 Chilolas or masked women participating in Viko Kasiki festivity. Photograph by the author, 2019

festivity, who wait for these days to return to the place where they were born. In this festivity, they have the opportunity to see each other, meet with their friends and families in the community and speak their mother language, the *Sa'an Savi*. This event is the union of extended families, who return to the community only on these dates. The *viko kasiki* gives us an identity by recognizing us as part of a community, which also forces us to provide what we can to those who come to dance at our houses. We offer breakfast, lunch or dinner to the masked men and women during the three days of the festivity. Migrants in the United States and Canada transfer money to support the cost of the festivity, they never forget their origin communities and they appreciate *viko kasiki* through Facebook LIVE. The *viko kasiki* gives us the opportunity to come back home, to our lands, to re-unite generations, to learn about life and death. Children of migrants learn about how to read the sky, the soil, when we have to cultivate corn and why we have to offer food and drinks to *Ñu'u Ndeyu*, Mother Earth, before planting and harvesting. *Viko kasiki* allows us to continue with Mesoamerican rituals and cultural values.

Conclusions

The Feathered Serpent is the most representative deity in Mesoamerica, known as Quetzalcoatl in Nahua culture; in Maya, as *Kukulkan* and in *Ñuu Savi* as *Koo Yoso*. As part of Mesoamerican peoples and cultures, the Ñuu Savi people have maintained their linkage with the past and, at the same time, integrated new values into those they already possessed. *Koo Yoso* is alive in Ñuu Savi, we can see him each year when the *tatyi-savi* comes and the chilolos dance in the houses and fields.

This chapter illustrates the dynamic cultural memory about the Plumed Serpent in *Ñuu Savi*; where, despite the effort by Spanish and religious authorities, Mesoamerican values continued, thanks to the language and the cultural synergy of the peoples themselves, as illustrated by *Viko Kasiki* in Santo Tomás Ocotepec, a reminiscence of the elaborate Mixtec ritual calendar in precolonial times. Thus, in this chapter, we show the importance of a living heritage that upholds the knowledge and cultural values of Mesoamerican civilizations for the well-being of the modern community.

Bibliography

Aguilar, S. O. (2019). Re-interpreting Ñuu Savi pictorial manuscripts from a Mixtec perspective. Linking past and present. In M. Jansen, V. M. Lladó-Buisán, & L. Snijders (Eds.), *Mesoamerican manuscrips. New scientific approaches and interpretations* (pp. 321–348). Brill. https://brill.com/view/title/36446?fbclid=iwar0qwahjknhwpnh00ejvvpodizwba39qxxkqjbsp_md5bywu7oluzxxptg

Aguilar, S. O. (2020). Ñuu Savi: Pasado, Presente y Futuro. Descolonización, Continuidad Cultural y Re-apropiación de los Códices Mixtecos en el Pueblo de la Lluvia (Tesis Doctoral). Archaeological Studies Leiden

University, Leiden University Press. https://scholarlypublications.universiteit
leiden.nl/handle/1887/138511

Anders, F., Jansen, M., & Pérez, J. G. A. (1992a). *Origen e Historia de los Reyes
Mixtecos. Libro explicativo del llamado. Códice Vindobonensis Mexicanus 1.*
Fondo de Cultura Económica.

Anders, F., Jansen, M., & Pérez, J. G. A. (1992b). *Crónica Mixteca. El
rey 8 venado, Garra de Jaguar, y la dinastía de Teozacualco-Zaachila.*
Libro explicativo del llamado Códice Zouche-Nuttall. Fondo de Cultura
Económica.

App "Códices Mixtecos". (2019). https://play.google.com/store/apps/details?
id=com.codice.celina.codicesmixtecos

Clark, J. C. (1938). *Codex Mendoza* (3 vols.). Waterlow & Sons.

Codex Mendoza, see Clark, J. C. (1938).

Codex Tonindeye (= Codex *Zouche*/Codex *Nuttall*), see Anders, F., Jansen,
M., & Pérez, J. G. A. (1992b). https://www.britishmuseum.org/collection/
object/E_Am1902-0308-1

Codex Yuta Tnoho (= *Códice Vindobonensis Mexicanus I*), see Anders, F., Jansen,
M., & Pérez, J. G. A. (1992a).

Filloy, N. L. (2019). "De la pluma y sus usos" en *Arqueología Mexicana*, Vol.
XXVII, núm. 159, septiembre-octubre (pp. 18–23).

Florescano, E. (2004). *Quetzalcóatl y los mitos fundadores de Mesoamérica.
Colección Pasado y Presente.* Taurus.

Hernández, D. J., & Ángeles, C. G. C. (2005). *Carnavales en la Mixteca.
Entre el culto a la fertilidad y el festejo católico.* Instituto de Investigaciones
Sociológicas de la UABJO, CONACULTA-FONCA.

Monaghan, J. (1995). *The covenants with earth and rain. Exchange, sacrifice, and
revelation in Mixtec sociality.* University of Oklahoma Press.

Jansen, M. E. R. G. N. (2012). *Monte Albán y la memoria mixteca.* Facultad
de Arqueología, Universidad de Leiden.

Jansen, M., & Pérez, J. G. A. (2007). *Encounter with the plumed serpent. Drama
and power in the heart of Mesoamerica.* University of Colorado Press.

Jansen, M., & Pérez, J. G. A. (2009). *Voces del Dzaha Dzavui. Mixteco Clásico.
Análisis y conversión del Vocabulario de Fray Francisco de Alvarado (1593).*
Gobierno del Estado de Oaxaca/Colegio Superior para la Educación Integral
Intercultural de Oaxaca/Yuu Núú A.C.

Jansen, M., & Pérez, J. G. A. (2011). *The Mixtec Pictorial Manuscripts: Time,
Agency, and Memory in Ancient Mexico.* Koninklijke Brill NV.

Pérez, J. G. A. (2008). Sahìn Sàu. *Curso práctico de lengua mixteca (variante de Ñuu Ndéyà) con notas históricas y culturales.* Universidad de Leiden/Colegio Superior para la Educación Integral Intercultural de Oaxaca (CSEIIO).

Witter, H. J. (2011). *Die gefiederte Schlange und Christus. Eine religionhistorische Studie zum mixtekisch-christlichen Synkretismus.* Doctoral dissertation. Leiden University.

12

Tlaneltoquilli tlen mochihua ica cintli ipan tlalli Chicontepec: tlamantli chicahualiztli ipan tochinanco

Eduardo de la Cruz Cruz and English translation by Brisa S. Zavala

Cintli Itlaneltoquil: Ce Tlamatli Chicahualiztli

Pan ni tequitl zaniloa ica tlaneltoquilli tlen cintli huan quenniuhcatzan monehpanoa ica cualli nemiliztli. Cualli nemiliztli mocuamachilia quence tlamantli yehyectzin huan cualli tlen nochipa neci ipan tonemiliz. Ce tlamantli tlen queniuhtzan techyolpactiah titequitizceh, timonohnotzazceh huan timotlahpalozceh pan ce canahya. Axtlahuel monequi ticpiyazceh ce tlamantli para timomachilizceh cualli. Yeca hueli niquihtoz, cualli nemiliztli ce tlamantli tlen moaxiltoc ica ce cualli tlatlepanittaniliztli, zanilli huan ce cualli tlaneltoquilli tlen totlalhui. Ipan ni capitolo nouhquiya zaniloa ica queniuhcatzan nemiyaya generaciones pan altepetzitzin nahuatl huan queniuhqui monextiltihualtoc ixtlamatiliztli.

E. de la Cruz Cruz (✉) · English translation by Brisa S. Zavala
IDIEZ, Zacatecas, Mexico
e-mail: xochiayotzin@gmail.com

University of Warsaw, Warsaw, Poland

© The Author(s) 2024
J. Olko and C. Radding (eds.), *Living with Nature, Cherishing Language*,
https://doi.org/10.1007/978-3-031-38739-5_12

345

Nimacehualli. Niehua huan niixtlapanqui Lindero2, Chicontepec, Veracruz. Niquillamiqui quemman nieliyaya nipiloquichpiltzin, nipiltelpocatzin, queniuhqui nitequitiyaya millah huanya nonanan, huanya totlayimeh. Nechpactiyaya huan nocca nechpactia cintli itequiuh huan itlaneltoquil. Quemman nicpiyaya 18 xihuitl nitequitito huan nimomachtito ipan huei altepetl. Naman nitlamachtia ica nahuatl ipan huei caltlamachtiloyan tlen Estados Unidos. Nimomachtiah nodoctorado ipan Huei Caltlamachtiloyan Varsovia huan nitlayecanquetl Zacatlan Macehualtlallamiccan (Instituto de Docencia e Investigaciones Etnológicas de Zacatecas). Iuhcatzan niccalactoc ipan tlamachtiliztli huan tlatehtemolizztli, tlahuel nitequiti para nicchicahuiliz tlahtolli huan macehualtlallamiccayotl nahuatl ica altepetzitzin tlen Chicontepec. Nicchihua notlatehtemoliz ica ce tlachiyaliztli emic quence zaniliztli huan tlamachtiliztli.

Ni tequitl moihcuiltoc ica nahuatl pampa nicnequi ma ahciti ica macehualmeh; nouhquiya pampa miac tlahtolli, ixtlamatiliztli ixpolihui ica ingles zo caxtillan quemman tlatehtemolizltli itzonyo huallauh ica nahuatl huan, necnequi nicmanextiz tohhuantin tlen nahuatl nouhquiya ticpiyah totozcac ipan caltlamachtiloyan. Mocalaquia ce tlahtolcuapaliztli ica ingles tlen quichihuaz ma tizanililocan ica cequin tlen quiamatih ni tlatehtemoliztli. Tlahtolcuapaliztli ica ingles quipiya achiyoc tlayolmelahualiztli tlen cequi tlahtolli ica nahuatl axmoihtoa.

Macehualmeh quiyehyecoah itlaneltoquil cintli ica macuilli campeca: Xinachtli, Miltlacualtiah, Elotlamanah, Cintlacualtiah huan Tlatlacualtiah. Xiquitta cuamecatlahtolpamitl 1.

Cintli itlaneltoquil motehtemohtoc ica cequi tlahcuiloanih, tlahuel neci ica ixtlamatinih macehualmeh tlen Huasteca Veracruzana. Pan inintlatehtemoliz zaniloah ica campeca tlen cintli quence

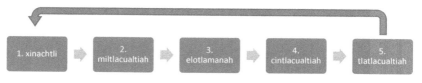

Ixcopincayotl 1 *Campeca tlen cintli* (Quimanextia campeca tlen cintli huan quen motecpana)

tlatlacualtiliztli. Ipan Argüelles Santiago (2012) ica itlahcuiloliz "El maíz en la construcción y transmisión de una identidad cultural de la Huasteca Veracruzana", zaniloa ica campeca tlen elotlamanaliztli, ce achi itlaneltoquil cintli. Quichihua ce cuecuetztzin tlapannextiliztli huan nouhquiya quiihtoa, ayocmocencuilia ni campeca pampa calaqui tlamantli politica tlen ce tlanahuatihqetl.

Nava Vite (2012) ipan itlahcuiloliz "El costumbre: ofrendas y música a Chicomexochitl en Ixhuatlán de Madero", Veracruz, zaniloa ica huentli tlen Chicomexochitl, tlahuel quimanextia quitl macehualmeh quichihuah temaihtolli ica tlaltepactli para quitetilizceh tlatoctli tlen oncah millah. Tlahuel monequi moihtoz tlahcuilohquetl quimanextia itlaneltoquil cintli: cintoquiztli (xinachtli), miltlacualtiliztli, miyahuacalaquiliztli, elotlmanaliztli, cintlacualtiliztli huan zaniloa tlahuel quentzin quence ce parrafo para cehce campeca. Zanpampa, tlahuel zaniloa huan techpanextilia ica atlatlacualtiliztli (tlatlacualtiliztli). Gómez Martínez huan Van't Hooft (2012) pan inintlahcuiloliz "Atlatlacualtiliztli: la petición de la lluvia en Ichcacuatitla, Chicontepec", zaniloah ica ce huei ixtlamatiliztli tlen atlatlacualtiliztli huan queniuhcatzan macehualmeh tlehcoh ipan tepetl malhuilli Postectli. Ipan Gómez Martínez (2002) pan iamox *Tlaneltoquilli La espiritualidad de los nahuas chicontepecanos*, quipiya ce tlahcuilolli tlen quitocaxtia Rituales Agrícolas, quichihua ce huei tlaixpannextiliztli ica campeca tlen atlatlacualtiliztli huan ce cuecuetztzin tlahcuilolli ica cequinoc campeca tlen cintli (pp. 180–113).

Tlayolmelahualiztli tlen quipiya ni capitolo huallauh ipan nonemiliz, tequitl tlen nicchiuhqui ipan 2011 huan 2019, huan tencopinanliztli ica nahui piltetahtzitzin ipan 2013 huan 2014. Peuhqui ninequiti millah quemman nicpiyaya chicome xihuitl. Tlacameh tlen niquinhuayatequitiyaya quiittaqueh nechyolpactia miac ni tlamantli tlen cintli, yeca nechmacatiyahqueh ixtlamatiliztli huan tlallamiquiliztli, cencah quen quichiuhtoyah ica inintatahhuan. Tlahtolpamitl tlen niccalaquia ipan ni capitolo ni ce tlahtolli tlen huahcapantl tlen mopouhtihualtoc ica nototatahhuan inintatahhuan, yeca axhueli niquihtoah acquiya mero quiihtohqui quence ce citah. Zan cualli nioquichpil niyohuiyaya ipan Xochicalli aztah ipehuayan xihuitl 2000, quemman noaltepeuh ayocquichiuhqui campeca tlen campa nochi macehualmeh yohuih. Ipan 2011, nitequipanoto ipan campeca tlen Elotlamanah ipan Lomas del Dorado, Ixhuatlan de Madero, huan nimomachtihqui nicchihuaz xochitl tlen ica tlachihchihuah. Ipan 2019 nitequipanoc ipan campeca tlen Tlatlacualtiah ipan Tecomate, Chicontepec.

Niquintencopinqui nahui macehualmeh tlen tlahuel cualli tequipanoyayah huan quiyecanayayah campeca tle mochihua ipan ininaltepeuh. Achtohui ce tepahtihquetl cihuatl tlen quipiyaya 60 xihuitl, ichan La Pahua, Chicontepec. Ompan, ni eliyaya itetah tlen nocca tequipano pan campeca Miyahuacalaquiah. Expan ce tlatzotzonquetl tlacatl tlen quipiyaya 50 xihuitl pan nopa cahuitl, ehua Tepecxitla, Chicontepec. Nauhpan ce tlacatl tlen quipiyaya 80 xihuitl, ehua Tecomate, Chicontepec, ya eliyaya ce huei tlayecanquetl tlen Elotlamanah huan Tlatlacualtiah, huahcauhqui quence 30 xihuitl pan ni tlaneltoquilli aztah ipehuayan 2000, quemman macehualmeh ayocquinecqueh tequipanoah Xichocalco.

Cintli quemman eli xinachtli, miyahuatl huan Chicomexochitl

Macehualmeh tlen Chicontepec, cintli quitechtiliah eyi itocah: xinachtli, miyahuatl huan Chicomexochitl. Xinachtli, quimanextia cintli tlen motocaz millah; miyahuatl, quimanextia ixochiyo toctli huan Chicomexochitl, quimantextia piltotiotztzin. Cehce itocah quipiya canin huan quemman moihtoa.

Xinachtli, quilliah quemman motechtia motocaz millah. Ce cintlancochtli mocuapaz xinachtli, tetatahhuan quitlapehpeniah, quiahhuiah huan quitlaliah ceyohual ipan atl. Ica yohuatzinco quiilliahya xinachtli; malhuilli huan monequi moicneliz quence ce conetzin quemman ontlacati. Quemman tiquitzquiz xinachtli achtohui timomaahhuiz, pampa ni ce tlamantli tlatlepanittaliztli huan tlaneltoquilli. Conemeh axhuelih quiihitzquiah. Anque chaneh huan iteixmatcahuan quitlatlalhuiliah; tlacameh contocah millah. Xinachtli quimanextia chicahualiztli huan tlacualiztli ica macehualmeh tlen itztoqueh pan nopa calli.

Miyahuatl, ni ixochiyo toctli tlen techillia huallohua yancui pixquiztli. Quemman pannihuetzin xilotl, eli elotl, mochicahuilia, huaqui huan nimantzin mocuapa cintli. Huehuehtlahtolli quiihtoa "quen mocahuato xinachtli millah monequi mocuiti". Ni quihtoznequi, axhueli mocuiz elotzin millah tlan axmotlacualtihtoc, axmotlaoniltihtoc yon axmopopochhuihtoc. Quemman motlacualtihtozza hueliya mocui huan hueli ticmacazceh ce acahya quence tlen nocalnechca para ma no quicua.

Chicomexochitl, ce tlahtolli tlen moihtoa quemman quitechtiah ce huei tlatlepanittaliztli ica nopa cintlancohtli zo quemman quimanextiah quichihuilizceh ce campeca ipan xochicalli. Macehualmeh tlen Chicontepec, cintli ne tlen ica mohmoztlah monehpanoah ipan cuamezah ica ce cuahcualtizn tlacualli, ni quipiya itecoh tlen quiilliah Chicomexochitl, zanpampa, quemman tetahmeh zaniloah ica ce campeca ya quiilliah Chicomexochitzin. Huacapameh quimanextihteuhtoqueh quence piloquichpiltzin huan pilcihuapiltzin. Altepemeh tlen Huasteca Veracruzana huan altepetzintzin tlen iihtico Chicontepec, zaniloah ica ce tlapohualiztli tlen moixmatih quence Chicomexochitl huan Tenantzitzimitl. Tlatempohualiztli tlen zan techpohuilia techmaca miac ixtlamatiliztli (De la Cruz Cruz, 2015; IDIEZ, 2010; Nava Vite, 2012). Nochi zaniloah ica ce tenantzin tlen quipiyaya ce iichpocauh huan ni quipixqui ce iconeuh oquichpil. Ni eliya ce oquichpil tlen nochipa zan yolpactoc huan axquemman tecualanmacac. Itonanan, axquemman quihuelittac iixhui. Ya quiyoltilia mahuiltiliztli para quimictiz iixhui, quence quemman mahuiltiah ica ornoh; zanpampa, axquemman hueli quimictia huan ya miqui ipan ne mahuiltiliztli. Chicomexochitl, nochipa cenyahtoc nemiyaya ica paquiliztli huan axtlen quicuezoyaya, nochi tlen quiilliyayah quichihuayayah.

Miac quimatih huan quipohuah tlapohualiztli tlen Chicomexochitl, quihtoah melahuac panoc huahcauhquiya. Naman quemman mochihua tlaneltoquilli quence tlatlacualtiliztli, zo elotlamanaliztli, nopayoh tiquittah Chicomexochitl, macehualmeh quipanmanextiah ica ome tlamantli:

Ixcopincayotl 2 *Chicomexochitl amatlatectli* (*Xiquitta* De la Cruz Cruz (2017)

1. Chicomexochitzin ipan tlatectli. Quitehtequih amatl, teipan yehyectzin quiyoyontiah. Xiquittah ixcopincayotl 2.
2. Chicomexochitzin ipan pilelotzin. Campa nouhquiya yehyectzin quicualtlaliah, quipiquih ica ce payoh zo tlaxcalyoyomitl, moilpia ica ce liztoh huan moxochiyotia ica xochitl tlen cempohualli. Xiquitta ixcopncayotl 3.

Nochi ome tlaixnextilli motlalia tlaixpan huan iuhquinon momactilia miac tlacualiztli huan tlaoniliztli.

Xinachtli

Xinachtli, tocayotl tlen campeca huan cintlancochtli tlen motlapehpenihtoc. Para mocuapaz xinachtli, ni pano pan miac piltequitzitzin tlen quichihuiliah. Tetatahhuan quitlapehpeniah petlayoh tlen onquizqui ipan toquiztli tlen onpanoc, quitlapehpeniah petlayoh tlen huehhuei huan tlen axcanah cualotoc. Zancehco tlaxipehuah.

Ixcopincayotl 3 *Chicomexochitl ica elotl* (*Xiquitta.* Huallauh ipan *Cenyahtoc cintli tonacayo: huahcapatl huan tlen naman*, quiihcuilohqui Eduardo de la Cruz, 2017, Varsovia, Huei Caltlamachtiloyan Varsovia, Facultad tlen "Artes Liberales." Ixcopincayotl quiquixtih Alan Sandstrom quemman quichihuayaya itlatehtemoliz ica tlaneltoquilli tlen tlatlacualtiah ipan altepetzin tlen itocah Cacahuatengo, Ixhuatlan de Madero, March 2007)

Teipan quitlahtlachiliah catlinya cualli huan yehyectzin, quicuih cintlancochtli tlen pahpatlactic, ehetic huan tlen axcanah ihtiyoltlapantoc. Tetatahhuan quicuih ce mahtzolli cintlancochtli huan quiahhuiah. Ticahhuiz xinachtli quihtoznequi ticmaca ce achi ihyotl tlen monemiliz huan quimanextia ce tlatlepanittaliztli tlen ticmacah ica toyollo. Tetatahhuan quiaquechiya huan quicahuah ceyohual ipan atl. Ica cualcan, tenanan quiquixtia xinachtli huan quitema pan ce chiquitl ma ayotemo.

Chaneh huan iteixmatcahuan monemiliah huan quitlatlalhuiliah xinachtli, ce yehyectzin tlacualli quimactiliah achtohui quemman tlacameh ayiquiontocah millah. Tenanan tlacualtlalia tlaixpan huan tlatlalia, quitlalia tlaxcalli, tecciztlapahuaxtli, pantzin, pan rozca, cafen, huan licor tlen aniz. Ya quicualtlaltozza ce tlacualli quemman tlaneztozza. Conemeh tlazahzacah, quence xochitl zo tlacualli huan quimactiliah tetatah para ma quitlali tlaixpan. Tenanan quitlatia cantelah huan quihuallican popochcomitl ica tlicolli tlen campa quitlalhuilizeh copalli. Ya quitlalia ce chiquihuitl ica xinachtli tlalchi, iixpan cuamezah zo inacaztlan tlen iicxitlan. Nimantzin quitlalia ce taza cafen ica pantzin iihtico zo inacaztlan chiquihuitl tlen quipiya xinachtli. Imah tazah monequi tlachixtoz ica tlapepecholli. Chicomexochitl mocehuia campa icuitlapan tlachiya ica tlapepecholli huan iixco ica chaneh huan iteixmatcahuan.

Quemman nochi tlatlaltozcehya tlaixpan, nochi mozancehcotiliah iixpan tlaixpan. Tetah zo tenan tlatennamiqui huan momaahhuia ica inahnacaztlan tlaixpan. Nican quemman momaahhuia, quihtoznequi tlatlepanittaliztli tlen quinextia ce ica cintli. Tetah zo tenan quicui popochcomitl huan tlapopochhuia tlaixpan huan quipanoltia iixco tlacualli huan inahnacaztla cuamezah huan tlalchi campa eltoc xinachtli. Iuhquinon cehce pano, tlapopochhuia huan momaihtoa quence ma yehyectzin eli xinachtli huan axcanah ma quiilcahua ichan campa quiztehua; ixhuaz campa quitocazceh, ixhuaz yehyectzin huan mochicahuiliz. Tlahtolli tlen tohuahcapahuan technextilteuhqueh axcanah ma timococolican, ma ticcahuacan xicoliztli, cualantli zo tequipacholli; quemman ce acahya quitlatlalhuilia xinachtli monequi quichihuaz ica iyollo. Tlaxque, xinachtli neci quence axquitlepanitta huan mocuezoa. Quemman xinachtli mocuezoa ni axcualli eli, huahhuaqui, pipilica huan axmonacayotia. Quemman xinachtli quicelia huentli, quiztehua tochan tlacuahtoc huan tlaonitoc. Eliz elotl huan ticcuazceh, huaquiz huan nimantzin mocuapaz cintli tlen ticahcohuizceh huan ticcuahtiyazceh noque ahci zampa ipohual pixquiztli.

Quemman xinachtli zo ceyoc tlamantli tlatoctli ixhua ipan tlalli, monequi momocuitlahuiz tlahuel cualli. Quen momocuitlahuia, ni monequi motlanempanchihuiliz, tictlahpihpiyalizceh para axma quitlanahuican totomeh, tecuanimeh. Quemman titequitih millah monequi ticnohnotzazceh nopa totlatochui. Tlan ce acahya quiilcahua quimocuitlahuiz, ichicahuiliz toctli moquetza, icuayo hueli huahhuaqui huan amo miyahuati zo mozcaltia huan amo xiloti. Quemmantzin piltoctzin mozcaltia, miyahuati huan xiloti, zanpampa, axyehyectzin monacayotia, zan huahhuahca motlalia xintlancochtli pan ioloyo, huan quemmantzin paniman axeli cintli pan olotl. Quemman pano ni tlamantli piltetahtzitzin quiihtoah: "Pilcintzin ayocquinequi eliz compah".

Cequi tlen motemachiah ipan ininmillah, cualcantzin mehuah huan yohuih ontlachiya. Nehnemi itehtenno milli, quitlahtlachiliah piltoctzitzin. Ica chiuhnahui huan mahtlactli cahuitl tlen yohuatzinco mocuapahya inchan, tlacuah huan zampa yohuih millah, quichihuah tequitl tlen monqui. Iuhquinon mohmoztlah quichihuah aztah quemman pixcah. Quemman pano ce zo ome metztli, anque chaneh quichihua ce campeca tlen itocah miltlacualtiah, quitlatlalhuiliah milli, piltoctzitzin tlen nopayoh mozcaltiah. Ni tlaneltoquilli naman ayoctlahuel neci.

Miltlatlacualtiah

Miltlacualtiah ni ce campeca tlen mochihuilia milli huan tlatocli tlen oncah nopayoh, quitetiliah, quichicahuiliah huan quipalehuiah para oncatih ce cualli pixquiztli. Ce tlaneltoquilli tlen campa axtlahuel ica tlatehtemotoqueh. Tetahmeh quimanextiah quence ce tlaneltoquilli tlen cintli. Momachtiltlatehtemoanih nahuatl Arturo Gómez Martínez huan Rafael Nava Vite quiihtoah pan inintequiuh (2002, p. 83; 2012, p. 38). Jesús Alberto Flores Martínez, ce huei momachtihquetl nahuatl quipannextiz ipan ce tlamahuizolli tlen quichihua. Iixtlapolihuiltil ni ce tlamantli tlen quichiuhtoc ma axoncah miac tlahcuilolli. Ipan notlanehnehuiliz quence macehualli, nicmanextia ni campeca quence tlahuel ipatiuh pampa yainon quihuicatiuh ce cualli pixquiztli.

Tetahtzitzin quipohuah quemman icuayo piltoctzin zancualli ce metro huan tlahco ihuahcapanca, chaneh pehua monemilia para quichihuaz campeca. Nican tlahuel monequi ̇ anque chaneh ma monemili huan ma quitemo ce huehuehtlacatl tlen quiceyacanaz campeca. Huehuehtlacatl quimanextia tlen tlamantli monequi mozancehcotiliz huan mohuallicaz quemman miltlatlacualtizceh. Quemman mochihua campeca, teteixmatcahuan huan huehuehtlacatl mozancehcotiliah tlatlahco milli. Nopayoh mochihua ce cuatlamapechtli quence cuamezah, nopayo tlatlaliah huan quimahuiztiliah totiotzin Chicomexohitl huan cequinoc totiotzitzin quence tonanan tlaltepactli, totiotzin tlen atl, totiotzin tlen axcualli. Inihhuantin quitlaliah pantzin huan cafen pan piltoctzitzin, zan yahyahualtic quitlalihtiyohuih.

Iuhquinon ni tlaneltoquilli quemman quence ixpoliuhtiyahqui, zan chaneh peuhqui quiyecana. Niquillamiqui quemman nieliyaya nioquichpil huan niquittac queniuhcatzan nonanan quichihuayaya ni tlaneltoquilli ipan tomillah, ya quinextilteuhqui itatah.

Tototatahhuan techilliyayah: "xiquitta queniuhcatzan mochihua pampa quemman tihueiyaz huan titlaahciz ya ni ticchihuaz, ni ticcencuiliz para ticpiyaz cuahcualli cintli, ticpiyaz tlen ticcuaz". Iuhquinon techtlaihilliyayah; huan naman tiquilcauhtoqueh inintlahtol. Tomacehualicnihuan tlen naman ayocquichihuah ni campeca.

Quemman macehualmeh quitlahuelcahuah ininmillah, quemman ayocquichihuah ni tlaneltoquilli, toquiztli axteyoti huan axtetiya. Quemman mochihua ni campeca toctli zancualli motlahtlantiyohuihya, miyahuati huan moxiloquixtia, teipan pehua monacayotia. Ipan ni cahuitl, tlahuel monequi momocuitlahuiz, iuhquinon axquicuah mapachimeh, coyotez huan cequinoc tecuanimeh tlen huelih quitlanahuiah. Mochihua miltlatlacualtiliztli nopa temachtli piltoctzin motetilia ipan tlalli iuhcatzan axmotequihuia ce tlamantli pahtli. Piltoctzin quiihyohuiz ce ehecatl, ce xopanal tlen huallauh ipan ce ehecayotl tlen tlatilana ipan ce atenno nechcatzin. Tecuanimeh nouhquiya axtlahuel quitlanahuizeh nopa tlatoctli. Naman tonatiuh, miltequitinih tlahuel motemachiah ipan pahtli. Pahtli nimantzin quichihua itequiuh, zanpampa axtlahuel cualli pampa tlaihtlacoa huan quitlanahuia tlalli.

Elotlamanah

Macehualmeh mozancehcotiliah xochicalli para quitlacualtizeh pilelotzin, ni quichihuah quemman ayimocuitoc yonce hueltah, iuhcatzan miac zo axmiac pilelotzin oncah millah. Axnochimeh tocah pan ipohual toquiztli, nochipa monapoh quence ome cemanoh zo ce metztli. Tlan ce acahya quipiya tlaxtlahuilli huan quinequi pixcaz achtohui zo teipan tlen quemman mochiuhtocca elotlamanaliztli tlen comuh, ni eli elotlamanah huan quinnotzah macehualmeh tlen cequinoc altepetzitzin para nouhquiya ma yacan. Tlan ce acahya axquipiya tlaxtlahuilli para campeca, zan miyahuacalaquiah (moittaz ipan tlahcuilolli tlen huallauh teipan). Macehualmeh elotlamanah pampa motlazcamatiliah ica nochi, quitlazcamatiliah Chicomexochitl, cualli quizato ipohual toquiztli huan quiceliah ipohual cintli tlen yancuic. Quinextiah tlatlepanittaliztli huan quiicneliah ica ce cualli huentli quence xochitl, tlamantli tlen mooni, tlacualli, piyomeh para tlacualli, tlatzotzontli huan mihtotiliztli.

Pan ni campeca tequipanoah miaqueh. Tlen chinancomeh motemacah tlayecanazceh huan quitemoah tlapalehuiliztli (tomin, tlamantli tlen ce acahya quinequi quitlaliz huan yohui ce acahya tlen moahxilia) ica ininteixmatcahuan huan tlen calnehnechca. Tlahuel ipatiuh quemman yehyectzin monemiliah.

Axipoliuhya itztoz huehuehtlacatl pampa ni ya quiyecana. Monemiliah huan quitemoah nochi tlen itechpohui: motemoa tlatzotzonanih tlen trio; motemoa coyolli huan cempohualli tlen ica tlachihchihuazceh. Mocohua ce yancuic popochcomitl, cantelah, copalli huan amatl tlen ica quichihuazceh tlatectli. Tlen ica tlacualchihuazceh quence piyonacatl zo pitzonacatl; cuatlacquetl quence cuaxilotl, camohtli tlaneuccalaquilli, alaxox, limah, cacahuatl, quemmantzin motlalia manzanah, tentlatzcayotl quence gayetah huan chocolatl; cequi tentlatzcayotl tlen mocualtlalia caltic quence tintinez, pemolez huan alfahorez; cequi tlamantli tlen mooni quence cafen, chocolatl, anizado, refrezcoh, atolli; papatzin tlachihualli quence etixtli, chiltlaxcaltzin tlen axtlahuel cococ. Quemmantzin mocohua ce quezqui tlatopontli tlen motoponia quemman mocalaquiaya pilelotzin.

Cihuameh huan tlacameh tlahuel monequi ma moxehxelhuilican tequitl. Tlen tlacameh, cequi pehuah tlachihchihuah tlaixpan ica ce xihuitl tlen quiilliah ilimonaria, cequi tlacameh quicualtlaliah arcoh caltenno ica iixpan calli campa quipanoltizceh elotl, campa macehualmeh huan Chicomexochitzin moixnamiquizceh huan motlahtlatizceh. Tlen anque xochiuhchihuah, quicualtlaliah xochitl quenne: coyolxochitl, maxochitl, xochicozcatl, xitlalimeh. Cequi tlacameh, yohuih millah concuih elotl ipan coxtalli zo cuachiquihuitl, zan cualli miac, nouhquiya quicuitehuah ce quezqui elotl ica icuayo tlen quipiya imiyahuayo. Quemman mocuapah, ni quitlaliah calmapan quence caltech noque tlamih mocualtlalia tlen motequihuiz.

Toahuimeh, nouhquiya moxehxelhuilia tequitl huan quichihuah tequitl tlen tlahuel monequi quence quintlaquentiah totiotzitzin, piyomictiah, ticih, quicualtlaliah tlacualli huan tetlamacah. Huehuehtlacatl quiyecana nochin ni tequitl huan quitlachiltiuh ma nochi cualli quiztiuh.

Tlatzotzonanih, ni ceyoc tlamantli tlen tlahuel monequi para mocencahuaz campeca tlen elotlamanaliztli. Inihhuantin pehuah tlatzotzonah zan quen ahcih huan moquetzah quemman tlami elotlamanaliztli. Quemman pehuahya tlatzotzonah, cequi macehualmeh pehuah mihtotiah tlaixpantenno, cequi tlen quence zanoc ahcih, tlatennamiqui huan nouhquiya pehuah mihtotiah huan iuhquinon mopatlatiyazceh. Tlatzotzonanih itechpohui pan ni paquiliztli tlen Chicomexochitl, quimatih miac zonez tlen quemmaniuhqui monequi motzotzonaz quence zon tlen ica piyomictiah, zon tlen campa ica tlatennamiquih, zon tlen ica tlatlaliah. Tlan inihhuantin mocuapoloa, huehuehtlacatl ya quinilltiz ce huan ce tlen monequi motzotzonaz.

Iuhquinon zancualli tlahcotona, huehuehtlacatl quiihtozza ya mocalaquizza pilelotzin pan arco, cequi tlacameh quitlalanahya elotzin tlen eltoc pan cuachiquitl zo coxtalli, cequi toahuimeh quicuih toccuahuitl huan moquetzah iican arco para quincelizceh, noque cequi cihuameh, tlacameh huan pipilmeh, moquetzah iixpan arco, ica tlatzotzontli nochi mihtotiah, quixochiyotiah pielotzin, quipopochhuiah huan quitennamiquih. Pan ni tlatoctzin, tlatzotzonanih cenyactoc tlatzotzonah huan macehualmeh tlen anque quipiyah cuachiquitl zo coxtalli mihtotiah ica yainon huan amo quitlaliah tlalchi para mociauhcahuazceh. Quemman huehuehtlacatl quiihtoz quence eltocca, nochi panoh ipan arco huan quicalaquihya elotzin calihtic.

Noque cequih quicencuiliah mihtotiah, tlacameh ica inincuachiquiuh zo inincoxtal tlen elotzin quicemoyahuah tlahco calli huan nopayoh, quence ce tlayohualolli pehuah quitechpochoa, zancualli tlahco metroh ihuahcapancauh eli, quence ce tzacualli yohualtic. Huacca cihuameh huan tlacameh pehuah quitlahtlalhuiliah. Achtohui quitlalhuiliah ce xochicozcatl huehueyac huan coyolxochitl quence pepechtli, teipan quitlalhuiliahya tlacualli, cuatlacquetl, tlamantli tlen mooni zanpampa axcanah huinoh. Noque cequi quicencuiliah mihtotiah, tlatlaliah huan tlapopochhuiah.

Quemman tlamih tlatlaliah nochi mihtotiah ce cualli ratoh, zan tlatoctzin, nochi tlen itztoqueh nopayoh motlamacah, huetzcah, camanaloah ica tlen quinpantiah ipan milli, pipilmeh tlacuah huan zanoliah ica mahuiltiliztli. Nochi paquih, yolpaquih, motlazcamatiliah ica Chicomexochitzin pampa oncazza cintli tlen yancuic.

Tlantihuetzih tlacuah, cihuameh huan tlacameh mopalehuiah quichihuah xamitl. Iuhquinon quiquixtiah elotzin ipetlayoh tlen ica moquipiz xamitl, cequin quixipehuah huan quixixintiyohuihya. Teipan moticih ipan metlatl, tlaxque pan molino moxamania, quemman notici huan moxamaniah pehua quiza iayo, huacca quinamiquih huan quichihuah atolli, ni atolli quiilliah eloatolli. Quemman eltocca tixtli tlen elotzin, nochi moyohualoah ipan cuamezan huan pehuah quipiquih. Teipan momana ipan chachapalli huan quitlalhuiliah ce quezqui elotl ma ica iucci. Pan ceyoc tlixictli ica chachapalli ya quiixcan eloatolli. Noque iucci xamitl, totlayimeh huan toahuimeh mozanilhuiah tlen tequitl quinpoloah quichihuazceh ininmillah noque iucci xamitl huan cualtia eloatolli. Zantlatoctzin quicuah xamitl huan quionih eloatolli, tlan axtlami ni xamitl, huacca momahmacah.

Elotlamanah ce campecah tlen quichihua ipan xochicalli naman pano huan mochihua pan ce techan, tlahuel neci ni tlamantli ipan Chicontepec. Macehualmeh Elotlamanah zan huanya ininteixmatcahuan. Elotlamanaliztli tlen moyehyecoa xochicalco ixpolihui pampa miac macehualmeh ayoctequipanoah, ayoctlapalehuiah huan ayocmoahxiliah. Cequi tlen quipiyah tlaxtlahuilli huan motemachiah ipan Chicomexochitl, quena Elotlamanah, ininchan quichihuah zan quemman oncah ce cualli pixquiztli, ayoctlahuel motetlanehuiah, zan ininteixmatcahuan quinilliah. Elotlamanah ce acahya quemman oncah ce cualli pixquiztli, neci quence ce tlamantli tlamahuizolli. Campeca tlen mochihua xochicalco huan pano pan techan quichihua ma cotoni paquiliztli ica pilaltepetzin huan ma moxelocan macehualmeh. Elotlamanaliztli, ce campeca campa motlepanitta huan motlazcamatilia Chicomexochitl, eli ce tlaneltoquilli campa ce cahya tlen quipiya tomin zan ya hueli eltlamanah. Cequin tlen axhuelih quichihuah elotlamanaliztli huan motemachiah ipan

Chicomexochitl, inihhuantin zan miyahuacalaquiah, ce campeca tlen campa quitlatlalhuiliah miyahuatl.

Quemman miyahuacalaquiah

Miyahuacalaquiah ni ce ixtlamatiliztli tlahuel cuecuetztzin tlen monextilia calihtic quemman chaneh quinequiya quicuiz elotl huan xochicalco ayicanah elotlamanah, elotlamanquehya zo cemeh axelotlamanazceh, huan chaneh axtlahuel tetiya ica tlaxtlahulli. Miyahuacalaquiah quihtoznequi cencah tlamantli tlen elotlamanaliztli. Chaneh quimanextia itlazcamatiliz pampa ahcic pixquiztli ipohual, yeca Chicomexochitl momactilia huentli.

Nican tequipanoah zan tlen itztoqueh pan ce calli huan cempauhcan mopalehuiah. Motemohtoc cequi tlamantli tlen motequihuiz quence pantzin, rozca, refrezco, cantelah, ce xochimantli huan ce yehyectzin copalli. Yohuatzincotzin quemman ayicanah cafenonih, ce acahya tlen calihtic, melahua tetatah, quinahuatiah ma quicuiti miyahuatzin. Ni moquechpoztequi toctli imiyahuayo tlen millah inahnacaztlan huan tlen tlahtlahco, tlan quipixtocca pielotzin, moquechcopina ce macuilli. Iuhquinon tlen anque yahtoc millah quichihua ome tlatzquintli miyahuatl huan tlan oncah cempohualli, no quixochiyotia. Quemman ahci caltic, ni mochiya caltenno zo moquetzah iixpan caltzauccayotl huan nopayoh mochiya, iuhquinon anque chaneh huan tlen itztoqueh calihtic nopayoh moixnamiquih, nopayoh quitlahtlahtiah, quitennamiquih huan quipopochhuia miyahuatl, quemman nochi panotoquehya tlacopalhuiah, mopanoltiaya calihtic ni miyahuatzin, huan quenne ipan cuamezah tlen tlaixpan nopayoh quinquehquetzah inahnacaztlan cuamezah.

Tlantihuetzi quicualtlaliah, pehuaya momactilia cafentzin ica rozca, piyocaltoh ica tlaxcaltzin (xiquitta ixcopincayotl 4), zan tlatoctzin nochi motlamacah huan nopayoh mocencahua ni campeca. Quemman mochiuhtocca tlaneltoquilli tlen miyahuacalaquiah, huelihya quicuih elotl zo quichiyah ma huaqui huan iuhquinon quipixcazceh cintli.

Quen miyahualacaquiah ce acahya axcencah ica tlen quichihua ceyoc macehualli.   Tetatahhuan   axquinpohuiliah   zo   quinilliah

Ixcopincayotl 4   *Miyahuacalaquiah* (*Xiquitta* De la Cruz Cruz (2017)

ininteixmatcahuan   tlen   quiihtoznequi   cehce   tlamantli,   zan
quintlanahnahuatiah. Pipilmeh monequi quiittazeh queniuhcatzan
mochihua huan quiillamiquizceh quemman tlaahcizceh, inihhuantin
quicencuilizceh huan quinmactilizceh ininconehuan. Tlaixpan ni ce
tlamanextilli tlen ni ixtlamatiliztli. Cehce techan quipiya ce tlaixpan
ipan huei calli, ni ce cuameza tlen motehtonihtoc ipan tlapepecholli.
Ihuexca cuamezah axcehca tlen oncah ipan cequinoc tetlaixpan, quence
1 metrohpan ica 0.5 azta ce cuameza tlen 2 metroh ica 0.75 tlen
ihuexca. Ipan miac techan quipiyah ce totiotzin tlen catolicoz
(cruztzitzin, santos), ce tlamantli tlaneltoquilli tlen monepanohtoc ica
pilaltepetzitzin. Pan ce techan ica inintlaixpan, hueli oncaz zo oncaz
miac ixcopincyotl tlen totiotzitzin. Tetatahhuan hueli quinnextilia
ininconehuan   queniuhcatzan   moixmachiyotizceh   quemman
monechcahuiah tlaixpan huan cequi tetatahhuan quinnextiliah
queniuhcatzan tlatennamiquizceh (quen moihtohqui ipan tlahcuilolli
tlen xinachtli).
        Naman tonatiuh, axmiac macehualmeh tlen Chicontepec quichihuah
ni campeca. Pan ipohual elotl, cintli, quicuih, pixcah huan quicuah;
yon axquimactiliah ce cantelah para motlazcamatlizceh quihzatoya

ipohual toquiztli. Miac piltetahtzitzin quihtoah: "Chicomexochitzin mocuezoa. Axquitlamacatoqueh yon axquitlaoniltihtoqueh, naman quinequi techtlahuelcahuaz". Piltetahtzitzin quiihtoah axteyoti cintli, nochi tlahuaqui huan oncah tequipacholli pan pilaltepetzitzin, pano ni tlamantli pampa quiilcauhtoqueh Chicomexochitl, quiilcauhtoqueh cualli nemiliztli huan quiilcauhtoqueh tlaneltoquilli tlen tohuahcapahuan technextilteuhqueh. Tlan axcanah timonemiliah, campeca tlen Miyahuacalquiah ixpolihuiz, cenca quen panotoc ica Miltlatlacualtiliztli huan Cintlatlacualtiliztli.

Cintlacualtiah

Cintlacualtiah ni ce campeca campa quitlahtlaniliah cintli tlen onmopixcac huan motecpichohtocca, motemachia ma quizati aztah quemman zampa ahciqui yancui pixquiztli. Ipan Chicontepec, axoncah ce cualli tlaixnextiliztli tlen queniuhcatzan mochihua ni campeca. Piltetahtzitzin tlen niquinteconpinqui quipouhqueh quitl inintototatahhuan zanilohqueh ica ni campeca, zanpampa axquiillamiquih tlan ipantequipanoqueh. Tlen niquintencopinqui zan quimanextihqueh anqueh chaneh monequi quitlacualtiz huan quitlaoniltiz campa tecpichtoc. Zan iuhquinon quiihtoyayah. Cintlatlacualtiliztli hueliz quipiya ce nahui zo macuilli generaciones ixpoliuhtoc ipan Chicontepec.

Hueliz ce tlamantli tlen nocca motzquitoc ica campeca tlen Cintlatlacualtiliztli, quiilliah cintetl (tetl tlen cintli). Ni ce tetl campa ihuehueyaca, itlapauhca huan itilauhca neci quence ce piquiz. Moahci quemman ce acahya tlamehua imillah. Iixpan tetl quipiya miac tlahuazantli quence pamitl, huan ica itzalan zan alaxtic. Moihtoa ni cintetl quimahcahuah tiotiotzitzin tlen *ahuahquez* huan quemman ce ahquiya quiahci, nopa quitl quihtoznequi tlahuel izoerteh, nochipa quipiyaz miac cintli zan monequi ma quichihua huan ma quimocuitlahui imillah.

Macehualmeh quemman pixcah, quiquixtiah cintli tlen petlayo huan tlen molcatl. Petlayo, quimanextia cintli tlen huei, etic, yehyectzin huan moquixtia ica totomochtli. Ni hueli motecpichoa calihtic, calmapan zo pan ce xahalli. Tlan anque chaneh quipiya tlen quillia cintetl, ni quitlalia zan tlatlahco tlatecpicholli tlen petlayo para ma teyoti huan zancualli quiahxiliti quemman zampa oncati cintli. Cintli tlen axtlami huan moixnamiqui ica tlen zampa mopixcazza quiillia xicintli. Molcatl, quimanextia cintli tlen cuecuetztzin huan cuahcualli, cualotoc zo palantoc huan miquixtilia itotomochyo. Cintli tlen molcatl yanni pehua motequihuia achtohui, quemman tlamiya, pehuaya mocui tlen petlayo.

Quemman nicpiyaya mahtlactli xihuitl ipan 1996, niquillamiqui niquittac queniuhcatzan nonanan quichiuhqui inon. Iuhcatzan nocca ticpiyah cintetl, ayocquiahcohui huanya cintli. Hueliz quipiya miac xihuitl, melahua motlacualtiyayah cintli quen ne campa motecpichohtoc petlayo huan campa tentoc molcatl, huan nictlalia quence quitlatlalhuiliah huan quitlazcamatiliah pampa itztocca caltic. Naman, ni tlaneltoquilli zan moixmatih ica itocah. Tlatehtemoanih tlen nahuatl quence Arturo Gómez Martínez huan Rafael Nava Vite nouhquiya quiihtoah pan inintequiuh (2002, p. 83; 2012, p. 39). Axoncah tlaixnextiliztli tlan mochiuhtoc.

Quemman tlatlacualtiah

Ni ce tlaneltoquilli malhuilli huan tlahuel ipatiuh para macehualmeh tlen Chicontepec, motlahtlani ce cualli ipohual atl huan yehyectzin ma mohuicacan macehualmeh pan ininnemiliz. Miac altepetzitzin mozancehcotiliah huan monemiliah para mochihuaz ni campeca tlen huahcahua nahui tonatiuh. Tlen achtohui eyi tonatiuh zan monemiliah huan pan nahui tonatiuh tlehcoh ipan tepetl malhuilli tlen itocah Postectli, nouhquiya quiiixmatih quence Ichcacuatitla. Piltetahtzitzin quihtoah monequi titlatlacualtizceh ohomeh xihuitl huan xihxihuitl monequi tictlalitih ce huentli ipan amelmeh tlen altepetzin para tictemachizceh ma oncah atl, ma tlaahuetzi.

Ce macehualli pan ialtepeuh monequi motlalanaz huan tlayecanaz. Ipan nechicoliztli tlen quichihua pan ialtepeuh, tlayecanquetl quinillamiquilia tlahuel monequi mochihuaz campeca tlen tlatlacualtiliztli. Chinancoehuanih monequih ica ininyollo quiiihtozceh tlan tequipanozceh. Miac macehualmeh monequi motemacazceh para quinemilizceh campeca, ni macehualmeh eli quence tequihuahmeh tlen ni campeca. Campeca axmochihua tlan axoncah miac tepalehuianih. Teipan, tlayecanquetl huan itlapalehuihcahuan, tlatitlanih ica cequinoc altepetzitzin tlan nouhquiya quinequih tepalehuizceh zo tequipanozceh. Cehce altepetl tlen tequipano motlalana ica itequihuah. Tequihuahmeh tlen altepetzitzin mozancehcotiliah para quitemozceh nochi tlamantli tlen motequihuiz, huan axipoliuhya quitemozceh huehuehtlacatl huan tlatzotzonanih.

Achtohui eyi tonatiuh tlen tlatlacualtiliztli, tlacualtlaliah huan tequipanoanih xochiuhchihua, quichihuah tlatectli tlen totiotzitzin, mihtotiah huan quizancehcotiliah nochi huentli (quence piyomeh/ cuapelechmeh tlen quinmictizceh, pantzin, tzopelatl, chichic, galletaz). Ipan nahui tonatiuh ica cualcan, nochimeh mozancehtoliah xochicalco. Cehce macehualli quicui ce huentli (paquete chichic, cuachiquihuitl tlen tlapihpiya, piyomeh) huan eli ininixcahuil quiahxititih iixco tepetl tlen malhuilli, quemman quihuicah ce huentli axcualli quitlalizceh tlalchi yon motlapatiltiyazceh zo mopalehuihtiyazceh. Zancehco quizah huan tlehcoh ipan tepetl. Quemman quiitzquiah ohtli nopa ce huei tlaihyohuilli. Tlen tlehcoh pan tepetl mozahuah huan tlacuah quemman mocuaptoquehya ipan tepetl quence ica tiotlac. Axtlacuah yon axyohuih cuatenno quemman tlehco ipan tepetl.

Achtohui para tlehcozceh ipan tepetl, panoh ipan ce amelli tlen eltoc itzintlan tepetl, nopayoh quicauhtehuah ce huentli para totiotzin tlen atl huan quicencuiliah ininohhuih. Iuhquinon monequi panotiyazceh ipan ome canahya malhuilli para ahcitih iixco tepetl. Canahya tlen achtohui eltoc itzintlan tepetl, ni ce cuamezah tlen campa tlatlaliah, tlen campa pehuah para tlehcozceh, ceyoc tlen mocahua zan tlatlahco tepetl. Inin quimanextia ce cuamezah tlahuel huahcapatl. Ce tlapechtli tlachihualli ica cuahuitl zo ohtlatl, campa quitlaliah ce huentli, mihtotiah huan quintlepanittah totiotzitzin. Tlatlaliah ipan cehce cuamezah tlen eltoc ipan tepetl, nochimeh mihtotiah noque momaihtoah para ma oncah ce cualli ipohual atl. Nemaihtolli quimanctiliah Chicomexochitl, totiotzin tlen tonatiuh, tlen tlitl, tlen atl, tlen tlalli huan tlahueliloc. Ontlamih tlatlaliah iixco tepetl, mihtotiah, momaihtoah huan nopayoh mocencahua campeca tlen tlatlacualtiliztli. Oncah miac tlamantli tlen axcanah momanextia ica tlaneltoquilli pan ni tlapohualiztli.

Huehuehtlacatl ipan achtohui tonatiuh tlen campeca nochipa quinhuehuehtlahtoltia macehualmeh tlen altepetzitzin ica queniuhcatzan cualli itztozceh. Quinillamiquiltia tlahuel monequi mopalehuizceh ica ce tequitl, catlinya itequiuh cehce macehualli huan quinhuehuehtlahtoltia queniuhcatzan quixolehuazceh ce cualantli quemman moixpanon. Quinillia tlen cualantli quiitta pan ininpilaltepeuh huan tlen monequi quichihuazceh para quixolehuazceh. Ni ce tlamanatli tlen itequiuh huehuehtlacatl huan monequi quiihtoz, para iuhquinon cualli huan yehyectzin yahtiyazceh.

Naman ipan miac pilaltepetzitzin ni tlaneltoquilli ayoctlahuel quichihuah. Quichiuhtihuetzih quemman quiittah quinpanhuetztoc ahuaquiztli, melahuac tonatiuh quipanhuetzi altepetzitzin huan ayocquinequi tlaeliz millah, pehuaya oncah cualantli huan cuezolli. Huacca nopayoh yoli tequimacholli. Ni campeca quence tlatepotzco mochiuhqui Tecomate ipan 2019, panotoyaya mahtlactli xihuitl axmochiuhtoya. Pan nopa xihuitl tlehcoqueh ipan tepetl tlen Tepecxitla huan amo ipan Postectli. Tlanahuatihquetl tlen nopa altepetzin ayoctemaca manoh ma zan nehnemican ipan Postectli pampa cequi macehualmeh micqueh noque nehnemiyaya quence zan paxaloanih. Pan ni campeca monequi ce huei tlatlepanittaliztli ica totiotzitzin, mozancehcotiliah ehelihuiz altepetzitzin para quichihuazceh ni campeca. Tlan axomochihua ni tlaneltoquilli, neci quence ixpoliuhtiuh inintlaneltoquil macehualmeh, quipatla comuntequitl ica tequitl tlen quichichua ce acahya icelti huan ixpolihui iteticcayo tlaneltoquilli huan macehualtlallamiccayotl (ixtlamatiliztli huan tlayecpamitl).

Cintli itlaneltoquil huan cualli nemiliztli

Cualli nemiliztli ni ce tlahtolli campa ce acahya yehyectzin momachilia, yolpactoc ica inemiliz huan nochipa quipiya ce tlachiyaliztli temachtli ica ce tequitl tlen mohmoztlah quipiya huan quichihua ica iyollo. Tlen anque quiyehyecoah cualli nemiliztli quinextiah ce huei paquiliztli ica macehualmeh, tlatlepanittah, quitlepanittah nochi tlamantli tlen oncah ipan tlaltepactli, huan mopalehuiah quemman oncah ce cualantli, ce tequipacholli. Campeca tlen cintli quinechicoa nochi ni tlamantl tlen cualli nemiliztli.

Ce acahya tlen pactoc huan tlaneltoca nopa ce chicahualiztli tlen eltoc iihtico cintli itlaneltoquil. Nochimeh tequipanoanih monequi axitztozceh ica cuezolli, mahmauhtli, xicoliztli, cualantli yon tequipacholli. Tlan ce acahya quipiya ni tlamantli zo axquineltoca tlaneltoquilli, totiotzin quinextiah ni tlamantli quence, ma huetzi ce tlahuilli, ma tlapani ce tlamantli zo ce acahya ma cuaixpoyahui; campeca axeli tlahuel ipatiuh. Pan campeca quemman xinachtli ipohual, quimanextia ni ixtlamatiliztli pan cehce techan. Nochimeh quipehualtiah toquiztli ica paquiliztli huan quiyehyecoa ce cualli nemiliztli. Nochimeh paquih huan zancehco motemachiah ipan xinachtli. Tlan ahachica mochihuaz campeca, quichihua macehualmeh ma quicahuacan cualantli huan ma quiilcahuacan cuezolli, iuhquinon oncah paquiliztli huan yehyectzin mohuicah. Quemman ixpolohui ce campeca quichihua ma axtimotlacaquilican, cualantli mocuapa ce tequipacholli huan quichihua ma mopazolocan toteixmatcahuan pan topilaltepeuh huan ma tlaihtlacahui pan totlalhui.

Cintli quinequi ma cualli timohuicacan huanya ya huan ma oncah tlatlepanittaliztli ipan tlaltepactli para cualli cueponiz. Nochi campeca quipiya chicahualiztli tlen monehpanoa ica tlaltepactli para yehyectizn eliz cintli. Iuhcatzan pan nochi campeca yohui para cintli, pan nemaihtolli tlen cehce campeca calaqui miac tlamantli tlen tlaltepactli huan tlen ce macehualli inemiliz. Miltlatlacualtiliztli ni ce campeca campa motlahtlaniah cualli ma mozcaltihtiuh piltoctzin huan cualli ma mochicahuilli. Quiihtoa tlahtolli quitl tecuanimeh quiihtlacoah tlatoctli. Huentli tlen quimactiliah millah ni quitl para tecuanimeh axcanah tlahuel ma quitlanahuican huan ma quiihtlacocan tlacualiztli tlen ica motemachiah chaneh ica iteixmatcahuan. Ipan campeca tlen tlatlacualtiliztli, macehualmeh tlahtlanih ma moxoxohuilli (xihuitl tlen mocua, xihuitl tlen pahtli, xihuitl) ininnentlan. Tlahtlani cualli atl, ma tlacehui, zampampa axcanah ma nelquipanohuili. Nochi nemaitolli yohui ica Chicomexochitl huan cequinoc totiotzitzin, quintlahtlaniliah cualli ma tlaelto. Quemman axmochihua campeca, macehualmeh zan quitemoah tlamantli (quence ce cualli calli, tomin torohmeh) tlen ica quipiyazceh quicahualiztli huan axquinehnehuiliah monohnotzazceh yehyectzin para cualli itztozceh pan ininnemiliz.

Pan cehce campeca oncah tlatlepanittaliztli ica ce chaneh, ica macehualmeh, ica altepetzin zo altepetzitzin. Ipan campeca, piltequitzitzin tlen tenanan huan tetatah cehca. Nouhquiya achi cehca ica campeca tlen mochihua xochicalco quence Elotlamanah, tequitl moxehxelhuiliah ica tlacameh huan cihuameh. Nochi tequitl cehca ipatiuh. Quemman altepetzitzin mozancehcotiliah huan monemiliah quichihuazceh Tlatlacualtiliztli, cehce alpetetl quitlalia itlahtol. Altepetzin tlen monechcahuia huan tequipanoa quipiya cehca tlatlepanittaliztli ica tlen axmonechcahuia xichucalco. Ni ome campeca quichihua chicancoehuanih ma monohnotzacan yehyectzin huan axma oncah cualantli; tlan axmochihua, huacca yoli tequipacholli huan mohueilia xicoliztli tlen quiihtlacoa tlatlepanittaliztli huan tlaneltoquilli.

Ce itequiuh Elotlamanaliztli huan Tlatlacualtiliztli, ya quitl monequi quichicahuiliz tomacehualtlallamiccauh huan totlachiyaliz. Quemman ce acahya tequipanoa ipan xochicalli, quichihua ce huei tequitl huan quichicahuilia quen timonehpanoah huanya ce acahya. Tequipanoanih quimachiliah chicahualiztli quemman quimatih oncah macehualmeh tlen ica huelih monechcahuizceh quemman quipiyah ce tlamantli cualli zo axcualli pan ininnemiliz, quence quemman ehua ce conetzin zo quemman miqui ce toteixmatcauh. Quemman ixpoliuhtiuh campeca, macehualmeh ayocquipiyah inintequixpoyohuan huan motemachiah zan pan ininteixmatcahuan, quence mopatla initlanehnehuiliz ica cequi.

Quemman ixpolihui cintli itlaneltoquil quichihua ma moahcomanacan altepetzitzin. Ica itlaneltoquil cintli tlen quipiya macuilli campeca, hueliz ce macuilli generaciones ticpolohqueh itlaneltoquil huan iixtlamatiliz Cintlatlacualtiliztli. Totatahhuan ayocmiltlatlacualtihqueh; itlaneltoquil tlen queniuhcatzan mochihua hueliz ixpoliuhqui nimantzin ica tototatahhuan. Togeneracion tlen naman, ayocquichihua Tlatlacualtiliztli yon Elotlamanaliztli. Ticchicoquixtiah inintlahtol huehuentzitzin huan ayoctinehnehuiliah ticchihuazceh campeca. Cequin tlen nocca quichihuah Elotlamanaliztli, quichihuah para quinextizceh tlen tlamantli quipiyah quence chicahualiztli huan quiilcauhtoqueh comontequitl huan itlaneltoquil zo itlacuamachiliz campeca. Naman zan ticchihuah tlen xinachtli, ce campeca tlen cintli itlaneltoquil. Tlan axtimotlalanah huan axtictlepanitta cintli itlaneltoquil, ipan ce omeh generaciones tlen huallauh, hueli ayocmochihuaz ipan Chicontepec. Naman quen moixnextihtiuh campeca, eliz quence zan ce tlamahuizolli. Monequi ticyoltilizceh cintli itlaneltoquil, tiquinnotzazceh tepolcameh, ichpocameh ma monehpanocan ica ni tlaneltoquilli.

Ixcopintlaltecpanalli

Amoxtlatecpanalli

Argüelles Santiago, J. N. (2012). El maíz en la construcción y transmisión de una identidad cultural de la Huasteca Veracruzana. In Pan A. van't Hooft & J. A. Flores Farfan (Eds.), *Estudios de lengua y cultura nahua de la huasteca* (pp. 11–29).

De la Cruz Cruz, E. (2015). *Tototatahhuan ininixtlamatiliz.* Amoxmecayotl Totlahtol (J. Olko & J. Sullivan, Eds.). Huei Caltlamachtiloyan Varsovia.

De la Cruz Cruz, E. (2017). *Cenyahtoc cintlo tonacayo: huahcapatl huan tlen naman.* Amoxmecayotl Totlahtol (J. Olko & J. Sullivan, Eds.). Huei Caltlamachtiloyan.

Gómez Martínez, A. (2002). *Tlaneltoquilli: La espiritualidad de llos nahuas chicontepecanos.* Ediciones del Programa de Desarrollo Cultural de la Huasteca/CONACULTA.

Gómez Martínez, A., & van't Hooft, A. (2012). Atlatlacualtiliztli: La petición de la lluvia en Ichcacuatitla, Chicontepec. In Pan A. van't Hooft & J. A. Flores Farfan (Eds.), *Estudios de lengua y cultura nahua de la huasteca* (pp. 100–118).

IDIEZ. (2010). Curso de Lengua Nahuatl Avanzado II. Instituto de Docencia e Investigación Etnológica de Zacatecas A.C.

Nava Vite, R. (2012). El costumbre: ofrendas y música a chicomexochitl en Ixhuatlan de Madero, Veracruz. In Pan A. van't Hooft & J. A. Flores Farfan (Eds.), *Estudios de lengua y cultura nahua de la huasteca* (pp. 31–65).

Ceremonial Practices Relating to Corn in the Region of Chicontepec: Local Aspects of Wellbeing

The Maize Ceremonies: An Act of Faith

This chapter discusses maize ceremonies and how they relate to *cualli nemiliztli*. *Cualli nemiliztli* is a way of life where one commits to keeping joy and beautiful things in one's life. We show our commitment daily in our attitude while working and interacting with others. Material possessions are accessories to a good life. Respect, friendly social interactions, and the local belief system ensure a good life. This chapter also touches on the relationships across different generations in the Nahua communities and the transmission of knowledge.

I am a *macehualli*.[1] I was born and raised in Lindero 2, Chicontepec, Veracruz. My childhood and adolescence are full of memories of working in the fields alongside my mother and other men from the community. I always showed interest in maize and its ceremonies. When I turned 18, I emigrated to larger cities in search of economic and educational opportunities. Currently, I teach Nahuatl at several universities in the United States. I am a postdoctoral candidate at the University of Warsaw and Director of the Institute of Teaching and Ethnological Research in Zacatecas (IDIEZ). Despite my heavy presence in academia, my primary focus is language and cultural revitalization through local initiatives that directly impact the villages of Chicontepec. I approach my research from an emic standpoint in dialogue with academia.

This chapter is primarily in Nahuatl because my target audience are Nahuas; we lose concepts in translation and show that we, too, have a voice in academia. Translation in English is included to allow for conversation with those interested in this topic. The English translation includes background information mostly as footnotes and is omitted in Nahuatl.

---

[1] Nahua; indigenous person.

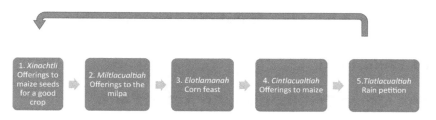

Fig. 12.1 Maize ceremonies. This figure shows the five maize ceremonies and the order they occur

Nahuas practice five maize ceremonies, *Xinachtli*, *Miltlacualtiah*, *Elotlamanah*, *Cintlacualtiah*, and *Tlatlacualtiah*. *For* this, see Fig. 12.1.

There are works pertaining to the maize ceremonies, most written by Nahua scholars from the Huasteca Veracruzana. Their research focuses on one ceremony rather than all five maize ceremonies. In the case of Jazmín Nayelly Argüelles Santiago (2012) in her article "Maize in the Construction and Transmission of a Cultural Identity of the Huasteca Veracruzana," talks about *Elotlamanaliztli* (Corn feast). She briefly describes the ceremony and argues that the discontinuity of the ceremony is caused by the influence of government policies. Nava Vite (2012) in his article "El Costumbre: Offerings and Music to Chicomexochitl in Ixhuatlán de Madero, Veracruz," talks about offerings to Chicomexochitl, emphasizing that Nahua families plea and pray to the elements of nature to strengthen the crops. In a short paragraph, Nava Vite defines maize ceremonies: *Cintoquiztli (Xinachtli)*, *Miltlacualtiliztli*, *Miyahuacalaquiliztli*, *Elotlmanaliztli*, and *Cintlacualtiliztli*. He then includes an extensive explanation about *Atlatlacualtiliztli (Tlatlacualtiah)*, rain petition ceremony. Gómez Martínez and Van't Hooft (2012) in their article "Atlatlacualtiliztli: Rain Petition in Ichcacuatitla, Chicontepec," speak of an extensive ethnographic description of rain petition and pilgrimage to the sacred mountain Postectli. Gómez Martínez in the book *Tlaneltoquilli: The Spirituality of the Nahauas from Chicontepec*, under the Agricultural Rituals section, makes a broader description of *Atlatlacualtiliztli* (*Tlatlacualtiah*) and a brief explanation of the rest of the ceremonies of the cycle (Gómez Martínez, 2002, pp. 180–113).

Information contained in this chapter is from personal experience, fieldwork in 2011 and 2019, and interviews with four elders in 2013 and 2014. I started to work in the fields at the age of seven. The men I worked alongside noticed my interest in maize and passed down their knowledge and wisdom as their fathers had. The quotes I include throughout the chapter are theirs, quoting from their fathers' words. These words are from our ancestors; thus, I cannot credit a person. As a child, I attended *Xochicalli*[2] regularly until early 2000, when our community stopped practicing communal ceremonies. In 2011 I attended an *Elotlamanah* celebration in El Dorado in the municipality of Ixhuatlan de Madero and learned how to make flower adornments. In 2019 I attended *Tlatlacualtiah* in Tecomate, Chicontepec. I interviewed four elders who actively participated in or organized communal ceremonies. First is a healer of 60 years old from La Pahua, Chicontepec. Second is her 60-year-old husband, who continues to practice *Miyahualcalaquiah*. The third is a 50-year-old musician from Tepecxitlan, Chicontepec. The fourth is an 80-year-old man from Tecomate, Chicontepec, who was the main organizer of *Elotlamanah* and *Tlatlacualtiah* for 30 years until early 2000, when there was no community support.

Maize as *Xinachtli*,[3] *Miyahuatl*, and *Chicomexochitl*

The people of Chicontepec have three terms to refer to maize in a ceremonial context. These are *xinachtli* (seed), *miyahuatl* (tassel/male inflorescence), and Chicomexochitl (maize deity). Nahuas use each term in different spaces.

Xinachtli represents the seed and planting season. For a dry corn kernel to become a seed, *xinachtli*, the parents select it, give it breath, and leave it to soak overnight. The following morning the kernels are

---

[2] House designated as a sacred place to give offerings to Chicomexochitl and perform other communal ceremonies.

[3] Maize kernels picked to give offerings to and then plant.

transformed into *xinachtli*; they are sacred and must be handled as you would a newborn baby. Before holding the *xinachtli*, you must cup your hands and exhale one breath into them. This is a way of giving reverence. Children are not allowed to handle *xinachtli*. The family gives the *xinachtli* offerings; the men head off to the field to sow it. The *xinachtli* holds the family's hopes and livelihood.

Miyahuatl is the corn's flower, tassel (male inflorescence), and announces the upcoming harvest season. Once it sprouts, the corn immediately silks and starts to mature. At this stage, the elders say, "just as one went to plant *xinachtli*, one must fetch it," meaning that corn should only be taken from the milpa if preceded with an offering; it must be watered and incensed. Only after observing local beliefs should it be harvested and shared with neighbors.

Chicomexochitl is a term that indicates respect for maize or implies that a ceremony will be held for the deity in a sacred place called *Xochicalli*. Chicomexochitl is the maize deity with whom the people of Chicontepec interact daily at the table while having a meal. When parents talk about a maize ceremony, they call him Chicomexochitzin. The ancestors describe Chicomexochitl as a boy and a girl. Chicontepec and the surrounding municipalities share Chicomexochitl's origin story, widely known as *Chicomexochitl huan tenantzitzimitl*, "Chicomexochitl and the angry grandmother." Narrations passed through oral tradition share the same plot (De la Cruz Cruz, 2015; Nava Vite, 2012; IDIEZ, 2010). The tale tells of a grandmother whose daughter gives birth to a son, Chicomexochitl. This boy is always cheerful and never causes any trouble. The grandmother is not fond of her grandson. She devises games to kill him, such as playing in an oven; however, she never succeeds and goes on to die during one of the games. Chicomexochitl lives happily, with no worries, and always does as he is told.

Many who know and narrate the story of Chicomexochitl say it is true, that it recounts what happened long ago. Today, when the Nahuas carry out a local ceremony according to their customs, such as the rain petition or the corn ceremony, they represent Chicomexochitl in two ways:

Fig. 12.2 Paper cut out representing Chicomexochitl. De la Cruz Cruz, 2017 [Chicomexochitl][Photograph])

1. A paper cutout. The paper cutout represents the deity and is clothed. See Fig. 12.2.
2. Ears of Corn. An upright bundle of three or four ears of corn wrapped with a handkerchief or embroidered napkin, tied with a ribbon, incensed, and decorated with marigolds. See Fig. 12.3.

Nahuas place both representations on the altar and present offerings such as food and drinks.

Xinachtli: Offering to the Corn Seed

Xinachtli is the ceremony's name, and the kernels are selected as seeds. To become *xinachtli,* the grain undergoes a process. The parents pick out corn from past season's *cintli tlen petlayoh,*[4] looking for those with a

---

[4] Dry mature corn harvested and stored without removing the husk.

Fig. 12.3 Ears of corn representing Chicomexochitl. From Cenyahtoc cintli tonacayo: huahcapatl huan tlen naman, by Eduardo de la Cruz, 2017, Warsaw, University of Warsaw, Faculty of "Artes Liberales." Alan Sandstrom took the photograph during his fieldwork about the rain petition ceremony in Cacahuatengo, Ixhuatlan de Madero, in March 2007

whole ear of large intact kernels. Together they husk and shuck the corn. Then, they examine the kernels closely, selecting and placing in a pile kernels that are wide, flat, heavy, and have an intact seed coat and tip cap. Both parents pick up a handful from the pile of kernels, cup their hands together, bring them inches from their mouths, and exhale a short breath over the kernels. Exhaling a short breath into one's cupped hands symbolizes sharing one's breath with the seeds and represents one's wholehearted commitment. The parents gently drop the seeds into a bucket half full of water. The kernels soak overnight. In the early morning, the mom pours the bucket over two woven baskets to catch the now *xinachtli*.

The family prepares to give *xinachtli* an offering, a meal before the men set off to sow the seeds. The mother clears the family altar[5] and

---

[5] It is customary for each family to have an altar in their homes. The altar is a wooden table, ranging in size, up against the wall, with a plastic or embroidered tablecloth. Due to religious syncretism most families have Catholic imagery and statues on the table or hung up on the

lays out the offerings, including tortillas, boiled eggs, bread, *rozca*,[6] coffee, and liquor made from anise. She prepares the food before dawn. Children help by fetching items such as flowers or food and handing them to their parents to place on the altar. The mom lights the candles and brings the *popochcomitl*[7] with embers to burn the copal incense. She places one basket with *xinachtli* on the floor, slightly in front of the table or next to the table leg. Then, she places a cup of coffee next to or inside each basket with a piece of bread. The cup handle must face toward the wall. Chicomexochitl sits with his back against the wall facing the family.

Once the altar is set, all household members gather in front of the altar. The head of household makes reverence by exhaling into their cupped hands and slightly bowing on the right front corner of the table and then the left. Here exhaling a short breath into one's cupped hands is an act of reverence representing one's wholehearted commitment. The head of household takes the *popochcomitl* in their right hand and passes it above and around all offerings to incense them. One by one, all family members incense the offerings, silently asking the *xinachtli* to grow well and not forget its people; it ought to sprout where they plant it; it must grow lush and mature. The ancestors' words taught us to let go of envy, anger, or resentment to give offerings to *xinachtli* wholeheartedly. Otherwise, the *xinachtli* feels our lack of respect and commitment; it may grow sad. A saddened *xinachtli* grows poorly, wilts, withers, and does not bear fruit. Upon receiving the offerings, the *xinachtli* leaves the house fed and watered. It will return, brimming with flavor as fresh corn for immediate consumption or maize to be stored and last until the following harvest.

When the *xinachtli*, or any crop, breaks through the soil one must tend to it with high frequency. The care includes weeding the field, scaring off thrushes, birds, or other animals, and even talking to the

---

wall. The altar is in the common room. Most families' homes are made of two structures. One structure has a opens into a large common space room, similar to a living room, that connects to several bedrooms. The other structure is a kitchen space where most meals are cooked and consumed. The bathroom is an outhouse, near the edge of the property.

6 Crisped bread, intertwined strands in a torus shape, can be salty or sweet.

7 A clay chalice.

plants as one works in the fields. If one forgets these responsibilities the plants' growth stunts; the stalk may dry before flowering, stop growing, or continue to grow and not bear fruit. At times, the stalks do flower, grow, and bear fruit, but the ears of corn partially fill or lack kernels altogether. When this happens, the elders say, "The crop no longer wants to grow compah."

Those who have faith in their milpas joyfully rise to check on their crops at dawn. They walk along the outer edge of the milpa, examining the plants closely. Between nine and ten in the morning, they return home, have a meal, and return to the milpa to complete that day's tasks. That daily routine starts on the day of sowing until harvest. A month or two after planting the maize, farmers make an offering, *Miltlacualtiah*, to the milpa and the maize plants growing in it. This practice is almost nonexistent today.

Miltlacualtiah: Offering to the Milpa

Miltlacualtiah is an offering to the milpa/crops that provides the corn plants with nutrition, strength, and protection to produce a good crop. This ceremony is not widely known compared to the other corn ceremonies. Elders mention it when listing corn ceremonies. Nahua scholars Arturo Gómez Martinez and Rafael Nava Vite mention it in their work (Gómez Martínez, 2002, p. 83; Nava Vite, 2012, p. 38). Jesus Alberto Flores Martínez, another Nahua scholar includes it in his forthcoming film. Loss of practice is one reason for the lack of documentation. As a Nahua, I consider *Miltacualtiah* one of the most important rituals because a good harvest depends on it.

Elders narrate that when the corn stalks are about a meter and a half tall, the head of the house starts the ceremony preparations. First, to lead the ceremony, they seek out a *huehuehtlacatl*.[8] The *huehuehtlacatl* specifies what items they will need to gather and bring on the agreed date. On the day of the ceremony, the family and *huehuehtlacatl* gather

---

[8] A wise old man with vast ancestral knowledge, also known as a healer, responsible for directing ceremonies such as weddings, the handwashing of godparents, among others.

in the center of the milpa. In the middle of the milpa, they make a table from bamboo that functions as an altar where they place offerings to worship Chicomexochitl, *Totonanan tlaltepactli*—mother earth—and the deity of water. They place bread and coffee on the surrounding corn plants.

As this practice started to decline, the father began to lead the ritual. As a child, I recall seeing my mother carry out this ritual in our milpa as her father taught her. Our parents and grandparents often told us, "Watch how it is done because when you grow up and get married, this is what you will do. You will continue this practice to have quality maize, and your food will suffice." That was their advice; we have forgotten their advice. Our generation does not practice this ceremony.

When people neglect their milpa, when they neglect to practice this ritual, there is a low crop yield. Around the time this ceremony takes place, the plants are close to reaching their final height, the tassel and ears emerge, and the kernels start to form. At this stage, keeping watch is crucial so raccoons, coyotes, or other animals do not harm the crop. Practicing *Miltlacualtiah* ensures that the corn obtains necessary nutrients from the soil without using fertilizers. The corn will withstand strong winds and rain if a hurricane occurs on a nearby coast. Animals will also not excessively take from or harm the crops. Today farmers rely on insecticides and fertilizers. The products were effective in the short term but are increasingly ineffective.

Elotlamanah: Blessing of the Corn and Feast

Nahuas gather as a community at the *xochicalli* to celebrate *Elotlamanah* before harvesting or consuming the season's corn, regardless of the crop yield. Not everyone plants corn at the same time; often, there is a difference of two weeks to a month. If a wealthy family needs to harvest before or after the communal celebration, they organize their *Elotlamanah* and invite the community to participate. If a family does not have economic capital for a large celebration, they practice *Miyahuacalaquiah* (explained in the following section). Nahuas celebrate *Elotlamanah* to thank Chicomexochitl for the season's yield and welcome the harvest season. They show respect and gratitude through offerings that include flower arrangements, drinks, food, sacrificial chickens, music, and dance.

This ceremony requires many hands. Community members volunteer as lead organizers and seek help (monetary, in-kind, and time) from relatives and neighbors. Planification is key. They must seek a *huehuehtlacatl* to guide the ceremony, hire a huapango trio,[9] and find *coyol* (palm-like) plantling leaves and marigold flowers for adornments. They purchase a *popochcomitl*, candles, copal, paper for cutouts, bananas, sweet potatoes (to make compote), oranges, limes, peanuts, apples, and sweets (e.g., cookies and chocolate bars). They preorder some food items. They make *tintinez, pemolez,* and *alfahorez*.[10] They prepare or purchase drinks: coffee, hot chocolate, a liquor made from anise, soda, and atole. They make two types of enchiladas with freshly cooked tortillas, one slathered in tomato sauce and the other with refried beans. Sometimes, they buy rocket fireworks (launched into the air, exploding with a loud bang) to light when they welcome the corn.

The labor division between men and women is essential for this ceremony. Men form several groups, each taking on a different task. One group decorates the altar with *limonaria*.[11] Another makes an arch in front of the house where they will welcome the corn. A third group makes adornments such as *coyol, maxochitl*,[12] necklace flowers, and star-shaped flowers with marigolds and *coyol*. A fourth group takes sacks or baskets to harvest enough corn for the ceremony and a few corn stalks with an ear of corn attached. They place these in the corridor outside the *xochicalli* while the rest wrap up the preparations to welcome the corn.

Similarly, the women divide into groups to complete tasks of equal importance. The duties include dressing the deities, preparing the sacrificial chickens, making tortillas, cooking, and serving food and

---

[9] A trio that consists of a violin, huapanguera, and jarana characterized by its rhythmic structure that mixes duple and triple meters. Ceremonial music includes only the melody with no lyrics.

[10] *Tintinez, pemolez, and alfahorez* are desserts whose principal ingredients include maize flour and *chancaca* (unrefined brown sugar sold in solid cone shapes).

[11] Small green leaves used only as decoration.

[12] hand-like adornment.

drinks to all participants. The *huehuehtlacatl* coordinates all these activities, ensuring everything is done correctly and on time.

The musicians are another vital element. Their presence is essential to appease Chicomexochitl. As soon as the musicians arrive, they begin to play and stop when the *huehuehtlacatl* concludes the ceremony of corn. When they start to play, those present, men and women, dance in front of the altar. Those who arrive later greet everyone and start to dance; they all take turns dancing. The musicians play a specific song when sacrificing the chickens, welcoming the corn, and placing the offerings on the altar. They know which piece corresponds to each stage of the ceremony. If they play the wrong melody, the *huehuehtlacatl* reminds them of the correct piece.

Around noon, the *huehuehtlacatl* announces it is time to welcome the corn at the arch. Men grab the corn and wait outside in front of the arch facing the *xochicalli*. Women take the corn stalks, stand facing the men, and welcome the corn. Women, men, and children gather by the arch; as the music plays, everyone dances. They adorn the corn with flowers and incense it. The musicians must not cease to play, even for a second— those carrying the corn dance continuously. At the *huehuehtlacatl's* signal, they conclude the first part of the ceremony by bringing the corn inside. Everyone stands by the altar.

As some continue dancing, the men empty the corn in the center of the room. They stack it to form a cylinder about half a meter high. The stacked corn's top layer serves as a table where women and men lay out the offerings. First, they place the necklace and *coyol* arrangements; then, they offer food, fruit, and drinks (non-alcoholic). Meanwhile, the rest continue dancing. Once they place all the offerings and incense, everyone dances for an extended period. Then they eat, laugh with joy, and share anecdotes that have happened to them in the milpa. The children eat and talk about games. Everyone is joyful. They are happy and grateful to Chicomexochitl; they will have maize again.

When everyone finishes eating, women and men prepare sweet tamales with tender corn (kernels in the milk stage). Some carefully remove the green corn husks they will use to make the tamales. Others cut the kernels from the cob and grind them using a metate or mill. They collect the corn dough in one container and the juice separately to make special atole, *eloatolli*. Everyone gathers around the table to make tamales. Once all the tamales are wrapped, they cook them in a clay pot and some fresh corn over the fire. Over another fire, they prepare and cook the atole. While the corn tamales and the atole are cooking, the adults talk about the work they still need to do in the fields. Once again, they share a meal; they eat tamales and drink atole. To conclude, they divide any leftovers equally among all.

Elotlamanah is transitioning from a communal practice to a private practice. The shift is significant in Chicontepec; Nahuas only celebrate *Elotlamanah* with their kin. The communal practice dropped as fewer community members were willing to organize the ceremony, contribute economically, and volunteer. Wealthy families who believe in Chicomexochitl practice *Elotlamanah* in their homes only when they have a high crop yield and limit invitations to their extended family. Celebrating it only when there is a high crop yield makes the practice folkloric. The switch from a communal to a private celebration weakens harmony in the community and creates divisions and rivalries. *Elotlamanah*, a celebration that honors and thanks Chicomexochtl, passes to signal family wealth and exclusion. Those who cannot afford to celebrate *Elotlamanah* yet maintain strong faith in Chicomexochitl celebrate *miyahuacalaquiah*, an offering to the tasseled plant.

Miyahuacalaquiah: Offering to the Tasseled Plant

The offering to the tasseled maize plant is a small ceremony practiced at home when a family needs to harvest before/after the communal *Elotlamanah*, they do not have the capital to organize their own, or there is no *Elotlamanah* celebration in their community. The meaning of *Miyahuacalaquiah* is the same as *Elotlamanah*. The family demonstrates their gratitude for the season's yield by giving offerings to Chicomexochitl.

All members of the household participate; everyone helps with the task. Some purchased items are bread, *rozca*, soft drinks, candles, a vase, and a copal. At dawn, before anyone has even a cup of coffee, a member of the family, usually the father, sets out to bring the corn stalks. They cut stalks from each corner and one from the center of the milpa. If the ears are ripe, they bring five. The father divides the stalks into two bunches and adorns them with marigold flowers. Upon arriving home, he waits outside the common room. The master of the ceremony, usually the mother, gathers everyone inside the house to welcome the maize. The family meets the maize at the door and greets it reverentially. Each person takes the *popochcomitl* and incenses the stalks. The father brings the stalks inside, placing one on each side of their alter.

To conclude, the family places food offerings on the table including coffee with *rozca* and chicken soup with freshly made tortillas (see Fig. 12.4). The family waits between 5 to 8 minutes for the corn deity to take the offering. The ceremony then ends, and the family eats their first meal of the day together. After *Miyahuacalaquiah*, the family can start harvesting or wait for the corn to dry.

Fig. 12.4 Miyahuacalaquiah. De la Cruz Cruz. 2017 [Miyahuacalaquiah][Photograph])

Domestic rituals' protocol varies by household. Parents do not narrate the process, give instructions, or explain the meaning of each action. Children must carefully observe and remember every detail to practice when they marry and pass it down to their children. The altar is an example of this. Each household has an altar, a table against the wall in the common room. Table size can vary from 1 meter by 0.5 meters to 2 meters by 0.75 meters. Most families have catholic imagery (cross, saints) due to syncretism. The amount of imagery varies from a few to having a whole wall/table plastered with imagery. Parents may teach their children to make the sign of the cross when approaching the altar and other parents teach them only to give reverence (explained in the *Xinachtli* section).

Today, few Nahuas of Chicontepec people practice this ceremony. When it is corn season, they harvest and eat it; they do not even offer a candle to the maize as gratitude for corn. Many elders say, "Chicomexochitzin is sad. They have not fed or given him a drink, and now he wants to abandon us." Elders express that the decline in corn production, droughts, and suffering in the region is because the people have forgotten Chicomexochitl, *cualli nemiliztli,* and the ways of our ancestors. If we take no action, *Miyahuacalaquiah* will be lost, as has happened with *Miltlacualtiah* and *Cintlacualtiah.*

Cintlacualtiah: Offering to Maize Grain

Cintlacualtiah is a ceremony to ask that the season's harvest lasts until the next harvest. In Chicontepec, there is no record of anyone actively practicing this ceremony. Elders I interviewed recounted that their grandparents spoke of this ceremony; however, they do not recall witnessing it. Interviewees stated that families should feed and water the maize where they store it. The explanation did not extend beyond that. *Cintlacualtiah* was likely lost in the region of Chicontepec four to five generations ago.

A possible remnant of *Cintlacualtiah* is the *cintetl*, maize stone. This stone is about the size (length, width, and thickness) of a *piquiz*.[13] A person can find one when they are tilling their milpa. The front of the stone has engraved grooves lengthwise, resembling furrows. The back is smooth. People say that the *Ahuahquez*[14] drop them when it rains. Only the chosen will stumble upon these rocks; when they do, so long as they tend to their milpa, they will always have an abundance of corn.

When Nahuas harvest maize, they collect dry whole ears of corn with and without intact husks. Larger, heavier corn that animals have not picked at are left with the husk. These are stacked and stored inside the house/shack or against an outside corridor. If the family has a *cintetl*, they put this in the center of the stack. The *cintetl* will help the corn last until the next harvest. Any dry corn left when the next harvest arrives passes to be *xicintli*, which means old corn. The dry corn harvested without husks is either small, rotten, or has birds pecked it. The family consumes this corn first, then the dry corn with husks.

In 1996, when I was ten, I last recall my mom doing this. Although we still have a *cintetl*, she no longer stores it with the maize. Perhaps several generations ago, families made offerings to the maize after storing it. They may have expressed gratitude for having corn to harvest and bring home and thanked Chicomexochitl for being in their home with them. Today, only this ceremony's name is known; Nahua scholars such as Arturo Gómez Martínez and Rafael Nava Vite also mention it in their work (Gómez Martínez, 2002, p. 83; Nava Vite, 2012, p. 39). There is no evidence of its practice.

Tlatlacualtiah: Rain Petition

Tlatlacualtiah is a sacred ceremony of great cultural value to ask for a good rain season and social and life relations. Several villages gather to organize and carry out this tradition over four days. The first three days are preparation of the offerings, and on the last day, they pilgrimage to

---

[13] Bean tamal.
[14] Water entities.

Postectli, a sacred mountain also known as *Ichcacuatitla*. Elders say we must practice *Tlatlacualtiah* every two years and give an offering to the village wells every year to ensure sufficient rain each year.

A person from the community must self-appoint as the lead organizer. During one of the community meetings, the led organizer steps up to remind everyone of the need to celebrate *Tlatlacualtiah*. The community must express willingness to help, and a handful of people must volunteer to organize the ceremony; this group becomes the ceremony committee. The ceremony is not held if there is not enough support. Then the led organizer and the committee invite nearby villages to participate. Each participating village forms its committee. The committees of participating villages gather to bring together all necessary items and participants, including the *huehuehtlacatl* and the musicians.

The first three days of *Tlatlacualtiah* include cleansing all participants, making flower arrangements, cutting paper representations of deities, dancing, and gathering all offerings (e.g., sacrificial chickens/roosters, bread, soda, beer, cookies). On the fourth day, before dawn, everyone gathers at the *xochicalli*. Each person takes an offering (pack of drinks, basket of bread, chicken) and is responsible for carrying it to the top of the sacred mountain without setting it down or asking for help. They head out together and hike up the mountain. Undertaking the hike is an immense sacrifice. Pilgrims fast from the previous night until the end of the ceremony (late afternoon). They must abstain from eating, drinking, and going to the bathroom.

When they arrive at the mountain base, they stop at a well at the foot of the mountain, leave an offering for the water deity, and start up the mountain. They pass two sacred sites on their way to the mountain's peak. The first site is near the start of the base, and the second is halfway up the mountain. Both are referred to as the old/ancient table, referencing the table there. The table is made from bamboo. They place part of the offerings at each table; everyone dances while praying for a good rain season. Prayers are directed to Chicomexochitl, the deity of the sun, fire, water, earth, and *tlahueliloc*.[15] They place the remainder

---

[15] A deity of both good and evil. It can bestow and take wealth, health, and overall wellbeing.

of the offerings at the summit, dance, pray, and conclude the rain petition. Various details and rules regarding this ceremony are omitted.

The *huehuehtlacatl* throughout the ceremony, starting on the first day, gives the community advice on how to live a good life. He reminds them of the importance of the family unit, the responsibility each person has, and advice on how to handle disagreements with each other. He points out problems he sees within the community and how they should resolve them. This is an integral part of his role during the ceremony to ensure the restoration of harmonious social relations.

Today the villages of Chicontepec only gather to practice this ceremony after multiple years of drought that cause extreme water rationing (a bucket of water per day per person), little to no corn production, and dying livestock. Only then do concerns arise. The last ceremony was in Tecomate in 2019, after over ten years. That year they hiked Tepecxitlan instead of *Postectli*. The local government banned anyone from hiking *Postectli* because people died while hiking due to a lack of maintenance. This ceremony requires an act of great faith, effort, and respect for the deities, bringing together different villages of the region to participate. Its lack of practice reflects declining faith in the local religion, the shift from community union to personal greed, and the loss of cultural capital (knowledge and environment).

Maize Ceremonies and *Cualli Nemiliztli*

Cualli nemiliztli is a Nahua concept of wellbeing rooted in being joyful across all aspects of one's life and projecting a good attitude when undertaking daily tasks. Those who practice *cualli nemiliztli* demonstrate a high degree of coexistence, respect people and nature alike, and unite in stressful situations. The maize ceremonies echo these principles.

A joyful and faithful heart is the center for any of the maize ceremonies. Everyone participating must be free from feelings of sadness, fear, envy, jealousy, anger, and resentment. If even one person holds such feelings or is doubtful of the practice the deities manifest their displeasure by causing candles to fall, items to break, and that person to feel dizzy; the ceremony is not as effective. The *Xinachtli* ceremony sets the tone for the season within each family unit. Everyone starts the agricultural cycle practicing the core values of cualli nemiliztli. They start joyfully placing their hope and faith in the seed together. The frequency of ceremonies pushes people to resolve and let go of negative feelings, thus maintaining social harmony. The loss of even one ceremony allows for disagreements and conflicts to cement into rivalries that start to weaken kin, community, and regional ties.

Corn needs coexistence and balance in nature to flourish. All ceremonies have elements plea for balance in nature so that the corn may flourish. Although the ceremonies are centered around maize, prayers during each ceremony include various aspects of nature and human life. *Miltlacualtiah* is a ceremony asking that the crop continue growing and reach maturity. There is an understanding that wild animals will eat from the crop. Offerings to the milpa are so that animals do not excessively damage the crop and jeopardize the family's food security. During *Tlatlacualtiah,* Nahuas ask that the surroundings return to their lush state (edible plants, medicinal plants, weeds). They ask for sufficient rain, for the heat to subside but not for it to be gone. All pleas to Chicomexochitl and other deities ask for a balance, not an excess. The lack of practice of these ceremonies results in individuals focusing their energy on accumulating goods (large houses, money, livestock) rather than reflecting on how they can live a more balanced life.

Respect among family, community members, and across villages is incorporated into each ceremony. In domestic rituals, the mother's and father's roles are of equal importance. Similarly in communal ceremonies, such as *Elotlamanah,* labor is divided among men and women. All tasks are of equal importance. When villages gather to organize *Tlatlacualtiah* each village has an equal voice. The hosting village is not privileged or prioritized. These two ceremonies allow community members to interact in a conflict-free space; the lack of practice allows conflicts to fuel rivalries that, in turn, deteriorate respect.

One of the functions of *Elotlamanah* and *Tlatlacualtiah* is to strengthen the sense of belonging and group identity. Participating in a group ritual develops and strengthens camaraderie. Participants feel a sense of security knowing that they have a network of friends to fall back on during stressful life situations such as the birth of a child or the loss of a family member. The decreased practice results in individuals feeling excluded, relying only on kin, and shifting toward an egotistical mentality.

The loss of these practices has resulted in the destabilization of communities. Out of the five maize ceremonies, about five generations ago, we lost practice and knowledge of *Cintlacualtiah*. Our parents stopped practicing *Miltlacualtiah*; knowledge of how to carry it out may soon be lost with our grandparents. Our generation stopped practicing *Tlatlacualtiah* and *Elotlamanah*. We ignore elders' advice to resume these practices. Those who practice *Elotlamanah* do so to display their wealth and have forgotten the community aspect and the ceremony's true meaning. We only practice *Xinachtli*, one out of the five maize ceremonies. If there is no intervention, over the next two to three generations, the corn ceremonies will cease to be practiced in Chicontepec. The celebration of these ceremonies will become folkloric at best. We need to try to revive these ceremonies and involve the younger generation.

Page 21 Figure 1. Maize
Page 24 Figure 2. Paper cut out representing Chicomexochitl
Page 24 Figure 3. Ears of corn representing Chicomexochitl
Page 30 Figure 4. Miyahuacalaquiah.

References

Argüelles Santiago, J. N. (2012). Maize in the construction and transmission of a cultural identity of the Huasteca Veracruzana. In A. van't Hooft & J. A. Flores Farfan (Eds.), *Estudios de lengua y cultura nahua de la huasteca* (pp. 11–29).

De la Cruz Cruz, E. (2015). *Tototatahhuan ininixtlamatiliz*. Amoxmecayotl Totlahtol (J. Olko & J. Sullivan, Eds.). Huei Caltlamachtiloyan Varsovia.

De la Cruz Cruz, E. (2017). *Cenyahtoc cintlo tonacayo: huahcapatl huan tlen naman.* Amoxmecayotl Totlahtol (J. Olko & J. Sullivan, Eds.). Huei Caltlamachtiloyan Varsovia.

Gómez Martínez, A. (2002). *Tlaneltoquilli: The spirituality of the Nahuas from Chicontepec.* Ediciones del Programa de Desarrollo Cultural de la Huasteca/ CONACULTA.

Gómez Martínez, A., & van't Hooft, A. (2012) Atlatlacualtiliztli: Rain petition in Ichcacuatitla, Chicontepec. In A. van't Hooft & J. A. Flores Farfan (Eds.), *Estudios de lengua y cultura nahua de la huasteca* (pp. 100–118).

IDIEZ. (2010). Curso de Lengua Nahuatl Avanzado II. Instituto de Docencia e Investigación Etnológica de Zacatecas A.C.

Nava Vite, R. (2012). El Costumbre: Offerings and music to Chicomexochitl in Ixhuatlán de Madero, Veracruz. In A. van't Hooft & J. A. Flores Farfan (Eds.), *Estudios de lengua y cultura nahua de la huasteca* (pp. 31–65).

Index

Printed in the USA
CPSIA information can be obtained
at www.ICGtesting.com
LVHW022237171223
766680LV00001B/3